Mine Fire Prevention and Control

Bo Tan · Haiyan Wang · Hongqing Zhu ·
Liyang Gao
Editors

Mine Fire Prevention and Control

Springer

Editors
Bo Tan
China University of Mining
and Technology-Beijing
Beijing, China

Haiyan Wang
University of Science and Technology
Beijing
Beijing, China

Hongqing Zhu
China University of Mining
and Technology-Beijing
Beijing, China

Liyang Gao
China University of Mining
and Technology-Beijing
Beijing, China

ISBN 978-981-96-0465-4 ISBN 978-981-96-0466-1 (eBook)
https://doi.org/10.1007/978-981-96-0466-1

This Springer imprint is published by the registered company Springer Nature Singapore Pte Ltd.
The registered company address is: 152 Beach Road, #21-01/04 Gateway East, Singapore 189721, Singapore

Preface

Coal plays a central role in China's energy mix. Although the development of clean energy has been vigorously promoted in recent years, coal is still the most important energy source in China, especially in the power generation industry, according to statistics, coal electricity accounts for 60-70% of China's electricity generation. At the same time, China is also a major producer of coal, accounting for almost half of the world's coal production. Coal mine safety has always been an important and sensitive issue in China's coal industry. Although the level of safety management has been improving in recent years, and the number of coal mine accidents and deaths have dropped significantly, the safety situation remains grim due to China's huge coal production. The main safety challenges include: gas, water damage, fire, dust, etc. Among them, mine fire is one of the main disasters in coal mines, which has the characteristics of complex occurrence mechanism, high danger degree and serious consequences, and is easy to lead to major accidents resulting in serious casualties. Although the current mine fire prevention technology has been a considerable development, but to completely prevent the occurrence of mine fire is still very difficult. In addition, there are not enough books on mine fire prevention. Therefore, the study of the causes, characteristics and preventive measures of mine fire is not only of great significance for the prevention and control of mine fire, but also provides a reference for the study and research of mine safety management personnel, coal mine practitioners, relevant scholars and other readers.

This book aims to sort out and compare the research and progress that has been made so far, and extract and summarize the key practical theoretical methods to provide theoretical reference for the prevention and control of mine fire. Firstly, the paper introduces the causes, key characteristics, complexity and influence of mine fire on underground ventilation system through practical cases. The relevant contents correspond to the first and second Chaps. 1 and 2 of this book. Then, according to the influence of mine fire on underground ventilation system, one of the key disposal measures in the fire period - air flow control technology in disaster period is introduced. The related content corresponds to the content in Chap. 3 of this book. At the same time, the corresponding monitoring and prevention methods are given for different places and different types of mine fires, and the commonly used fire extinguishing methods and equipment are systematically combed and analyzed. The relevant contents correspond to the fourth, fifth and sixth Chaps. 4–6 of the book. The research and analysis results of this book have certain significance for improving and strengthening the relevant technical measures of mine fire prevention and control, ensuring personnel safety and production safety. This book is also characterized by actual accident cases, narrated from the experience and lessons of the accident, so as to facilitate readers to more easily understand the basic theory of mine fire and related countermeasures.

This book was written by me and three other authors (Haiyan Wang, Hongqing Zhu, and Liyang Gao). Their profound insights on the mechanism, disposal technology, monitoring and prevention of mine fire provide a solid theoretical basis for the completion of this book. With their support, the content of this book has been continuously developed and improved. Special thanks to Hongqing Zhu for his professional advice during the writing of this book. His patient guidance and careful proofreading helped to avoid many irregularities and imprecisions in this book.

This book was jointly supported by the National Key Research and Development Program of China (2023YFC3009101) and the National Natural Science Foundation of China (52074156).

In addition, I would like to thank the students of the Fire Monitoring and Control Laboratory of China University of Mining and Technology (Beijing) for their efforts in data analysis and field research, which made the conclusions of this book more rigorous and credible.

Finally, I would like to thank all the participants for their continued support and selfless contributions, and the book could not have been completed without their joint efforts.

Beijing, China
September 2024

Bo Tan

Contents

Chapter 1
Overview of Mine Fire

Bo Tan, Haiyan Wang, Hongqing Zhu, and Liyang Gao

Mine fire refers to any uncontrolled burning that occurs on the surface or underground of a mine that threatens the safe production of the mine and creates a disaster. Mine fire is one of the major disasters in coal mines. When a mine fire occurs, the fire develops rapidly. It is not only complex, but also affects a wide range of areas, and its often affects normal production. In serious cases, it can burn coal resources and mine equipment, or in more serious cases, it can ignite gas and coal dust explosions or poison the mine with fire smoke, which can lead to major accidents with casualties. In addition, the complexity of fires can make it extremely difficult to make disaster response decisions. The political and economic damage caused by a major fire accident is often incalculable. The emotional toll on miners is not short-lived and can have political repercussions [1]. Although current mine fire prevention technology has developed considerably, it is still difficult to prevent mine fires from occurring. Therefore, it is important to ensure that mine fires are prevented and extinguished in order to ensure that production is carried out safely. As one of the major disasters in coal mines, mine fire is a serious constraint to the safe production and development of coal mines. Therefore, it is crucial to study the causes of fire, fire characteristics and develop corresponding fire prevention measures to predict and prevent fire. In

B. Tan (✉) · H. Zhu · L. Gao
China University of Mining and Technology-Beijing, Beijing, China
e-mail: tanbo709@126.com

H. Zhu
e-mail: zhq@cumtb.edu.cn

L. Gao
e-mail: gaoliyangcc@163.com

H. Wang
University of Science and Technology Beijing, Beijing, China
e-mail: b2146431@ustb.edu.cn

B. Tan et al. (eds.), *Mine Fire Prevention and Control*,
https://doi.org/10.1007/978-981-96-0466-1_1

the process of coal mine production, we should strictly do our best to prevent disasters and continuously improve and strengthen prevention and control techniques and measures to ensure the personal safety of coal mine personnel and normal production.

1.1 Conditions and Causes of Mine Fire

Mine fire is combustion. Combustion must satisfy the "three elements" of combustion, which also known as the "triangle" of combustion. It includes combustible, ignition source and supporter of combustion, as shown in Fig. 1.1a.

1.1.1 Combustible

In underground coal mines, coal itself is a large and widespread combustible. Secondly, the large quantities of coal dust and gas produced in the course of carrying out production, as well as the pitwood, electromechanical equipment, explosives, oil and other substances used, are all combustible. Their presence is a fundamental element in the occurrence of fires.

1.1.2 Ignition Source

Only a heat source with sufficient heat can cause a combustible mass to burn. In coal mines, fires can be caused by spontaneous combustion of coal, combustion and

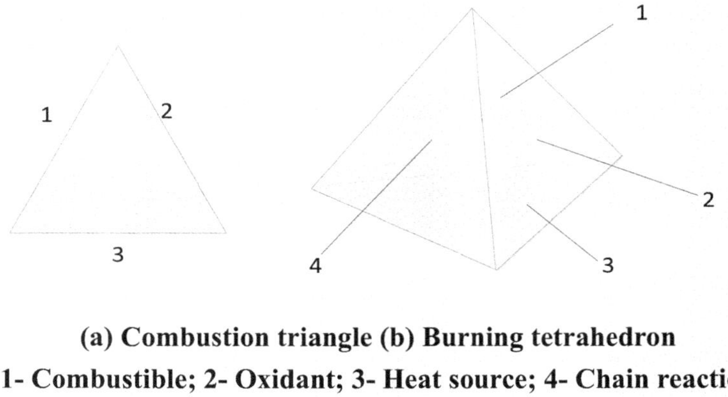

Fig. 1.1 Combustion triangle and burning tetrahedron. **a** Combustion triangle **b** burning tetrahedron. 1—Combustible; 2—Oxidant; 3—Heat source; 4—Chain reaction

explosion of gas and coal dust, blasting, heat from mechanical friction; overheating from short-circuiting of electrical currents, poorly functioning electrical equipment; smoking, welding and other open flames.

1.1.3 Supporter of Combustion

Combustion is defined as a violent oxidation reaction that produces heat and light, accompanied by the production of smoke. Although any combustible has a heat source with sufficient heat, if it lacks sufficient concentrations of O_2, combustion is unsustainable. Therefore, the supply of O_2 is also essential for combustion. Experiments have shown that no combustible combustion can be maintained in an air environment with an O_2 concentration of 3%. Gas loses its explosive properties in air with an O_2 concentration of less than 12%. Candles also do not burn at O_2 concentrations below 14%. Therefore, the source of O_2 supply in this case is air with a normal oxygen content, not oxygen-poor air.

Not only must combustion have the necessary (three elements) and sufficient conditions, but the above conditions must act in conjunction with each other for combustion to occur and last. Otherwise, combustion will not take place.

In recent years, the concept of burning tetrahedron has also emerged, i.e. the addition of a chain reaction element to the three basic elements of combustion mentioned above. Chain reactions are present in certain combustion processes, in particular the free radical chain reaction at the flame front, which controls the growth rate of the fire due to its extremely fast reaction rate. This is shown in Fig. 1.1b.

New fire extinguishing and flame retardant agents based on the principle of interrupting chain growth reactions in combustion have been invented to provide better fire extinguishing effects, thus supporting the new view that chain reactions are a fundamental element of the combustion process.

1.2 Coal Spontaneous Combustion Process and Theory

1.2.1 Coal Spontaneous Combustion Process

Coal spontaneous combustion is the phenomenon of coal catching fire on its own without ignition. It is a process in which coal with a tendency to spontaneously combust comes into contact with oxygen in the air, and the heat generated by oxidation is greater than the heat lost to the surrounding environment.

Coal spontaneous combustion has been studied extensively in the field of coal spontaneous combustion mechanism, among which the theory of coal-oxygen complex action is generally accepted. Coal spontaneous combustion occurs mostly due to the continuous oxidation of the coal remains in the mined-out area under the

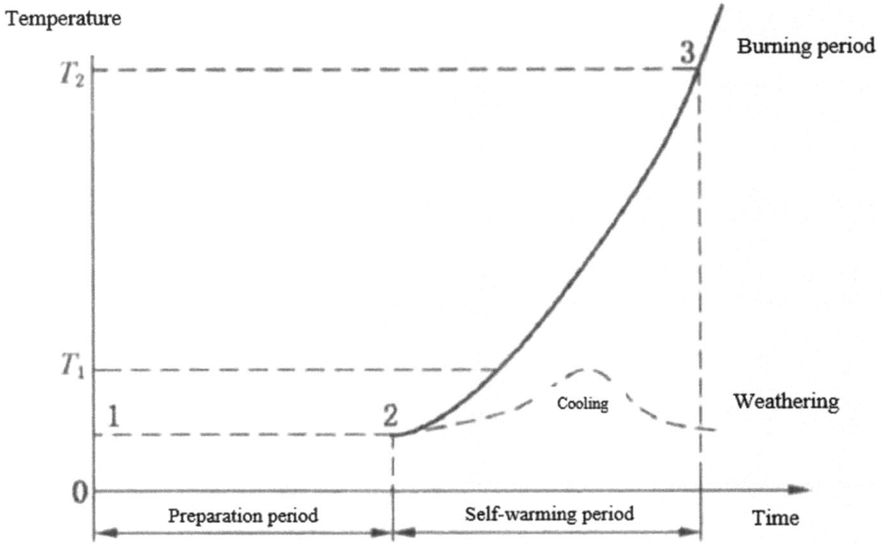

Fig. 1.2 Schematic diagram of the coal spontaneous combustion process

condition of air leakage in the mined-out area. The coal spontaneous combustion phenomenon is caused by a combination of factors, and the physical and chemical reactions occurring within the coal spontaneous combustion are also very complex. The coal spontaneous combustion is affected by internal factors such as coal type and particle size, porosity and specific surface area, as well as external factors such as environment and air leakage. It is also influenced by external factors such as the environment and air leakage.

The coal spontaneous combustion process is divided into three phases: the preparation period (latent period), the self-heating period and the burning period. The spontaneous combustion process can be represented in Fig. 1.2 [2].

(1) **Preparation Period (Latent Period)**

The oxidation rate of coal is slow and little heat is generated, leaving the temperature of the coal almost unchanged, but the chemical activity of the coal is increased. At the same time, the preparation period of the coal spontaneous combustion process can be prolonged if the conditions of coal crushing and stacking, heat dissipation and ventilation and oxygenation are improved.

(2) **Self-Heating Period**

The oxidation of coal gradually accelerates and the unstable oxides begin to decompose into H_2O, CO_2 and CO. If the heat generated is not dissipated in time, the accumulated heat is able to cause the coal temperature to rise gradually. When it rises to a certain critical value (generally considered to be 60–80 °C), dry distillation of coal occurs. Under different coal temperature conditions, different kinds and concentrations of landmark gases appear, such as CO, C_2H_4. C_2H_2, H_2 etc. The appearance of these gases presents a strong regularity. The use of these gas change laws can predict the forecast of natural coal ignition and carry out the judgment of the burning state of the fire area. However, if the oxygen supply and heat dissipation conditions of the coal suddenly change before the coal temperature rises to the critical temperature, the heating process of the coal will automatically slow down. Then, it will enter the cooling stage, and the coal will continue to oxidize slowly and finally enter the weathering state [3].

(3) **Burning Period**

When the temperature of the coal reaches its ignition point temperature, (generally considered anthracite greater than 400 °C, bituminous 320–380 °C, lignite 210–350 °C) it begins to ignite and burn, accompanied by the appearance of open flames, high temperature smoke, CO, CO_2 and various combustible gases and other ignition phenomena.

1.2.2 Thermal Theory of Coal Spontaneous Combustion

The process of spontaneous combustion is very complex and the theory of thermal spontaneous combustion is based on an analysis of the balance between heat production and heat dissipation in the system to determine whether spontaneous combustion is possible. Some of the heat is used to heat the combustible loose media itself to accelerate the oxidation and continue to produce heat. Some of the heat is dissipated to the outside environment through the combustible loose media boundary. Under the influence of certain factors, the rate of heat production is greater than the rate of heat dissipation, the system thermal equilibrium is broken, the temperature in the system continues to rise, thus prompting the automatic acceleration of the process of self-ignition in the system, spontaneous combustion occurs. Conversely, when the rate of heat dissipation is greater than the rate of heat production, the system temperature will continue to fall and the spontaneous combustion process will end.

In combustible processes, the rate of reaction increases exponentially in the later stages, especially in the flame burning phase. It is difficult to analyse the combustion process in detail using thermal theory (free radical). The acceleration of combustible spontaneous combustion can now be explained by the chain reaction mechanism after extensive experimental verification.

There are two reasons for the increase in the amount of free radical during combustion reactions. The increase in temperature leads to an increase in the rate of thermal movement of the molecules, which produces more free radical, but the increase in the amount of free radical produced in this way is very small. The chain reaction leads to more free radical during spontaneous combustion, which is the main reason for the production of free radical. The chain reaction theory suggests that there is a free radical in the combustion process and that as long as a small amount of free radical is produced under certain conditions, the chemical reaction will continue until the reaction is complete. The free radicals are produced and disappear at the same time, mainly because they collide with each other to form stable neutral molecules or lose their activity due to collisions, resulting in a decrease in concentration.

The chain reaction can be divided into a straight chain reaction, where the amount of free radical remains constant during the reaction, and a branched chain reaction, where the amount of free radical increases significantly with time during the reaction. The chain reaction can be divided into three stages—chain initiation, chain transfer and chain termination reaction.

(1) **Chain Initiation**

The process of free radical production is known as chain initiation, which is the initial stage of the chain reaction and requires sufficient external energy to complete the internal chain breakage reaction. Therefore, it is the most difficult stage of the whole chain reaction. This is the most difficult stage of the whole chain reaction. In the initial stage of spontaneous combustion, the coal gradually builds up heat through the oxidation of the primary free radical contained in the coal itself with oxygen, resulting in spontaneous combustion and speed, as in the initial stage of coal oxidation reaction.

(2) **Chain Transfer**

In the chain transfer process, the amount of free radical regenerated by the chain reaction is greater than the amount of free radical consumed, and the free price is conserved during the reaction. In spontaneous combustion, the amount of free radical is continuously increased through the chain transfer process, making the chain reaction speed up gradually and thus releasing more heat. When the heat release reaches a certain level, the coal will naturally ignite.

During spontaneous combustion of coal, gases such as CO, CO_2, alkanes, olefins, formaldehyde and methanol are produced. Since different temperature stages will produce different signature gases in large quantities, the pattern of signature gases can be used to predict the spontaneous combustion of coal. The mechanism of the chain transfer process can be used to analyse the pathways of the signature gases. For example, aldehyde free radicals can be pyrolysed to form CO, acyloxy free radicals can be pyrolysed to form CO, alkoxy free radicals can be broken to form formaldehyde, peroxides can be pyrolysed to form methanol, etc.

(3) **Chain Termination Reaction**

The chain termination reaction is the gradual disappearance of the free radical in the chain reaction. The chain termination reaction requires very little activation energy. In the spontaneous combustion phase of coal. The chain termination reaction consists of two main types of reaction: the recombination reaction between free radicals and the destruction reaction between free radicals in the gas phase. In the case of recombination reactions, the reaction between two free radicals produces a free radical, e.g. the reaction between hydrocarbon free radicals.

$$\mathrm{R} + \mathrm{R'} = \mathrm{R} - \mathrm{R'} \tag{1.1}$$

The recombination reaction of free radicals is an important part of the chain reaction, which produces many important free radicals, such as methyl, benzyl, etc. Gas phase destruction reaction is the reaction between free radicals to form a gas molecule, this reaction is the main termination reaction of the coal spontaneous combustion process by the free radicals. For example:

$$CH_3 + H \rightarrow CH_4 \tag{1.2}$$

$$2CH_3 \rightarrow C_2H_6 \tag{1.3}$$

$$CH_3 + CH_3CH_2 \rightarrow CH_3CH_3CH_2 \tag{1.4}$$

$$CH_3 + CH_3CH_2 \rightarrow CH_4 + C_2H_4 \tag{1.5}$$

$$H + H \rightarrow H_2 \tag{1.6}$$

The chain termination reaction process is easier than the chain growth process, but the free radical concentration for the termination of the chain reaction is relatively small in the high temperature phase. So the coal spontaneous combustion process is still dominated by the chain growth reaction.

1.3 Classification of Mine Fire and its Characteristics

In order to correctly analyse the causes of fires, their development and to formulate targeted countermeasures to prevent and extinguish them, it is necessary to classify mine fires. At present, mine fires are generally categorized into several types according to the location of the fire, the ignition source, the oxygen affluence and other circumstances.

1.3.1 Superficial Fire and Underground Fire

Fires can be classified as superficial fire or underground fire depending on where they occur.

(1) **Superficial Fire**

Fires that occur on the ground within the industrial square of a mine are called superficial fires. Superficial fires can occur in administrative offices, shaft buildings, coal processing buildings, coal storage yards, gangue hills and other locations.

The characteristics of a superficial fire: the external signs are obvious, it is easy to detect and the air supply is adequate. It burns completely and there are fewer sources of toxic gases from incomplete combustion; the ground space is wide and the smoke spreads easily, facilitating relief.

(2) **Underground Fire**

Fires that occur underground, and fires that occur near the wellhead and threaten the safety of the shaft, are collectively referred to as underground fires. Underground fires can occur in the wellhead building, shaft, underground yard, electromechanical chambers, gunpowder store, air intake and return alleyways, mining and mining working face and mined-out area, coal pillars, etc.

The characteristics of underground fire: due to the limited supply of air underground, it is difficult to burn completely. Toxic and harmful fumes occur in large quantities, spreading everywhere with the wind flow, poisoning the mine air and threatening the lives of workers. There is a risk of gas and coal dust explosion in mines, which may also cause an explosion, resulting in a major accident.

1.3.2 Ascensional Airflow Fire, Descending Airflow Fire and Inlet Airflow Fire

Underground fires can be further divided into three categories according to their location and impact on the mine ventilation system.

(1) **Ascensional Airflow Fire**

Upward airflow is the bottom-up flow of air along an inclined or vertical shaft, mining working face, i.e. the airflow flows from the low point of the elevation to the high point. When a fire occurs in this type of tunnel airflow, it is referred to as an ascensional airflow fire. When a fire occurs in the upstream airflow, the direction of action of the fire air pressure due to heat is the same as the direction of the airflow, i.e. the same as the direction of action of the main ventilation fan in the mine. In this case, the main features of its effect on the mine ventilation system are as follows.

The direction of the air flow in the main air path (from the intake air shaft flowing through the fire source to the air-return shaft) is generally stable, i.e. it has the same direction as the original air flow and the smoke flow air flow is discharged. In contrast, all other side branches that are connected in parallel with the main air path or converge at the rear of the fire source in the main air path have an unstable direction and may even be reversed, resulting in a turbulent air flow accident. Therefore, fire protection measures should be taken to avoid reversals of bypass flow.

(2) **Descending Airflow Fire**

Downward airflow is the top-down flow of air along inclined or vertical shafts, mining working faces such as intake air shafts, inlet air downhill and downward ventilated working faces, i.e. the flow of air from the high point of elevation to the low point.

When a fire occurs in this type of airflow, it is called a descending airflow fire. When a fire occurs in the downstream airflow, the direction of action of the fire wind pressure is opposite to that of the main ventilation fan in the mine. Therefore, as the fire develops, it is difficult to maintain the normal flow direction of the airflow in the main air path. When the fire wind pressure increases to a certain level, the wind flow in the main wind path will reverse and the smoke flow will then recede, resulting in a disrupted wind flow.

In the event of a fire in the downstream airflow, the change in the state of the airflow of the ventilation system due to the action of the fire wind pressure is much more complex than in the case of an ascensional airflow fire. Therefore, it is more hazardous and technically difficult to prevent.

(3) **Inlet Airflow Fire**

Fires that occur in the intake air shaft, inlet air shaft or in the inlet airway of a mining area are known as inlet airflow fires. The main reason for distinguishing this category of fire is the characteristics of its development, the hazards to underground workers and the technical measures required to extinguish it, which are to a greater extent different from upper and descending airflow fires. A coal spontaneous combustion fire that occurs in the incoming airflow is generally not easily detected in the early stages. The fire is not easily detected at an early stage, but after it has occurred, it develops rapidly due to the adequate oxygen supply and is not easily controlled. Underground workers are mostly in the downwind stream, which makes them vulnerable to high temperature fires and smoke, resulting in injuries and fatalities. For this kind of fire, in addition to using the corresponding control technical measures according to the structural characteristics of the fire-emitting wind path (upstream or downstream), technical measures adapted to the prevention and control of this kind of fire should be used according to the characteristics of the incoming wind stream, such as mine wide, regional or local counter-wind.

1.3.3 Exogenous Fire and Internal-Caused Fire

In coal mines, mine fires are usually divided into two categories depending on the source of the ignition heat: exogenous fire and internal-caused fire.

(1) **Exogenous Fire**

This refers to fires caused by external heat sources such as gas and coal dust explosions, blasting operations, mechanical friction, poorly functioning electrical equipment, short-circuiting of power supplies and other open flames, smoking, welding, etc.

Exogenous fires are characterized by their sudden and rapid onset. If they are not detected and controlled in time, they can often lead to major accidents. Although the proportion of exogenous fires in the total number of mine fires is relatively small (4–10%), it cannot be ignored. But in recent years, with the increase in mechanization, the proportion of exogenous fires is on the rise. According to statistics, more than 90% of recorded major malignant fire accidents in China belong to exogenous fire. Exogenous fires mostly occur in wellhead buildings, shaft shafts, electromechanical underground chambers, powder magazines and in roadways or working faces where electromechanical equipment is installed. The flames of a fire are usually on the surface of the burning material, which can be easily extinguished if detected and extinguished in time.

(2) **Internal-Caused Fire**

Internal-caused fire or spontaneous fire of coal (spontaneous combustion) is a fire formed when coal under certain conditions, such as broken coal pillars, coal walls, concentrated piles of floating coal, mined-out area leftover coal, etc., under the condition of having the right amount of wind supply, itself undergoes physical and chemical changes, oxygen absorption, oxidation, heat generation, and heat gathering leading to a fire.

The onset of an internal-caused fire is often accompanied by a process of heat build-up, which can be detected early based on early warning. However, the source of the fire is hidden and often occurs in mined-out areas or coal pillars that are difficult to access, making it difficult to locate the exact source of the fire. As a result, it is difficult to extinguish and fires can last for months, years or even decades. Sometimes, the fire spreads gradually, destroying large amounts of coal and freezing large resources.

1.3.4 Oxygen-Enriched Combustion and Fuel-Rich Combustion Fire

Mine fires are classified according to the combustion environment (oxygen and fuel), the composition and concentration of the combustion products, which are classified as oxygen-enriched combustion or fuel-rich combustion fire.

Underground fires occur in confined spaces, where the characteristics of the fire are closely related to the ventilation conditions. Depending on the amount of air supply, fires in confined spaces can be divided into two types: oxygen-enriched combustion and fuel-rich combustion.

(1) **Oxygen-Enriched Combustion**

Oxygen-enriched combustion is combustion that is adequately oxygenated and has a similar combustion and spread mechanism to superficial fire. It is also known as unrestricted combustion or fuel-controlled combustion.

Due to sufficient oxygen, the volatile gases from the combustion of the fire source are largely depleted in the combustion and no excess incandescent volatile gases converge with the main wind stream and preheat the larger area of combustible on the downwind side. The flames from the combustion heat the adjacent combustible to the ignition point in the form of thermal convection and thermal radiation, maintaining the continuity and development of the combustion. The small extent of the fire, the low intensity of the fire, the low rate of spread and the low oxygen consumption result in a considerable amount of oxygen remaining. The oxygen concentration on the downwind side is generally kept above 15% (by volume), hence the name oxygen-enriched combustion.

Fuel-rich combustion is under-oxygenated combustion, also known as restricted combustion or ventilation-controlled combustion. When the fire is burning, the fire is hot, the temperature is high and the fire generates a large amount of hot volatile gases, which are not only consumed by the combustion zone, but also merge with the main air flow heated by the hot fire to form a hot smoke stream, which preheats a large area of combustible downwind of the fire and makes it continue to generate large amounts of volatile gases. On the other hand, the flame at the combustion location heats the immediate combustible by thermal convection and thermal radiation to bring it up to its ignition point. Due to the persistence and development of the two factors that keep the combustion going, such fires allow combustion to take place on a wider scale and spread at a greater rate resulting in the main air stream being almost completely depleted of oxygen with a residual oxygen concentration of less than 2%. Therefore, the spread of such fires is limited by the amount of oxygen supplied to the main air stream. In superficial fires, these fires are also called restricted fires because they only occur in situations where space is restricted or the aisle section is small. They are generally referred to as fuel-rich or oxygen-poor fires due to their downwind flue gas oxygen concentration being close to zero. The downwind side of the smoke flow is often high temperature premixed combustible gas, and the side of the fresh air flow intersection, easy to form a new source of fire, the formation of multiple regenerative sources of fire called the phenomenon of fire development "jumping frog". That is multiple intermittent source of fire like a frog jumping landing point. The appearance of regenerative fires increases the probability of explosions caused by premixed gases entering the fire source and accelerates the rate of fire spread.

In a well-ventilated environment, most fires have an initial supply of oxygen greater than the demand for combustion, which is oxygen-enriched combustion. As combustion continues, combustible gases such as methane, hydrogen and water vapour are released from coal, wood and other combustible materials as the heat from combustion builds up and raises the temperature to a certain value. However, as there is not enough oxygen to support the combustion of these combustible gases, the combustion becomes fuel-rich combustion. The development of combustion from oxygen-enriched combustion to fuel-rich combustion is an important process. Once it has changed to fuel-rich combustion, it indicates a large fire and insufficient air supply. It signals a significant increase in the risk and severity of a disaster. All personnel, including ambulance crews, must be evacuated promptly in this environment.

Table 1.1 summarizes the basic characteristics of oxygen-enriched combustion and fuel-rich combustion fires.

Table 1.1 Basic characteristics of the two types of fire

Classification	Fuel-rich combustion (restricted combustion)	Oxygen-enriched combustion
Basic features	Lots of fuel and insufficient oxygen supply	Low fuel and high oxygen supply
Features	Large fire area, high fire intensity and rapid spread	Small fire area, low fire intensity, slow spread
	High oxygen consumption, low residual oxygen (around 2%)	Low oxygen consumption, high residual oxygen (around 15%)
	Large amount of combustible volatiles remaining	Combustible volatiles largely depleted
	Susceptible to regenerative ignition and explosion	Not prone to re-ignition and explosion
	More dangerous	Slightly less dangerous

1.3.5 Other Classification Methods of Mine Fire

In order to make a more in-depth investigation of mine fires, they are sometimes classified according to the location of the fire, the burning material and the nature of the ignition.

Depending on where the fire occurs, mine fires can be classified as shaft fires, roadway fires, coal pillar fires, face fires, mined-out area fires and chamber fires.

Due to the different combustibles they can be classified as electrical and mechanical equipment (tapes, cables, transformers, switches, dampers etc.) fires, gunpowder combustion fires, oil fires, pit wood fires, gas combustion fires, coal dust combustion fires and natural coal fires (spontaneous combustion).

By state of combustion, they can be classified as smoldering fire and open flame fire. Smoldering fire refers to a fire with no visible flame, open flame fire refers to a fire with a longer flame.

Depending on the nature of the ignition, it can be divided into native fire and secondary fire. Secondary fire refers to fires caused by native fire. In the burning process of native fire, there are high temperature smoke streams which are not yet combustible. Once they converge with the wind stream in the smoke discharge channel, they may burn again when they are supplied with oxygen. Especially if the convergence point is located in the dry wood support area, it is more likely that a secondary fire will occur and expand the fire area.

1.4 Hazards of Mine Fire

1.4.1 Causing Injury or Death

(1) The high temperatures generated by the fire source directly cause casualties. The casualties occur mainly in the vicinity of the fire source and in the immediate area.
(2) Mine fires produce large amounts of hot fire smoke and the temperature near the fire source is often above 1000 °C. The large amount of smoke and heat generated by a fire cannot be easily dissipated underground. In the case of shaft ventilation, the high temperature smoke stream can spread rapidly with the wind over a large area, exposing personnel in these areas to high temperatures. In addition, the spread of high temperature toxic and harmful gases from the fire can also cause poisoning and casualties on the downwind side. It also makes it more difficult to evacuate and rescue people and poses a threat to safety. Statistics show that most people killed in mine or building fires are not killed by heat, but by poisoning. Therefore, during a fire, the control of the flow of toxic and hazardous fumes is one of the main measures of mine fire relief.
(3) When mine fire develops to a certain intensity, there is a fire wind pressure, it may cause a change in the direction of the wind flow in the mine ventilation network, causing a disturbance in the mine wind flow, thus making the flow of smoke out of control and even allowing toxic and harmful airflow to enter the inlet area, further expanding the disaster area and exposing more underground personnel to the poisonous effects of fire smoke. At the same time, this poses great difficulties and dangers to the safe evacuation of the underground.
(4) Fire-induced accidents can also cause mechanical injuries to personnel, impacts, collisions and injuries from exploding flying rocks.

[Case 1]

On 23 May 1998, a natural fire occurred in the zero piece mined-out area of the eighth floor of a mine, which was handled by opening the dampers without authorization, causing the reversal of the wind flow in the mined-out area and causing the back-flow of gas accumulated in the mined-out area to flow through the fire area causing a gas explosion with 27 casualties [4].

A coal mine has a production capacity of 200 × 104t per year and it produces Taixi coal. It is a high gas mine and the coal seam is at risk of spontaneous combustion. In October 2003, a fire and explosion occurred in the mine. The accident first occurred near the intersection of the mined-out area and the four cross-currents at the working face. As the fracture pathway connecting the mined-out area to the surface of the mine was open, the surface leakage was large, and the gas gush rate was high. There were sufficient sources of oxygen and gas recharge in the fire area to constantly replenish the consumption of oxygen and gas by the explosion, resulting in the fire area being closed to create an asphyxiating environment within the fire area. As a result, the fire zone closure was repeatedly extended, resulting in multiple interchanges between fire and gas explosions, causing more than 100 explosions. The mine is divided into the South Flank production area and the North Flank preparation area.

The ambulance team arrived at the affected area after receiving the alarm. It was found that there was a lot of smoke from the explosion inside, and decided to observe 24 h. A day later, it was found that the smoke did not decrease, so a forced closure was carried out. The temporary confinement after the closure was destroyed by the gas explosion. Then, the confinement was expanded to close the area again after the confinement was destroyed by another explosion in the fire area, and the fire area was expanded to play 7 permanent confinements to confine the whole fire area. Two days later, in the ambulance team for containment reinforcement, because of the fire area connected with the transport alley two cross river gas explosion, the old confinement destroyed. The explosion in the second cross-channel destroyed the confinement in the second cross-channel. But fortunately, it did not cause the collapse of the transport roadway and the ambulance team that was working on the confinement was able to withdraw safely.

The General Relief Command decided to construct a burst containment in the 1660 Middle roadway beyond the intersection of the 1.5 and 2 cross-currents of the transport roadway A. During the construction of the containment, conditions were created for future grouting and the relief personnel on site dismantled the gas extraction line. The line was to be used as a future grouting pipe. As the gas drainage stops a large amount of gas gushing retrograde, the gas concentration in the closed place gradually increases from 0 to more than 2%. After being informed that the closure could be completed within half an hour, the General Command decided to fight for time to complete the closure as soon as possible under the premise of closely monitoring the dynamic changes in gas concentration. However, the observation into the mouth of the Second Cross River had turned into an open fire. Upon being informed of this, the General Command immediately ordered the 160 people who were transporting materials and building the containment in various parts of the mine to retreat. After a few minutes, the gas concentration at the confinement was found to be 0. At this point, there was a dispute between two opinions. One opinion was that the retreating crew should be asked to return in time to get the confinement in place. The other opinion is that caution should be exercised. Because in general, only a change in the direction of the wind flow to the incoming wind can make the gas concentration at the site 0. A change in wind flow can only cause a change in gas concentration, not make it 0. And changes in the direction of wind flow often foreshadow the possibility of a gas explosion. Therefore, no personnel were notified to return. 4–5 min later, a strong gas explosion occurred, causing the blast doors of the south and north wing wind shafts to burst open. Fortunately, the retreating personnel were not allowed to return to play the seal, otherwise very serious damage would have been caused.

This explosion caused the solution of making a confinement in the transport roadway to be impossible to implement, whether to make a surface closure in the production area of the south wing (including the stone gate confinement B connected to the north and south wings), or to make a confinement underground. The Command held a meeting to analyse the situation and two opinions emerged. One being that the decision was made to further extend the confinement underground out of a desire to reduce the impact on production, the other being that such a confinement would require the drilling of 11 permanent confinements, which would be extremely unsafe in the face of constant explosions. In the end, the command decided to close the two pairs of vertical shafts in the south wing from the surface and to close the stone doors connecting the north and south wings of the shaft.

During the period, when the main and secondary wells on the surface were closed and the stone gate was closed, explosions were constantly occurring in the affected areas, taking into account that the affected areas were more than a thousand meters from the wellhead and stone gate locations. At the time, all the explosions were transmitted to the wellhead and stone gate locations, and the blast winds were already near room temperature. According to the analysis of the situation at that time, it was safer to play confinement in these places. So the use of the explosion interval to grab hit the seal, and finally all closed. 3 h after the gas explosion, the destruction of the south two transport on the wellhead seal and stone door explosion-proof seal B. Stone door explosion-proof seal B above the formation of high 600 mm, across the entire width of the tunnel space, sandbags all thrown out, indicating that the explosion-proof seal above the roof is not secure. The command decided to use the interval between the blast to grab the seal and finally close it all with an explosion-proof seal. Considering that the friction coefficient of woven sandbags was too small to resist the lateral impact in the direction of the original roadway, the sandbags in the stone door explosion-proof closure B were all changed to sacks and the stone door explosion-proof closure was reinforced with support above. About 3 h later, a fairly violent explosion occurred, destroying the newly constructed wellhead explosion-proof closure and breaking down the explosion-proof dampers and wind chambers, and the sandbags of the explosion-proof closure washed out of the wellhead, but the stone door explosion-proof closure was not damaged, indicating that the sandbags of the stone door explosion-proof closure were effectively reinforced with sacks and support above. The command analyzed the reason for the multiple closures and explosions. Because the mine had a large surface air leak, a large influx of coal seam gas and a fire zone closure that could not form an asphyxiating environment.

Finally, the command decided to take measures to flood the south wing by reinforcing the stone door explosion-proof containment B connected to the north and south wings to prevent leakage. The fire and explosion inter-conversion disaster at the mine was one of the most complex in recent years, with over 100 gas explosions, and no casualties were sustained throughout the relief process.

1.4.2 Existence of Potential Gas Explosion Hazard

There are two main types of accident caused by gas in mines. High gas concentrations lead to a drop in oxygen concentration resulting in asphyxiation. Gas concentrations within the explosive limit can burn and explode under high temperature heat source conditions, resulting in injury or death and damage to the mine production system.

The high temperature gas volatiles generated by the fire and not yet burned out are mixed with the gas, which may mix with the fresh air flow of the associated intake duct during the flow process to form a gas mixture. The presence of large amounts of flammable material everywhere in the underground coal mine makes it extremely easy for fires to develop and spread. High temperature fire smoke will form a new source of fire at the infiltration site when it infiltrates the fresh air flow on the way through the tunnel. Because of wind turbulence flow through the source of fire or secondary sources of fire caused by gas explosions. This type of accident has historically occurred in both high gas and low gas mines.

[Case 2]

At 21:56 on 29 March 2013, a particularly significant gas explosion occurred at Babao Coal Company, Tonghua Mining Group Company, Jilin Province, China, which resulted in 36 people being killed (7 people were concealed by the company and verified after being reported by the public) and 12 people being injured. The direct economic loss was RMB 47,089,000.

On April 1, the mine did not implement the Jilin Provincial People's Government's directive prohibiting personnel from working down the shaft, and illegally arranged for personnel to enter the shaft for construction of the confinement, which resulted in another gas explosion at 10:12 a.m., killing 17 people and injuring eight.

① The occurrence of the "3·29" accident and the rescue situation

Babao Coal Mine is a former ballast coal mine of Tonghua Mining Bureau, renamed as Pine Town Coal Mine Babao Mining District in 2004, which is renamed as Jilin Babao Coal Company Limited when it underwent renovation and expansion in 2007. The mine's industrial and commercial business licence, coal production licence, safety production licence, mine manager's qualification certificate and mine manager's safety qualification certificate are all valid. The enterprise name of the mining licence has not been changed and remains as Tonghua Mining Bureau Pine Town Coal Mine Babao Mining District.

Babao Coal Mine has six mine-able coal seams, all of which have a propensity for spontaneous combustion rating of Class II, which is a spontaneous combustion seam. It is a high gas mine with an explosive risk of coal dust. The mine is a vertical shaft with 5 shafts and 5 production areas (1 integrated mining area and 4 water mining areas) prior to the accident. The mine's deepest elevation had reached the −780 m level, exceeding the −600 m level permitted by the mining licence. The accident occurred in the mined-out area of the −4164 east water mining workings in the −416 mining area, where the roof was managed by the natural collapse method and the mined-out area gas was extracted by buried pipes.

At approximately 16:00 on 28 March 2013, a gas explosion occurred in the mined-out area near the −416 quarry area, and the mine took two measures to add another closure to the −380 stone door closure in the −416 quarry area and to construct a new −315 stone door closure. At 14:55 on the 29th, a second gas explosion occurred in the mined-out area near the −416 quarry, and the newly constructed containment was broken. The carbon monoxide sensor alarm of −416 mining area-250 stone gate, the mining area personnel withdrawal.

After receiving the report, Ning Lianjiang, Chief Engineer of Tonghua Mining Company, and Chen Weiliang, Deputy Chief Engineer, rushed to Babao Coal Mine and decided to construct five confinements in the −315, −380 stone gate and the East 1, East 2 and East 3 stratified down-slots. At 16:59, Ning Lianjiang and Chen Weiliang led the ambulance crew and workers to the −416 mining area for mined-out operations. At approximately 19:30, a third gas explosion occurred in the mined-out area near the −416 mining area and the workforce retreated to the bottom of the shaft in a panic (six of the confinement workers ascended the shaft and firmly refused to risk further operations). The above three gas explosions all occurred in the mined-out area of the −4164 East water mining workings in the −416 mining area, which did not cause any casualties.

The mine not only failed to report and evacuate the operators as required, but also decided to continue the construction of the mined-out area. At 21:00, the underground site commander forced the construction workers to return again to implement the mined-out construction work. At 21:56, a fourth gas explosion occurred in the mined-out area, before the mine notified the underground to stop production and evacuate people and report to the relevant government departments, at which time the mine had a total of 367 people underground and a total of 332 people ascended on their own and by rescue. As of about 13:00 on the 30th underground search and rescue work ended, the accident caused a total of 36 deaths.

② "4·1" accident and rescue situation

After the search and rescue work of the "3·29" accident, in view of the fact that there were no more people underground and the disaster was serious, the working group of the Jilin Provincial People's Government and the State Administration of Safety Supervision requested the JI Coal Group to hire experts from inside and outside the province to conduct a serious analysis of the underground disaster area and formulate a safe and reliable fire-fighting plan, as well as decided that no one should go down the well without the consent of the Provincial People's Government. At 07:50 on 1 April, monitoring personnel detected a rapid increase in carbon monoxide concentration in the −416 mining area below the Babao Coal Mine shaft through sensors. Wang Shengyu, Executive Vice General Manager of Tonghua Mining Company, convened Li Chengmin and Wang Li, Vice General Managers, and Wang Qingfa, Deputy Mine Manager of Babao Coal Mine, to discuss the matter and then decided to send personnel down the shaft without permission, in defiance of the Jilin Provincial People's Government's directive that all personnel are strictly prohibited from working down the shaft. At 09:20, Wang Yubo, Director of Safety Supervision of Tonghua Mining Company in the mine, and Wang Qingfa led rescue team members down the shaft to the −400 roadway and −315 stone gate respectively to implement measures to hang wind barriers to block the wind flow and control the fire. At 10:12, a fifth gas explosion occurred in the mined-out area near the zone. A total of 76 people were working down-hole at this time, 59 survived the rescue (8 of whom were injured) and 6 were found killed and their remains removed from the shaft. Eleven people were still missing from the shaft, resulting in a total of 17 deaths and 8 injuries.

The fire area under the mine was gradually expanding. There was a risk of another gas explosion, which was repeatedly discussed by the expert group. The Jilin Provincial People's Government decided to adopt a disposal plan of putting out the fire before searching for it. A sixth gas explosion occurred at approximately 0810 h on 3 April. No new casualties were caused as no further personnel went down the shaft [5].

[Case 3]

On 15 January 1993, a gas explosion in the Xieer Mine B93342 mining working face upwind roadway of the Huainan Mining Bureau was followed by a fire, the spread and expansion of which led to 51 consecutive gas explosions.

Xieer Mine is located in the transition zone between the southern flank of the Bagong Mountain monoclinic structure and the Li Chengz oblique structure, with complex geological structure and high gas pressure. The mine is explored by a vertical shaft with a central boundary ventilation. In 1991, the mine was identified as a coal and gas protrusion mine with an absolute gas outflow of 40.949 m^3/min and a relative outflow of 15.36 m^3/t. There have been many gas explosions in the history of the mine. On July 30, 1985, there were three consecutive gas explosions caused by coal spontaneous combustion in the transportation roadway of coal mining team 1 in mining area 31.

The B93442 mining working face is located between the F13-8 and F13-7 faults in the north flank of the 34 mining area. The upper limit of recovery is a 401 m and the lower limit is a 456 m. The working face has a strike length of 330 m and an average mining length of 140 m. As of 15 January 1993, there was still 40 m of recovery remaining. The absolute gas outflow from the working face is 15 m^3/min. In order to solve the problem of gas. In order to resolve the impact of gas on the recovery, the tail-race was ventilated. The other end of the B10 tracked tunnel connected to the tail-race and a 400 m return stone door leads to the 31 mining area, which is a service tunnel for mining the C13 and B11b coal seams and is more than 400 m long. It is a long closed gas reservoir, from which the mine is pumping out gas.

After a gas explosion in the B93342 mining working face upwind roadway, a second gas explosion occurred on track 33 up hill at 23:47 on 15 January. A third gas explosion occurred in the affected area at 00:50 on 16 January, a fourth at 02:05 and a fifth at 03:25. This was followed by more than 40 gas explosions [6].

1.4.3 The Existence of Coal Dust Explosion Hazard

Coal dust is explosive. In general, in addition to a few anthracite coal, all other types of coal are explosive coal dust, the higher the volatile content of coal, the more explosive coal dust. Mine fire provides a high temperature heat source for coal dust. Once the concentration of coal dust suspended in the air in the explosive limit concentration range, coal dust will explode.

1.4.4 Damage to Underground Equipment and Facilities

The high temperature generated by mine fire causes damage to the roadway support, collapse of the roadway, burn damage to electrical and mechanical equipment and facilities, and burn a large amount of coal. While the closed fire area makes some areas of coal can not be mined, the management and opening of the closed fire area is difficult and becomes a hidden danger for safety production.

1.4.5 Severe Impact on Production

For underground fires, especially internal-caused fires that occur in the mined-out area or in the coal column are often difficult to eradicate in the short term. In such cases, the fire area is usually closed, resulting in a large amount of frozen coal and strained mine production continuity. For intensive production mines with one mine and one shaft, such closure can result in a total mine shutdown. In the first four years of China's Seventh Five-Year Plan period, there were more than 240 fire zones with a total of more than 70 million tonnes of frozen coal [7].

Mine fire seriously interferes with the normal production order. For a certain period of time, it causes deterioration of the production environment, increased psychological pressure on workers, reduced labour efficiency, decreased production and even production stoppages. At the same time, mine fire affects the normal succession of mining work due to the possible closure of the working face.

1.4.6 Pollution of the Environment

Mine fires produce large amounts of toxic and harmful gases such as CO, CO_2, SO_2 soot, etc., which can cause environmental pollution. In particular, coal seam outcropping fires in places like Xinjiang are not extinguished for a long time due to the large fire area, deep burning depth, high temperature of the fire area and lack of sufficient funds and advanced fire extinguishing technology, which not only burns a large amount of coal resources, but also causes the harmful gases in the atmosphere to seriously exceed the standard, resulting in widespread acid rain and greenhouse effect.

References

1. Zhou X, Wu B. Theory and practice of mine fire disaster relief. Beijing: Coal Industry Press; 1996.
2. Wang D. Mine fire science. Xuzhou: China University of Mining and Technology Press; 2008.
3. Yu M. Mine fire prevention and control. Beijing: National Defense Industry Press; 2013.
4. Mine Rescue Command Centre, State Administration of Work Safety. Cases and analysis of emergency rescue in mining accidents. Beijing: Coal Industry Press; 2006.
5. Jia Y. Compilation of typical cases of coal mine accidents. Xuzhou: China University of Mining and Technology Press; 2012.
6. Information Research Institute of the State Administration of Work Safety. Coal mine safety warning education cases. Beijing: Coal Industry Press; 2014.
7. Wang X. Coal mine accident rescue guide and typical cases. Beijing: Coal Industry Press; 2014.

Chapter 2
The Effect of Mine Fire on the Flow Regime of Air Flow

Bo Tan, Haiyan Wang, Hongqing Zhu, and Liyang Gao

Fires can also cause disruptions to the ventilation system in mines. This not only leads to difficult to control reversal of the normal air flow, but also poses a serious threat to the safety of the mine and damages the natural environment, resulting in significant resource losses and environmental pollution. Understanding the changes in air flow status and hazards caused by mine fires is an important guide to air flow control and fire suppression during mine fires.

2.1 Airflow Disorder from Mine Fire

The so-called airflow disorder means that in the event of a fire in a shaft, the direction of air flow and the distribution of air volume in the roadway under normal circumstances is disrupted by the action of fire and smoke, and the toxic and harmful smoke from the fire enters the air flow, causing the accident to extend further and causing a large number of casualties.

B. Tan (✉) · H. Zhu · L. Gao
School of Emergency Management and Safety Engineering, China University of Mining and Technology-Beijing, Beijing, China
e-mail: tanbo709@126.com

H. Zhu
e-mail: zhq@cumtb.edu.cn

L. Gao
e-mail: gaoliyangcc@163.com

H. Wang
School of Civil and Resources Engineering, University of Science and Technology Beijing, Beijing, China
e-mail: b2146431@ustb.edu.cn

B. Tan et al. (eds.), *Mine Fire Prevention and Control*,
https://doi.org/10.1007/978-981-96-0466-1_2

When a mine fire occurs, the effect of the air flow state, i.e. the effect of fire heating air pressure, is expressed as a "throttle effect" and a "buoyant effect". The increase in air flow mass and volume flow due to the addition of combustion products and water vapour generated by the fire and the further increase in air flow volume flow due to the effect of changes in air flow temperature is known as the throttle effect. throttle force, i.e. thermal resistance, increases the air flow in the opposite direction to the air flow. The throttle force, or thermal resistance, increases the resistance to air flow. Because it is always in the opposite direction to the direction of the air flow. The fire causes an increase in air flow temperature and a decrease in air density, causing the air flow to float upwards on its own, a phenomenon known as the buoyant effect. Buoyant effect is applied in a roadway with a height difference. The buoyant effect is the result of hot smoke in a chimney rising to the atmosphere without any other external power.

The buoyancy and throttle effect of a mine fire causes a turbulent change in the state of the mine airflow. This change can be divided into airflow (smoke) inverting, smoke return and smoke rollback.

2.1.1 Airflow (Smoke) Inverting

The combined effect of buoyancy and the throttle effect, which resists the influence of mechanical wind pressure, results in a change in the direction of air flow in certain roadways of the mine, which is known as air flow reversal. Reversal occurs mainly in the side branches where the reverse heat wind pressure is greater than the forward mechanical wind pressure (the main airway is the path from the inlet shaft to the return shaft via the fire origin. A bypass is a branch other than the main airway).

As shown in Fig. 2.1, fire origin is on branch 2–4 and air flow flows out of the network via trunk airway 1–2–4–5–6 and branches 1–2–3–5–6, where 3–4 is a side branch connecting the trunk airway to the branch. In a normal ventilation network, this branch air flow flows from 3 into 4 and then joins the trunk airway. When a fire has occurred to a certain extent, the effect of the catastrophe will cause a disturbance in the air flow field and the branch 3–4 may be reversed so that the branch air flow flows from 4 to 3 and then joins the other branch and exits. After the fire, if the side branch does not reverse, the fire will only affect the left main branch 4–5–6. If the side branch reverses, harmful gases will enter the other branch 3–5–6 along the branch 4–3, resulting in a wider area of fire impact and a greater disaster.

2.1.2 Smoke Return

Under the buoyancy or throttle effect respectively (depending on the inclination of the roadway), and the influence of temperature and pressure gradients in the longitudinal and cross-sectional directions of the roadway, fresh air flow continues

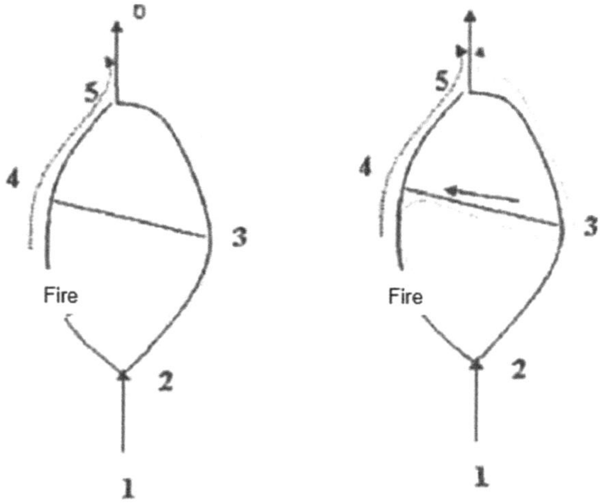

Fig. 2.1 Airflow (smoke) inverting diagram

to supply air along the upwind side of the fire origin of the ignition roadway while smoke continues to flow out along the bottom of the roadway. Air flow return may occur on the ignition roadway and its connecting trunk airway. Mine fire's smoke return is shown in Fig. 2.2.

As shown in Fig. 2.3, clean air flow in a normal ventilation network flows out of the network via trunk airways 1–2–4–5–6 and 1–2–3–5–6. When a fire breaks out, assuming that the fire origin is on branch 2–4, if the fire is rapid and the smoke generation is high, the downwind side of the fire origin is blocked from exhausting smoke and the smoke generated at the fire origin exits along the return air system 4–5–6 of the trunk airway. On the other hand, it fills the whole section of the roadway and flows against the smoke of the main airway to node 2 into branches 2–3 and 3–5, which causes the return smoke to attack a larger area with the air flow, thus expanding the hazard.

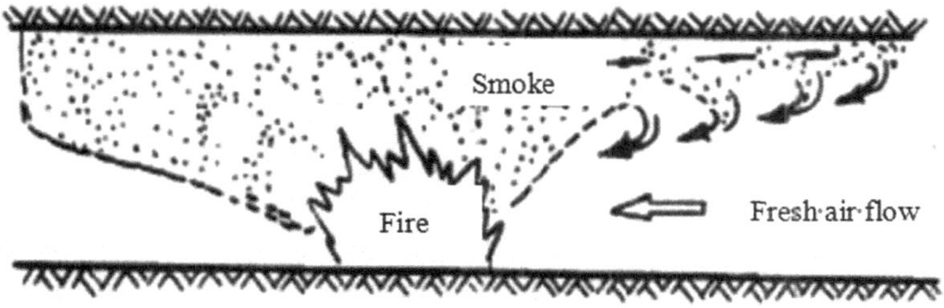

Fig. 2.2 Mine fire's smoke return diagram

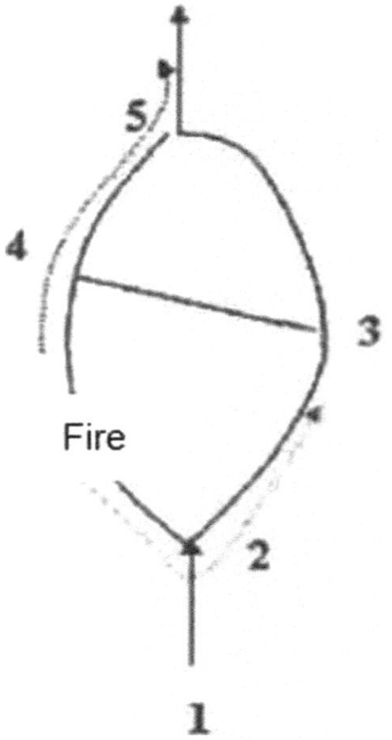

Fig. 2.3 Smoke return diagram

2.1.3 Smoke Rollback

Under the influence of the throttle effect on the downwind side of the fire origin as well as the temperature and pressure gradient at the roadway section, while the fresh air flow flows into the fire origin along the bottom of the roadway in the original wind direction, the smoke generated by the fire origin flows backwards along the top of the upwind side of the roadway and rolls back towards the fire origin. Under certain conditions, this phenomenon may also occur on the downwind side.

Reversal is the unidirectional flow of the same fluid, return is the an-isotropic flow of different fluids (smoke and fresh air flow), rollback is the an-isotropic flow of both fresh air flow and smoke in the same section, and the an-isotropic flow of the same fluid caused by the rollover of smoke. Rollback is the precursor to both return and reversal. Reversal is a precursor to the occurrence of return and reversal.

Figure 2.4a shows the change in the distribution of the heat effect force in the road, uphill and downhill ventilation. The high temperature smoke generated by the fire origin rises to the top of the roadway under the buoyant effect as well as flows up and downwind side respectively, where the downwind side flows easily, while the upwind side is more difficult to flow back upwind. When the smoke rises and dissipates to the main downwind side, a low pressure area is formed in the lower part of the fire origin, and the fresh wind from the upwind side flows into the fire origin along the bottom of the roadway to replenish it. Under certain conditions, smoke from the downwind side may also rollback into the fire origin.

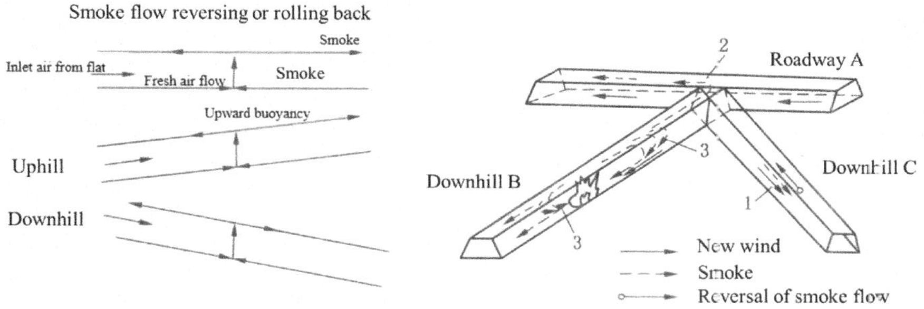

(a) Variation of heat distribution in different dip angle alleyway

(b) Schematic diagram of turbulent wind flow conditions near the fire source

Fig. 2.4 Airflow disorder caused by mine fire

As shown in Fig. 2.4b, air flow enters downhill B and C via roadway A. Downhill B catches fire, rollback smoke appears on the upwind side of the fire origin, and smoke returns as the fire strengthens, causing smoke to enter roadway A and downhill C. Under certain conditions, air flow in downhill C also appears to be reversed. The phenomenological characteristics of the turbulent changes in the air flow state of the mine are specified in Table 2.1.

Table 2.1 Airflow disorder phenomena during mine fire

Forms of disorder	Location of occurrence	Cause of occurrence	Representation
Air flow reversal	Mainly found in bypass branches	Reverse heat wind pressure > its forward mechanical wind pressure in the side branch	Smoke flowing in the opposite direction of the original air flow, generally full section reversal
Smoke return	Ignition roadway (trunk airway), upward, road ventilation; downwards ventilation	Fire origin of downwind side of throttle effect, reverse heat wind pressure + roadway cross section temperature, pressure gradient effect	Upwind side of fire origin with different fluids flowing in opposite directions in the same roadway section
Smoke rollback	A few upwind sides of the ignition roadway's fire origin can occur on the downwind side	Throttle effect on the downwind side + temperature and pressure gradient effects on the roadway cross-section	Both an-isotropic and smoke reversal flow in the upwind side (and sometimes in the downwind side) of the roadway section of the fire origin

2.2 Hazards of the Airflow Disorder Phenomenon

2.2.1 *Reduced Airflow*

The reduction in roadway airflow may not be a threat in mines where there is no gas or a small gas surge. However, in mines with large gas surges, explosive gas mixtures may form and pose a potential explosion hazard. Especially when the explosive gas mixture passes through the fire zone, it can easily cause a gas explosion.

2.2.2 *Air Flow Reversal*

Air flow reversal causes disturbances in the flow of air, which can make evacuation and relief work more difficult and dangerous

(1) Reversal air flow carries a large amount of toxic and harmful gases, spreading to a larger area and even contaminating the incoming air area, extending the affected area and even threatening the whole mine.
(2) Air flow reversal goes through the process of air reduction—air stoppage—air reversing. During the wind reduction and stoppage phases, the gas concentration in the air flow increases due to the drastic reduction in air volume and the reduction in air velocity creates conditions for the formation of localized gas accumulation.
(3) Air flow reversal increases the likelihood that volatile-rich air flow from the downwind side of the fire origin or foul air from local gas accumulation zones will re-enter the fire zone, thereby increasing the likelihood of an explosion. This is the reason why combustible gas explosions can also occur in metallic and non-metallic mine fires.

2.2.3 *Smoke Return*

The smoke return poses a direct threat to direct firefighters on the upwind side of the fire origin. As smoke mixes with the incoming air and re-enters the fire origin, a gas explosion may be induced under certain conditions. Smoke return causes smoke to enter other roadways, which may have similar results to air flow reversal.

2.2.4 *Smoke Rollback*

The rollback phenomenon causes smoke from the upwind side of the fire origin to mix with fresh air flow and then flow back into the fire origin. Under certain conditions, this can lead to a gas explosion. The smoke rollback also poses a direct threat to firefighters on the upwind side of the fire origin.

Therefore, maintaining the stability of the air flow and especially the wind direction during the mine fire period is one of the most important tasks in the relief work.

2.3 Effect of Fire on Air Flow Status in Different Roadways

In order to simplify the analysis and make it easy for the reader to understand, only a qualitative analysis is presented here to investigate the consequences of the buoyancy, throttle force and mechanical forces acting on the air flow direction in the road, uphill and downhill.

2.3.1 Horizontal Roadway Fire

Fires occurring in the horizontal roadway, the fire zoning diagram is shown in Fig. 2.5. A is the roadway inlet air section, D is the roadway outlet air section, and the fire occurs at P. B is the front face of the burning zone, and C is the rear face of the burning zone. The firing roadway can be divided into three zones-fresh air flow zone (A–B), fire burning zone (B–C), fire smoke zone (C–D). If you ignore the influence of temperature changes in the adjacent tilted roadway, it is generally believed that there is only a throttling effect, no buoyant effect. Throttling increases the resistance to air flow flow, the result of which leads to ignition roadway air volume reduction (air volume reduction of up to 30%). The throttle effect is related to the size of the burn and the size of the air velocity. When the burn size is less than 250 kg (wood), or the wind speed is less than 1 m/s, the throttle effect is not obvious. When the size exceeds 4500 kg (wood), there will be a significant throttle effect. In the case of a road fire, the ignition roadway and the roadway in series with it will not experience air flow reversal. But the wind direction in the corner roadway between the two parallel roads may reversal.

2.3.2 Upward Ventilation Roadway Fire

The fire in the upward ventilation roadway produced both buoyancy and throttling effects. As shown in Fig. 2.6a, the uphill airflow is increased due to the buoyant effect of the buoyant effect being greater than the wind reducing effect of the throttle effect. If there is sufficient fuel, the increased wind will increase the oxygen supply and enhance the fire, making the heat wind pressure effect stronger. As a result, uphill fire winds generally do not reversal, and the increase in uphill winds is accompanied

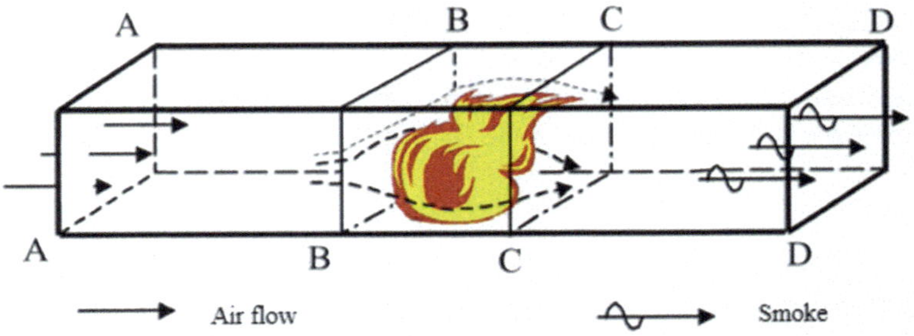

Fig. 2.5 Schematic diagram of the horizontal roadway fire zone

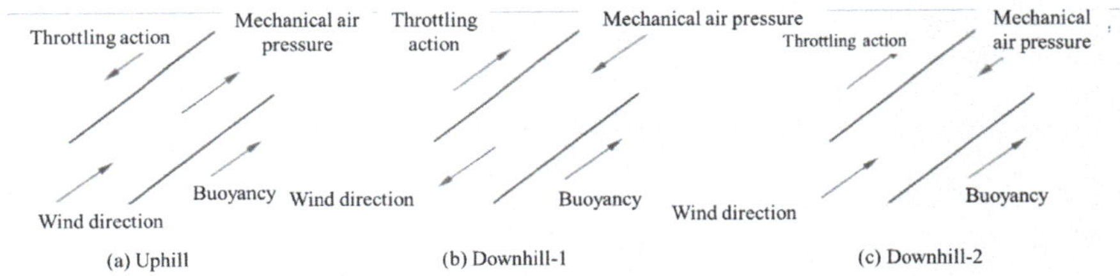

Fig. 2.6 Variation of wind direction in the roadway under buoyancy and throttle force

by a reduction in adjacent parallel roadway winds. If the original wind pressure in the adjacent parallel roadway is low, air flow stagnation or reversal in the adjacent uphill may occur. If the air flow in the adjacent parallel roadway is to be kept stable, the air intake side of the firing roadway should be increased to reduce the oxygen supply to the fire zone and to control the fire; or the ventilation pressure of the parallel roadway should be increased, such as increasing the fan air pressure or reducing the pressure of the inlet and return roadway in series with the parallel roadway, so as to increase the amount of fan wind pressure allocated to the parallel roadway.

2.3.3 Descentional Ventilation Roadway Fire

As shown in Fig. 2.6b, the descentional ventilation roadway fire produced throttling and buoyant effect with the original ventilation pressure acting in the opposite direction of the wind, which tends to reduce the air volume of the roadway, and even appear air flow reverse, smoke return phenomenon. In the case of reduced air flow, the oxygen supply decreases and the fire weakens, thus weakening the throttling and buoyant effect produced by the fire. There is a tendency to increase the air flow again. As shown in Fig. 2.6c, the descending air flow becomes upward in the air flow reversal case. During the reversal period, the air volume is generally much smaller than the normal downward air volume. Because the ascending air flow at this time is different from the normal ascending ventilation roadway situation, and its buoyant effect has to overcome the effect of the wind pressure and throttle effect of the fan acting in that roadway. The buoyant effect is not sufficient to overcome the fan pressure and throttle effect as well as causes the air flow to reverse downwards again. This frequent change in air flow rate and direction of flow occurs when a downwards ventilation roadway is on fire, unless the descentional ventilation roadway has a small difference in elevation. Or there is insufficient fuel, or the air pressure in that roadway is very high. Then, this phenomenon is significant and often results in air flow rate and even dramatic fluctuations in the direction of flow. Therefore, in the case of a fire in downhill, the wind direction is likely to be reversed and there may be frequent changes in wind direction, which is something to be particularly aware of when responding to a disaster. The ability to maintain a continuous direction of air flow reversal after a fire depends on several factors.

(1) Effect of fire origin in downhill position

The fire origin should have sufficient elevation difference on both sides. After the air flow reversal, the fresh air goes upwards and generates hot smoke directly into the top road, which does not generate enough natural air pressure to overcome the ventilation pressure. The air flow may revert back to the original direction of downward flow.

(2) Effect of oxygen concentration in smoke

The oxygen concentration in smoke, especially on the downwind side of a very large fire origin, is much lower than in fresh air. In the case of a reversal of the wind direction caused by a large fire, the incoming air from the fire origin is replaced by a return smoke with a low oxygen concentration, which inevitably weakens the fire. If the section of smoke with a low oxygen concentration returning to the fire origin is sufficiently long, it will reduce the fire heating air pressure generated by the fire origin over a longer period of time than is necessary to overcome the ventilation pressure, making it impossible to maintain this state of air flow reversal.

(3) Effect of original ventilation wind pressure

A continuous air flow reversal may be maintained if the descentional ventilation roadway has low wind pressure and the fire develops rapidly, reducing, stagnating and reversing the downward wind speed.

(4) Effect of fire size

The fire origin of the fire in the descentional ventilation roadway is so large that the buoyancy effect is greater than the combined effect of the mechanical force and the throttle force, so the wind direction is less likely to be reversed after the reversal.

(5) Effect of blending fresh air flow

When a fire lasts for a long time, pressurized air from a damaged pressurized air duct or air flow from another roadway mixes with reversal air flow on the downwind side of the fire origin, which increases the oxygen concentration in the air flow reversing into the fire origin, resulting in a higher fire heating air pressure. heating air pressure. It may then maintain the air flow reversal condition.

Fires in the descentional ventilation roadway are generally not as strong as those in the upward ventilation roadway. Because the upward and throttle effect of a fire in the descentional ventilation roadway reduces the airflow and oxygen supply by resisting the ventilation pressure.

2.4 Approximate Calculation of the Throttle Effect and the Buoyant Effect

The fire heating air pressure is formed during a mine fire due to the combined effect of the throttle effect and the buoyant effect. The fire heating air pressure is the thermal wind pressure caused by the fire, which can produce a strong additional wind pressure. It is also known as fire negative pressure or thermal negative pressure. The combined effect of fire on roadway wind pressure is obtained below by approximating the throttle effect and the buoyant effect respectively.

2.4.1 Calculation of Throttle Effect

The pressure drop Δh_L due to the throttle effect of air flow through a roadway is calculated as below

$$\Delta h_L = h_{La}(F^2\frac{T_m}{T_a} - 1) \tag{2.1}$$

In the Equation:

h_{La}—pressure drop from air flow flow before the fire, Pa.

T_m—the average absolute temperature of that roadway air flow after the fire, K.

T_a—the average absolute temperature of that roadway air flow prior to the fire, K.

F—coefficient of increase of air flow mass after a fire.

2.4.2 Calculation of Buoyant Effect

The pressure drop hN due to the buoyant effect of the air flow through the roadway is calculated as below.

$$h_N = g\rho_a\left(1 - \frac{T_a}{T_m}\right)L\sin\beta \tag{2.2}$$

In the Equation:
ρ_a—density of air flow before the fire, kg/m^3.
L—the length of the roadway, m.
β—the angle of inclination of this roadway, the original air flow direction is positive in the upper row and negative in the lower row.

Combining (2.1) and (2.2), the combined effect of the buoyant effect and the throttle effect on the change in pressure drop across this roadway air flow h_T:

$$h_T = h_N - \Delta h_L = g\rho_a\left(1 - \frac{T_a}{T_m}\right)L\sin\beta - h_{La}\left(\frac{F^2T_m}{T_a} - 1\right) \tag{2.3}$$

A positive value of h_T tends to increase the airflow; a negative value of h_T tends to decrease the airflow.

2.4.3 Example of Calculation

(1) Questions

Roadway section is S = 8 m^2, perimeter is U = 12.4 m, roadway length is L = 100 m, airflow is Q = 10 m^3/s, $\alpha = 0.01$ kg/m^3. Air flow density before the fire is $= \rho_a = 1.2$ kg/m^3, roadway inclination is $\beta = 60°$, average absolute temperature of the air flow before the fire is $T_a = 20$ °C. The average absolute temperature of the roadway after the fire is $T_m = 300$ °C, and the increase in air flow mass after the fire is F = 1.2. When this roadway is uphill, road and downhill (Fig. 2.7), the increase or decrease of the throttle effect and the buoyant effect on the air flow flow pressure drop is calculated.

(2) Solution

Roadway wind resistance:

$$R = \alpha UL/S^3 = 0.0242\text{kg/m}^7$$

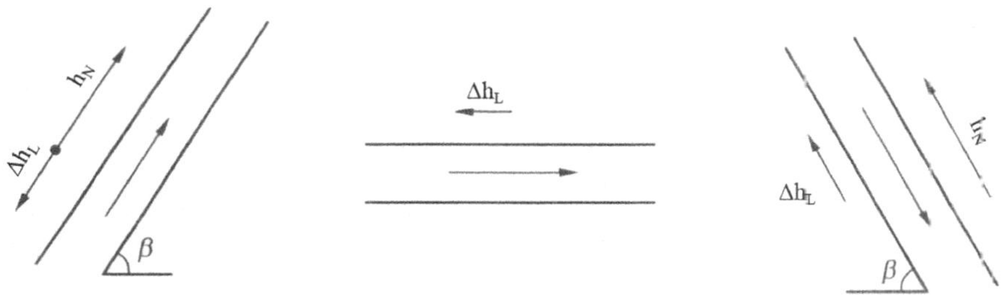

(a) Uphill (b) Flat roadway (c) Downhill

Fig. 2.7 Effect of throttling and buoyant effect on pressure drop variation in roadway air flow

Pre-fire roadway pressure drop:

$$h_{\mathrm{La}} = RQ^2 = 0.0242 \times 102 = 2.42\mathrm{Pa}$$

When the angle of the roadway is 60°, the combined effect of uphill wind-fire heating air pressure is as below.

$$h_{\mathrm{T}} = 9.8 \times 1.2 \times (1 - 293/573) \times 100 \times \sin 60° - 2.42 \times \left(1.2^2 \times 573/293 - 1\right) = 493.27\mathrm{Pa}$$

When the angle of the roadway is 60°, the combined effect of fire heating air pressure from downhill winds is as below.

$$h_{\mathrm{T}} = 9.8 \times 1.2 \times (1 - 293/573) \times 100 \times \sin(-60°) - 2.42 \times \left(1.2^2 \times 573/293 - 1\right) = 502.01\mathrm{Pa}$$

When the roadway is road, there is no buoyant effect, only a throttling effect.

$$h_{\mathrm{T}} = 2.42 \times \left(1.2^2 \times 573/293 - 1\right) = 4.39\mathrm{Pa}$$

(3) Analysis of results

① From Eq. (2.3), the value of the $(1 - T_{\mathrm{a}}/T_{\mathrm{m}})$ term is generally 1–1/5 of $(F^2T_{\mathrm{m}}/T_{\mathrm{a}} - 1)$, and they are in the same order of magnitude, while $g\rho_{\mathrm{a}}\ L\sin\beta$ is often more than 100 times the value of h_{La}, so the buoyant effect is much greater than the influence of the throttle effect. The greater the angle of inclination of the roadway, the greater the difference between the effects of the two effects. The direction of the throttle effect is always opposite to the direction of the air flow, while the direction of the buoyant effect depends on the direction of the smoke and the environmental conditions.

② In the ascending ventilation roadway, the buoyant effect acts in the same direction as the fan wind pressure and in the opposite direction to the throttle effect, with a general tendency to increase the air volume.

③ On road, the buoyant effect on the air flow flow pressure drop can be ignored and only the throttle effect exists.

④ In the downwind roadway, the throttle effect works in conjunction with the buoyant effect to resist the fan wind pressure. The tendency of their action is to reduce the air volume.

⑤ As T_{m} increases, both $(1 - T_{\mathrm{a}}/T_{\mathrm{m}})$ and $(T_{\mathrm{m}}/T_{\mathrm{a}} - 1)$ increase, but the latter increases more rapidly. Therefore, the difference between the throttle effect and the buoyant effect decreases slightly as T_m increases.

2.4.4 Other Calculation Methods

There are a number of ways to calculate the buoyant effect and the throttle effect, each of which is listed below and compared with the results of the above example.

Related literature proposes a formula for the post-fire roadway equivalent R_{equ}:

$$R_{equ} = \frac{h_{Lm}}{Q_a^2} = F^2\left[R_a\frac{T_m}{T_a} + \frac{\rho_a}{2S^2}\left(\frac{T_m}{T_a} - \frac{1}{F^2}\right)\right] \tag{2.4}$$

In the Equation:

h_{Lm}—the total value of the change in wind resistance of the roadway after a fire, due to the buoyant effect and the throttle effect.

From the example it follows that:

$$R_{equ} = 1.2^2\left[0.0242\frac{573}{273} + \frac{1.2}{2*8^2}\left(\frac{573}{273} - \frac{1}{1.2^2}\right)\right] = \frac{0.085\text{kg}}{\text{m}^7}$$

For road:

$$\Delta h_T = h_{Lm} - h_{La} = \left(R_{equ} - R_a\right)Q_a^2 = (0.085 - 0.0242) * 100 = 6.08\text{Pa}$$

For the uphill roadway, the empirical formula commonly used in the Federal Republic of Germany to calculate the natural wind pressure is as below.

$$\text{H}_\text{n} = 1.95 * \Delta\text{Z}/100\,(T_m - T_a) \tag{2.5}$$

From Eq. (2.5):

$$h_\text{n}=1.95 \times \frac{100 \times \sin 60°}{100} \times (573 - 293)=506.62\text{Pa}$$

2.5 Prevention and Control of Airflow Disorder and Examples

Once an accident has occurred in a coal mine, especially a major explosion or fire, one of the most urgent tasks is to effectively control the underground air flow and prevent airflow disorder from occurring. If airflow control measures are reasonable, they can not only effectively limit the uncontrolled spread of smoke in the shaft and prevent the scope of the disaster from expanding, but also minimize the damage and avoid casualties.

2.5.1 Control Methods of Air flow

Common methods of air flow control include maintaining normal ventilation and stabilizing air flow; stopping fans; air reversing; air flow short-circuiting; etc.

(1) The conditions for using the method of maintaining normal ventilation and stabilizing air flow are as below. The fire origin is located in the mining section, the smoke has spread over a large area and the personnel are widely distributed; the ventilation network is complex for high gas mines; the fire origin is located in the sole head roadway and the local fan cannot be stopped; the fire origin is located in the main return duct.
(2) The conditions for using the stop fan method are as below. The fire origin is located in the inlet shaft or shaft and cannot be fed into air reversing; the fire in the lone head roadway is long and the gas concentration exceeds the upper explosion limit; the fan becomes the ventilation resistance.
(3) The conditions for the use of the air reversing method are as below. The use of air reversing equipment and facilities to change the direction of the fire smoke so that the downwind side of the fire origin personnel are in the fire origin fresh air flow in the whole mine air reversing, regional air reversing, localized air reversing.
(4) The air flow short circuit method is used when the fire origin is located in the main air intake system of the mine and cannot be air reversing in time or cannot be air reversing.

Therefore, different forms of airflow disorder correspond to different prevention and control measures, which are shown in Table 2.2.

2.5.2 Airflow Disorder Control Measures for Roadway Fires

Mine fires occur in a variety of locations, including shaft and borehole fires, underground fires, fires in underground chambers, ventilation roadway fires, fires in extraction workings and solo roadway fires. Depending on the location of the fire, there are different methods of fire suppression and control. The following is a list of measures to control and extinguish a fire in a ventilation roadway that results in airflow disorder

(1) In the event of a fire in the inclined air inlet roadway, measures such as air flow short-circuiting, local air reversing and area air reversing can be taken to prevent the intrusion of harmful gases into the workplace.
(2) Fires occurring on the inclined upward return air flow roadway, the normal air flow direction must be maintained and the air supply reduced as appropriate.
(3) When a fire occurs on an inclined descending air flow roadway, it is necessary to reduce the return air resistance and prevent air flow reversal, but never to stop the fan operation.

Table 2.2 Measures to prevent and control the phenomenon of airflow disorder

Forms of airflow disorder	Prevention and treatment measures
Air flow reversal occurs in the bypass branch of ascending air flow	① Hanging wind curtains on the air intake side of the fire origin, constructing temporary confinement and taking direct fire fighting measures to stop the development of the fire, thus reducing the local fire heating air pressure in the inner part of the system and at the same time increasing the wind resistance of the inner part of the system ② To keep the main fan of ventilation in the burdened fire area in normal operation, never stop the operation of the main fan, let alone put down the main fan gate ③ Maximizing the air pressure in the outer part of the system by adjusting the fan blades ④ To reduce the wind resistance of the outer part of the system by lifting the gates in the main fan duct, opening the regulating dampers on the exhaust path, preventing localized roofing accidents in the exhaust path and maintaining the return air system roadway free from blockages
The main airway of the descending air flow undergoes air flow reversal	① If the fire occurs on the air intake side, air reversing shall be carried out at the beginning of the fire by means of a regulating fan and a dedicated roadway for air reversing, at least with the main fan stopped ② If the fire occurs on the return air side, you should ensure that the roadway on the return air side is not blocked; adjust the fan to increase the air pressure of the fan and try to make the fan pressure greater than the local fire heating air pressure generated by the fire
Return of smoke in ascending air flow airway	① Reduce the air supply and control the development of the fire ② Do not take measures to stop fan or reduce pressure easily, and at the same time increase the wind resistance of the bypass branch as appropriate to increase the main fan to the main airway ③ If the tendency for smoke return from the main airway is unavoidable, find a short-circuiting path on the air intake side of the fire origin to discharge the returned smoke back into the return air system, if possible
Return of smoke in descending air flow airway	The return of smoke in the descending air flow airway is a precursor to air flow reversal in the main airway of the descending air flow. Therefore, in addition to measures to prevent air flow reversal in the main stem of the descending air flow, the smoke that has already returned is mainly discharged back into the return air system by means of short-circuiting and local air reversing

(4) When fighting fires from the bottom up in an inclined roadway, protective cranes, protective partitions and other body protection should be provided to prevent injuries from falling objects.

(5) In the case of fire fighting in the inclined roadway, the middle roadway, small chute, contact roadway and pedestrian roadway should be used to approach the fire origin. If the fire origin cannot be approached, the mine car or skip can be used to lower the water sprinkler into the roadway to extinguish the fire, or fire the high-capacity foam to extinguish the fire from a distance.

(6) In the event of a fire in a road, stone gate and other horizontal roadway located in the main intake duct of a mine or a wing, the most effective ventilation method to

save lives and extinguish the fire is to be selected. The ventilation methods that can be adopted include air reversing, air flow short-circuiting, area air reversing in a multi-wind shaft and normal ventilation. When short-circuiting ventilation is used to prevent a fire from expanding, it is important to ensure that the fire's hazardous gases do not reversal.

(7) When fighting fires in the horizontal roadway of the mining section, normal ventilation is generally maintained, or in the case of coal mines, the air supply to the fire area is increased or reduced depending on the gas situation.

2.5.3 *Example of Airflow Disorder Prevention and Control*

A mine in Jiangsu uses vertical shafts with multiple levels of development, with production levels of −300, −700 and −1260 m levels and −520 m as an auxiliary level. The old main shaft and secondary shaft are up to −300 m level and the mixed shaft is up to −700 m level. The ventilation method of the mine is extracted and the ventilation method is central parallel, with the secondary shaft and the mixed shaft feeding the air as well as the central shaft returning the air.

During the mine's annual air reversing exercise, there was an unusual situation where the air flow of a key branch (the roadway) did not reverse, even though the air reversing facilities, air reversing airflow and other air reversing conditions were all met. Through field observations, the air flow direction of the north limb extension transport downhill was not reversed in the lower section and reversed in the upper section at the beginning of air reversing. As the air reversing time increased, the air flow of the whole roadway was not reversed. The sketch diagrams of the ventilation system during normal ventilation and air reversing are shown in Figs. 2.8 and 2.9. The network diagram of the ventilation system during air reversing in the mine is shown in Fig. 2.10 (the solid arrows in the diagram are the actual air flow direction and the dashed arrows are the theoretical air flow direction).

By analyzing the ventilation system in Figs. 2.8 and 2.9, it can be seen that the reason why the air reversing of the downhill air flow in the north limb extension is not reversed may be due to the small pressure difference between the front and back of the branch 5–4 and the air density difference between the north limb extension transport downhill and the north limb extension track downhill. When the differential pressure between the front and rear of the dampers becomes small, it often results in the lower section of the north limb extension transport downhill not reversing the air flow and the upper section reversing the air flow, i.e. the direction of the lower section air flow is 4→8 and the direction of the upper section air flow is 4→3. The difference in air density between the north limb extended transport downhill and the north limb extended track downhill can lead to the north limb extended transport downhill not reversing the direction of air flow throughout the roadway.

The following control measures are proposed.

(1) Before air reversing, check the main ventilation facilities in the mine to ensure that the ventilation adjustment facilities between the inlet and outlet roadways

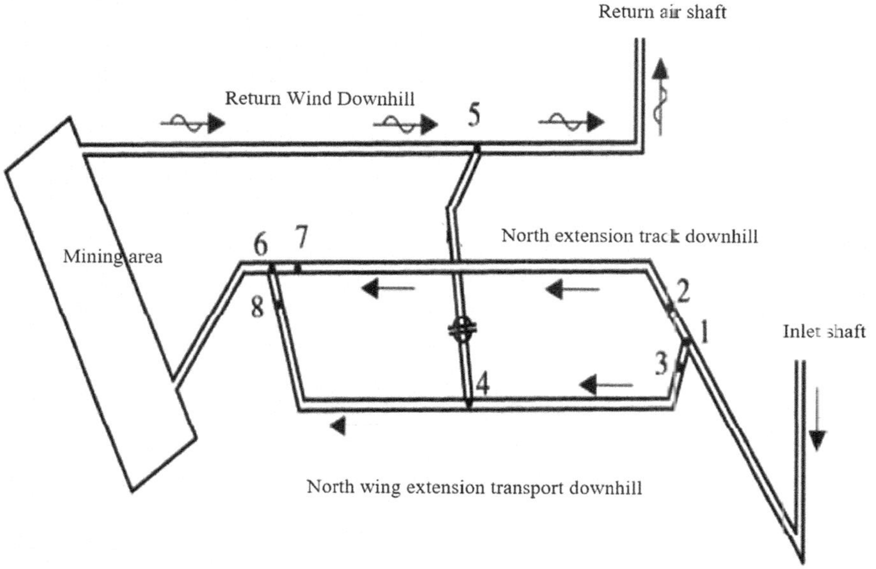

Fig. 2.8 Sketch of the normal ventilation system in a mine

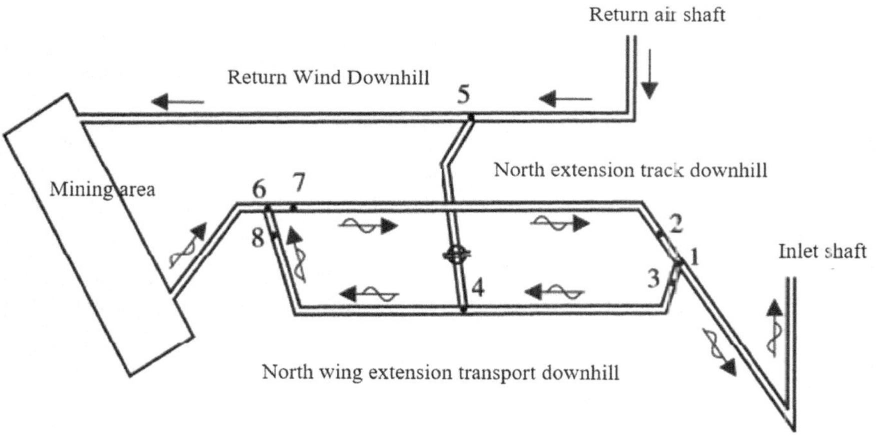

Fig. 2.9 Sketch of the ventilation system during air reversing in a mine

are intact, and plug the large air leakage facilities to reduce the amount of air leakage during air reversing.

(2) Reduce the temperature of the return air flow during air reversing. Install cooling facilities at hot operating locations down-hole to lower the temperature of the

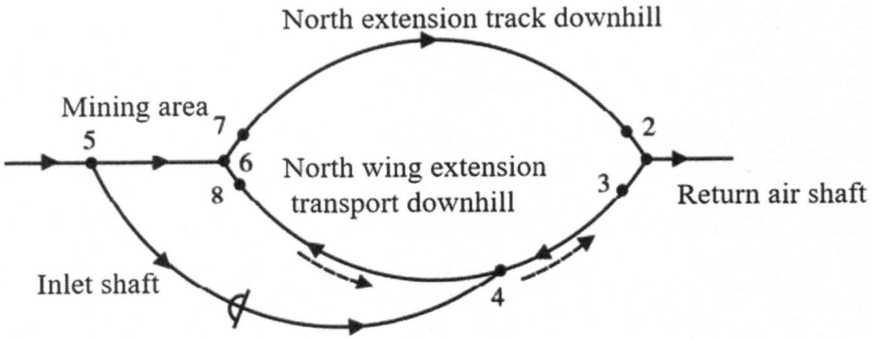

Fig. 2.10 Mine ventilation network diagram during mine air reversing

air flow and reduce the temperature difference between the north limb extension transport downhill and the north limb extension track downhill on the two roadways.

(3) Increase the number of liaison roadways between the north limb extension transport downhill and the north limb extension track downhill to reduce the temperature difference between the two roadways and ensure that the air density in the two roadways is basically the same.

2.6 Airflow Disorder Patterns and Examples during Fires

A simplified schematic diagram of a ventilation system for a mine is shown in Fig. 2.11. When the fire occurs in the uphill of the mining section, i.e. in the upstream airway a, a local fire heating air pressure hi is generated due to the high temperature smoke flowing through the upstream airway. Under the action of the local fire heating air pressure, the pressure distribution relationship of the original wind network will be changed. Since the action direction of the local fire heating air pressure is the same as the main fan, the wind direction of the main airway 1–2–a–3–4 will generally not be reversed; while the air flow direction of the bypass branch 2–b–3 may be reversed. In order to examine the wind direction variation pattern of the bypass branch, as shown in Fig. 2.12, the whole ventilation system is divided into two parts: the inner part of the system (denoted by the subscript i) and the outer part of the system (denoted by the subscript 0). h_0 represents the sum of the fire heating air pressure and the main fan pressure occurring in the outer part of the system; R_i and R_0 represent the synthetic wind resistance of the inner part of the system and the outer part of the system, respectively.

Using Kirchhoff's circuit theorem for networks, the discriminant for b-branch air flow reversal can be established as $\frac{h_i}{h_0} < \frac{R_i}{R_0}$. In the event of a fire in the descending air flow, the direction of the fire heating air pressure is opposite to the direction

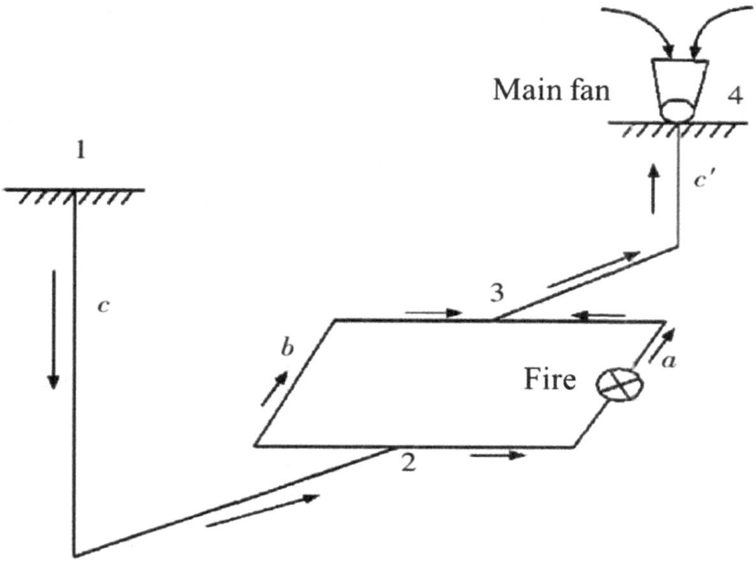

Fig. 2.11 Schematic sketch of the ventilation system for ascending air flow in a mine

Fig. 2.12 Closed loop diagram for ventilation system

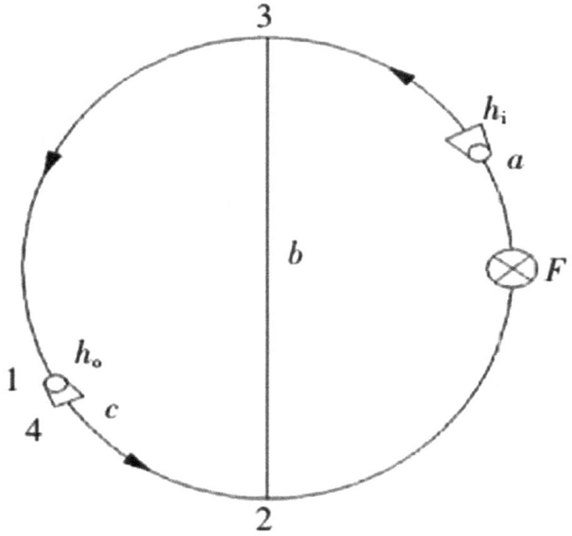

of the main fan air pressure, and the situation in the down-hole airflow disorder is fundamentally different from that of a fire in the ascending air flow.

2.6.1 Case of Air Flow Reversal in a Fire Side Branch of Ascending Air Flow

[Case 1] Fire in the uphill of the mining section of a mine

In one mine, due to weak daily management of mine ventilation and a lack of awareness of the changing pattern of air flow during a fire and improper handling of the situation, the mining section on the upwind side of the fire area, away from the origin of the fire, was attacked by smoke, resulting in the death of 25 people.

As shown in Fig. 2.13, fire origin F was located in the lower part of uphill F-6 in the mining section, and the old fire area F_0 reignited, igniting the wooden supports of the uphill in the adjacent mining section, which generated a high local fire heating air pressure hi the inner part of the system. The dampers in road 5-F were previously removed for track repair. This reduces the air resistance R_i in the inner part of the system branches 3–4–5–6–7 and soon satisfies the condition (2.6). In this case, the air flow in branch airway 3–7 is bound to reversal. The air flow reversal condition is particularly favourable because of the poor maintenance of the return airway and the high air resistance R_0 of the outer part of the system. After the air flow reversal in jack shaft 3–7, the fire smoke entered the stone door from point 3. As a result, almost the entire mine shaft was exposed to fire smoke, except for the air flow that split from point 2.

$$\frac{h_i}{h_0} > \frac{R_i}{R_0} \tag{2.6}$$

In most cases, normal mine ventilation should be maintained in the event of a fire to stabilize the air flow and provide a reliable guarantee of fire extinguishment, while also protecting underground operations [1].

[Case 2] A fire at the working face of the western mining section of a mine

As shown in Fig. 2.14, a coal spontaneous combustion fire occurred at the working face of the mining section in the western part of a mine. After the fire had occurred, several containment walls were constructed in the inlet airway to the fire area (T_1 in Fig. 2.14), during which no airway air flow reversal occurred. In order to fully enclose the fire area, it was decided to construct a containment wall on the return air side of the fire area and to choose to enter the return air side of the fire area via the 5–6 road. Road 5–6 had a dampener D to isolate the air flow. But when dampener D was opened, the 2–4 road air flow reversed, resulting in the sudden appearance of fire smoke at the bottom of the inlet shaft, which then spread throughout the mine.

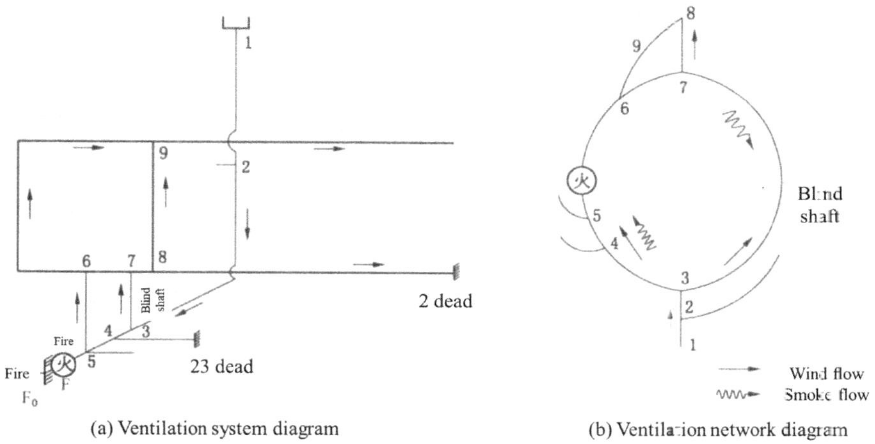

Fig. 2.13 Mine air flow reversal and fire smoke diffuse ventilation system diagram

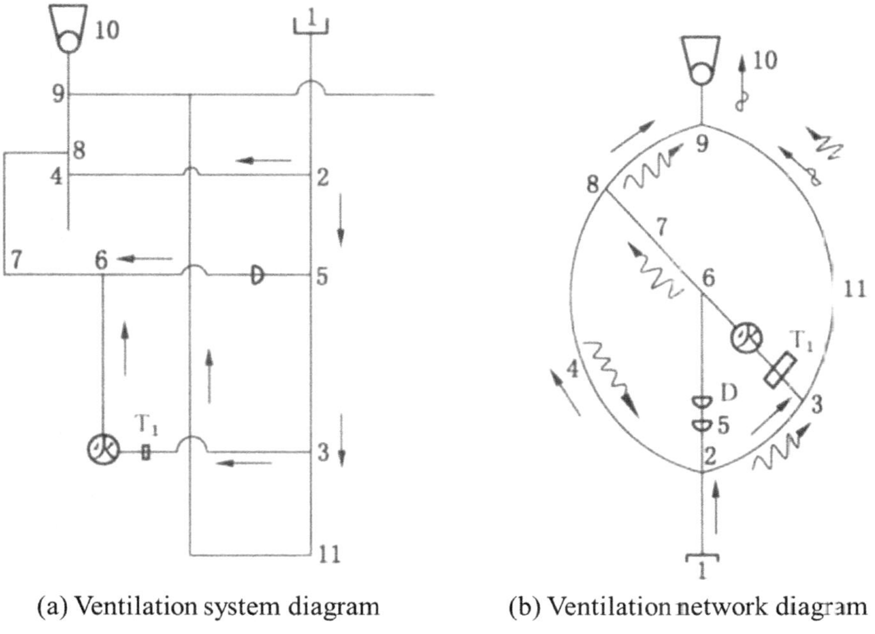

Fig. 2.14 Diagram of a mine air flow reversal with fire and smoke filled shafts

As can be seen from Fig. 2.14, the airtight wall T_1 constructed on the inlet side of the fire zone increased the wind resistance R_i of the internal system of airway 2–5–6 and air flow 2–5–6 failed to reversal. At the same time, air flow 2–4–8 also failed to reversal. Because the other branch 5–6, which forms the inner part of the system, is also damped, it creates a larger air resistance R_i in the inner part of the system. However, when the dampers D were opened, the air resistance R_i of the inner part of the system suddenly decreased, so that under the action of the local fire heating air pressure hf, branch air flow 2–4–8 reversal occurred, thus causing the accident of fire and smoke filling the mine.

When adjusting the ventilation system, someone familiar with the ventilation system should be responsible for ensuring that the ventilation system is stable and that the air flow is not disturbed after the ventilation system has been adjusted [2].

[Case 3] Fire accident in a coal mine at uphill

An exogenous fire occurred in a coal mine at uphill 3–4 near node 4. The heat wind pressure generated by the fire caused air flow reversal. Figure 2.15 shows a diagram of the ventilation system in the mine area. Figure 2.16 shows its corresponding ventilation network.

The air flow reversed along airways 8–7–5–2, 7–6–3 and 6–5 respectively (reversal air flow is indicated by dashed lines), resulting in the death of 48 miners. As the fire originated and spread rapidly towards node 3, the above airway winds recovered, but the air flow reversed along airway 4–9–1.

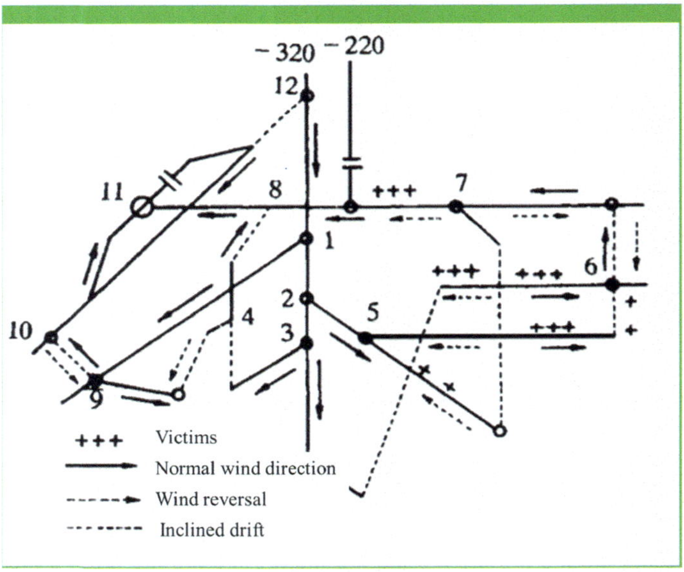

Fig. 2.15 Diagram of the ventilation system during a fire

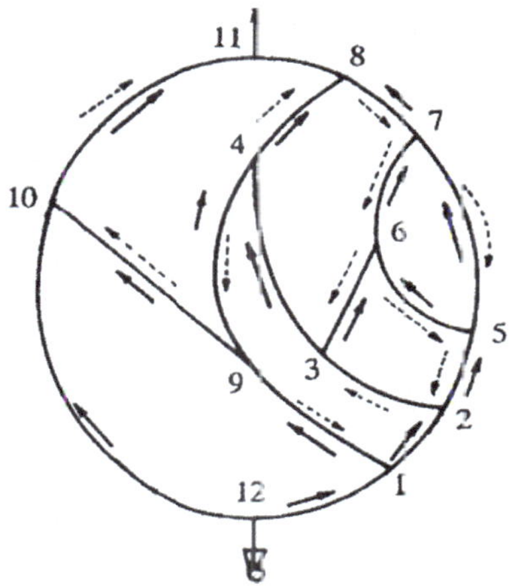

Fig. 2.16 Ventilation network diagram for the fire period

The above fire is divided into two phases. The first stage of the fire closed loop diagram is shown in Fig. 2.17 and the second stage of the fire closed loop diagram is shown in Fig. 2.18.

The first stage is when the fire origin is at the top of the uphill 3–4 roadway (close to node 4) and the second stage is when the fire origin spreads to the lower part

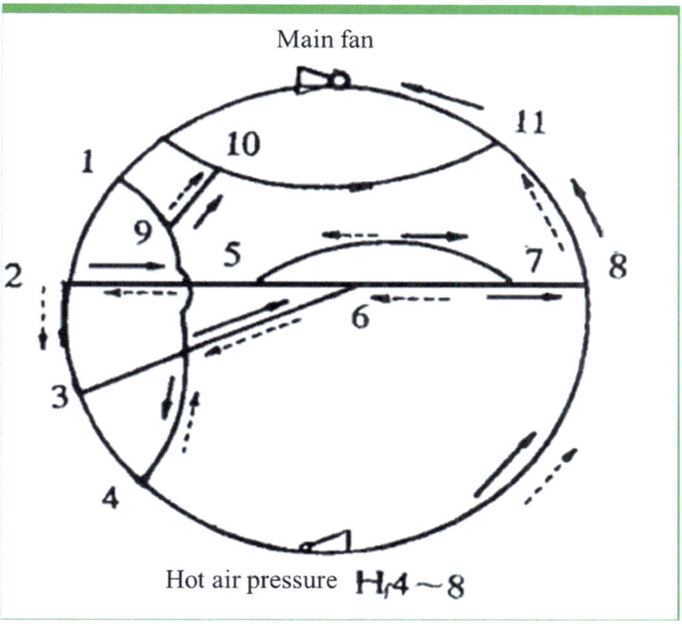

Fig. 2.17 Closed loop diagram of the first stage of the fire

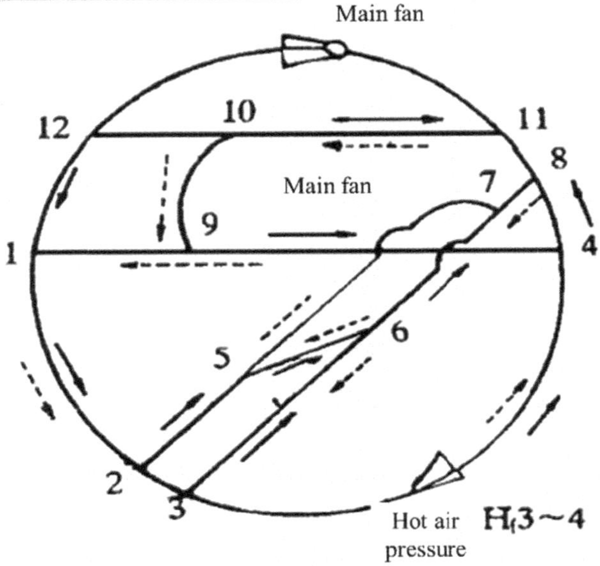

Fig. 2.18 Closed loop diagram of the second stage of the fire

of uphill 3–4 (close to node 3). In the first stage of the fire, the heat wind pressure generated by the high temperature smoke at uphill 3–4 is negligible as the fire origin is close to the top of uphill 3–4. When the high temperature smoke flows through uphill 4–8, uphill 4–8 will produce a larger heat wind pressure H_f 48, heat wind pressure H_f 48 in the wind network is like an auxiliary fan with gradually changing wind pressure value, prompting roadway 4–8 air volume to gradually increase, while roadway 2–5, 5–7, 5–6, 6–7, 3–6 air flow will gradually decrease. When the heat wind pressure H_f 48 reaches a certain critical value, the above mentioned roadway air flow will equal to zero. When the heat wind pressure H_f 48 exceeds the critical value, the above mentioned roadway air flow will reversal. With the development of air flow reversal, when the high temperature smoke along 8–7–5–2, 7–6–3, 6–5, they also generate a heat wind pressure (they are uphill roadways). But because the direction of their heat wind pressure is the same as the direction of the fan wind pressure, there is a tendency to make the above reverse air flow recover again. At the same time, the oxygen concentration of the reverse flow of high temperature smoke is lower. When they flow back to the fire origin again, the fire origin fire will be weakened because of the lower oxygen concentration. As a result, the above reverse air flow also has a tendency to recover. However, after a further period of time, the air flow will again reverse along the above airway and then recover again.

When the fire enters the second stage, i.e. the fire origin spreads from node 4 to node 3, the heat wind pressure H_f 34 generated by uphill 3–4 is first to reverse the side airway 1–9–4 air flow. The tendency for the air flow to reverse along 4–9–1 is stronger in the case of a severe collapse of the 4–8 airway through the first stage fire zone. When high temperature smokes flow along 2–5–7–8, 3–6–7, 5–6, they also generate heat wind pressure, the direction of their heat wind pressure is the same as the direction of the fan wind pressure, which has the tendency to make the air flow along the side airway 4–9–1 continuously and steadily reverse. Therefore, once the fire enters the second stage, the reversed air flow from the first stage will recover, but the air flow will reverse in the direction of airway 1–4–9.

The above fire is an ascending ventilation roadway fire, then

The side branch air flow reverse condition equation is as below.

$$h_i/h_0 > R_i/R_0 \tag{2.7}$$

The side-branch air flow stagnation condition equation is as below.

$$h_i/h_0 = R_i/R_0 \tag{2.8}$$

The side-branch air flow maintains the in-situ condition equation as as below.

$$h_i/h_0 < R_i/R_0 \tag{2.9}$$

In the Equation:

h_i—the sum of the local heat wind pressures occurring in the internal sub-network, Pa.

h_0—the sum of the local heat wind pressure and the main fan wind pressure occurring in the external sub-network, Pa.

R_i, R_0—are the synthetic wind resistance of the inner and outer sub-networks respectively; Ns^2/m^8.

Analyzing the wind control measures during the fire period, the reversal of branch air flows 2–5–7–8, 3–6–7 and 5–6 in the closed loop diagram shown in Fig. 2.17 was one of the main causes of casualties during the first phase of the fire. If the above air flows had not been reversed, most of these people would not have been killed. According to Eq. (2.9), one of the main measures to maintain the stability of the air flow in these branches is to increase the synthetic wind resistance R of the internal sub-wind network. For this purpose, it is possible to choose to add resistance in branches 3–4 or 4–8. If it is concluded that the fire origin fire will not spread towards node 3, it is better to add resistance at uphill 4–8, otherwise resistance should be added at uphill 3–4. This is because blocking at uphill 4–8 leaves a backlash to the stability of the second stage of the fire 1–9–4 airway air flow. As shown in Fig. 2.17 and formula (2.9), increasing the uphill 3–4 air resistance is not only beneficial to stabilizing the internal air network air direction, but also reduces the air supply to the fire origin, which serves the purpose of suppressing the fire. At the same time, it also increases the ventilation pressure under axial fan ventilation conditions. If the fire origin fire is large, increasing Ri may not be sufficient to stabilize the above branch air flow. Consideration should also be given to reducing the synthetic wind resistance Ri of the external sub-network, as shown in Fig. 2.17. Adding resistance to the side branches 1–9, 9–10 and 10–12 of the external sub-network can reduce the synthetic wind resistance R0; reducing resistance to the main branches 8–11 and 11–12 can reduce R0; thus achieving the purpose of stabilizing the air flow.

In the second stage of the fire, as in closed loop diagram 2–18 and Eq. (2.9), in order to keep the side branch 1–9–4 air flow in the same direction, the fire branch 3–4 should be blocked, while the branch 4–8, which is located in the main branch, should be blocked less. Therefore, in the first phase of the fire, it is also beneficial for the second phase of the fire to stabilize the 1–4–9 airway air flow if blocking is added to the fire branch 3–4. In addition, in the first phase of the measures to reduce R_0, blocking is added to the external sub-network in the bypass branches 9–10 and 10–11, which is still effective for the second phase of the fire to stabilize the 1–9–4 air flow. But airway 1–9 has become the branch to be stabilized.

In summary, in order to control the direction of air flow and to maintain air flow stability, the best air control solution is to add resistance at branches 3–4, 9–10, 10–11 etc. The most effective measure to prevent air flow reversal during a mine fire is to install resistance devices at the branch on fire. In the case of a rapidly spreading fire origin, it is necessary to analyse the effect of air control measures in the early stages of the fire on the stability of the air flow in the later stages of the fire in order to choose the best air control solution [3].

2.6.2 Descending Air Flow Fire Case of Air Flow Reversal on Main Airway

[Case 4] Major fire accident in a belt conveyor tunnel in a mine

In 1986, a major fire broke out in the −380 m horizontal belt conveyor roadway of a mine in Zaozhuang, which lasted for 3 days and killed 24 miners. A sketch of the ventilation system is shown in Fig. 2.19.

The first belt conveyor head funnel at the −380 m level was burned and welded during the mid-shift while the mine was closed for maintenance. A latent fire at the end of the burn ignited the conveyor belt at 2am the following morning. Due to the night shift, the belt conveyor roadway and return air roadway were unoccupied. By the time, it was discovered, the fire was uncontrollable.

The air flow reversal during the disaster is shown in Fig. 2.20, the fire origin point is in the upper warehouse belt machine tail alley and −380 level first belt machine head articulation at the diagonal alley belt burning. The generated fire heating air pressure makes the alley and the passage air flow reversal. Zouwu wind shaft main fan stop running about 10 min later, a dark diagonal shaft air flow occurred reversal, then through a horizontal contact alley to two two dark slant shaft winch room into two dark slant shaft, until −380 horizontal alley, invaded 223 track uphill.

Analysis of the causes of 223 uphill roadway airflow disorder: mine fire produces fire heating air pressure action results. Under normal ventilation (i.e. before the fire disaster), the local ventilation system of the Shanjialin mine can be analyzed in Fig. 2.21, which has three airway lines, namely: 1→2→3→4→Zouwu wind shaft;

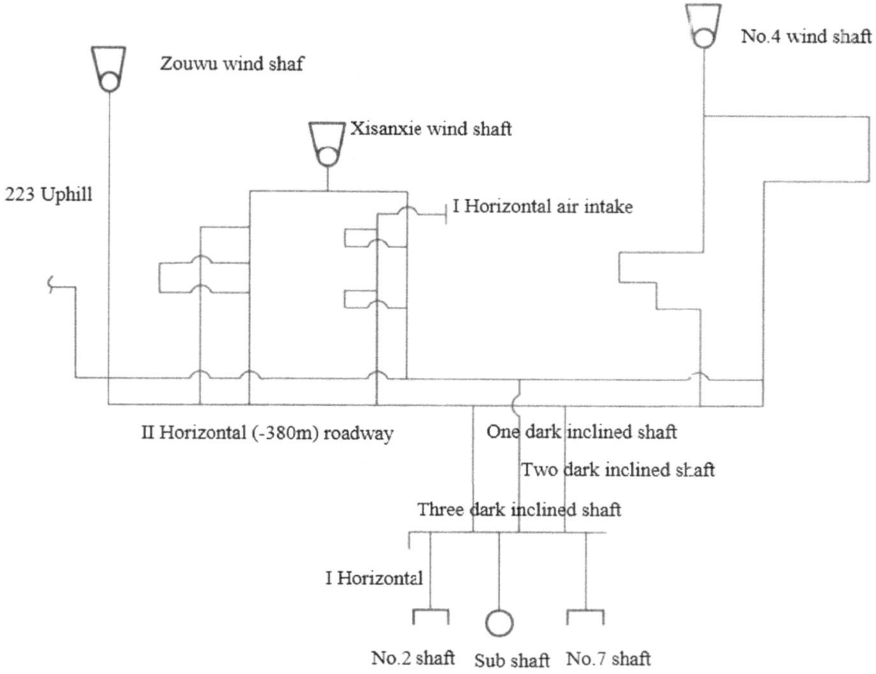

Fig. 2.19 Sketch of the mine ventilation system

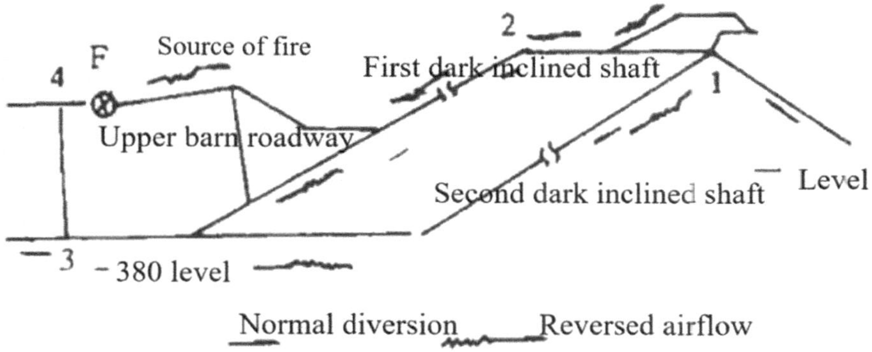

Fig. 2.20 Schematic diagram of the air flow path transfer in the mine

1→6→5→4→Zouwu wind shaft; 1→6→5→Xisan slant shaft. After a descending air flow fire, the high temperature smoke flows with the air flow into the 220 uphill and −380 sub-corridor, causing a reduction in air density in the high temperature smoke in the roadway. A localized fire heating air pressure is formed at the connection between the 220 uphill and −380 sub-corridor.

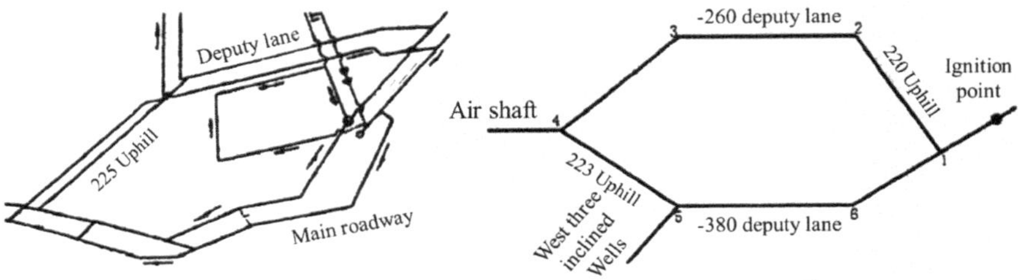

(a) Diagram of local ventilation system (b) Equivalent diagram

Fig. 2.21 Mine local ventilation system diagram

Fire heating air pressure produces the following effects respectively.

(1) 220 uphill is an upward ventilating area where the fire heating air pressure is generated in the same direction as the air flow. Therefore, the fire heating air pressure contributes to the ventilation in the 220 uphill roadway, which causes the high temperature smoke to spread more rapidly in the roadway during the fire period and further extends the disaster.
(2) For the −380 sub-roadway junction, under normal ventilation, this is a downwards ventilation and the resulting fire heating air pressure is in the opposite direction to the air flow. So the fire heating air pressure acts as a barrier to ventilation in the −380 sub-roadway. As the fire disaster progresses at different fan revolutions, it eventually leads to an airflow disorder in the 223 uphill and −380 sub-corridors. Airflow disorder was observed, resulting in localized circulating smoke poisoning of underground personnel. During the fire period, the air flow into the −260 mainline was significantly higher than during normal ventilation.

The fire is located in the descending air flow. The fire heating air pressure generated is in the opposite direction to the main fan. As a result, it tends to cause the main airway air flow to reverse, thus spreading the disaster throughout the mine.

A total of three air flow reversals occurred in this accident.

The first occurred at 04:15, the following morning in the upper silo belt conveyor roadway and its passage to the first dark inclined shaft. As shown in Fig. 2.22, the point of origin of the fire was at the point where the end of the upper silo belt conveyor met the head of the first belt conveyor in the −380 m horizontal alley, where the air flow reversal between the alley and the access road was caused by the fire heating air pressure generated by the burning of the conveyor belt in the inclined alley. The reversal of the air flow was split into 3 branches from the upper conveyor belt through the access road, a dark incline (the lower section reverses the flow by 100 m) and then merges to the −380 m horizontal roadway via the third dark incline shaft. The reversal of the air flow caused a dramatic extension of the disaster to all production and preparation mining sections fed from level II.

The second occurred at approximately 05:10, the following day in the first dark shaft. After the first air flow reversal, the high temperature air flow descended through the lower section of the first concealed shaft and the temperature in the upper section

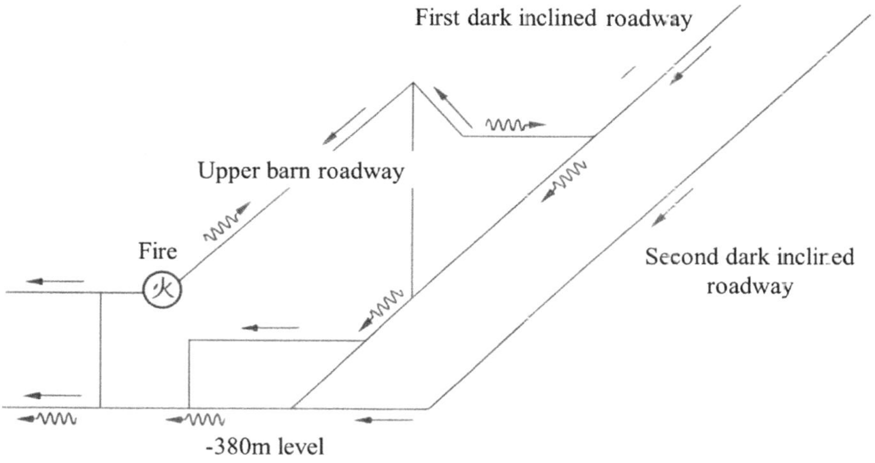

Fig. 2.22 Diagram of the first air flow reversal

was gradually increased by it. At approximately 05:00, the total mine air discharge dropped after the wind shaft fan stopped operating, and the air flow down the first dark inclined shaft also dropped. As a result of the fire heating air pressure, the first dark slope air flow reversal occurred at approximately 05:10. The route of the reversal is shown in Fig. 2.23.

The third place occurred at the 220 secondary uphill (upper section) and 221 positive uphill off-site. Reversal smoke harmful gas from a dark slope upward, through the I level across the second dark slope shaft winch room into the second dark slope

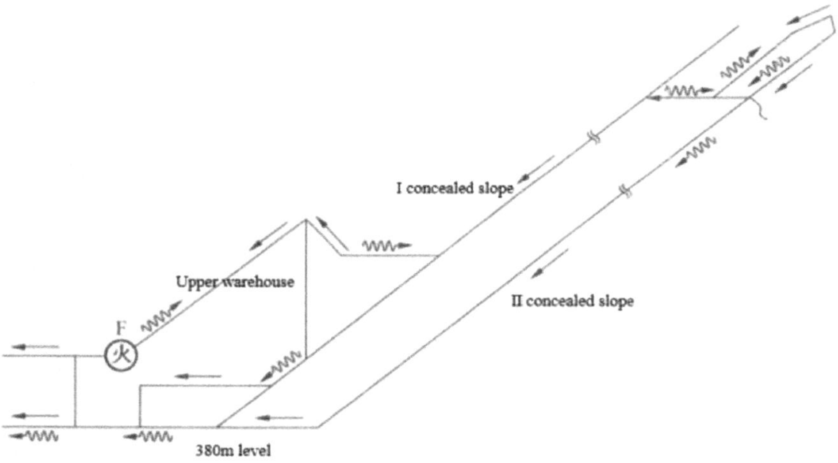

Fig. 2.23 Diagram of the second air flow reversal

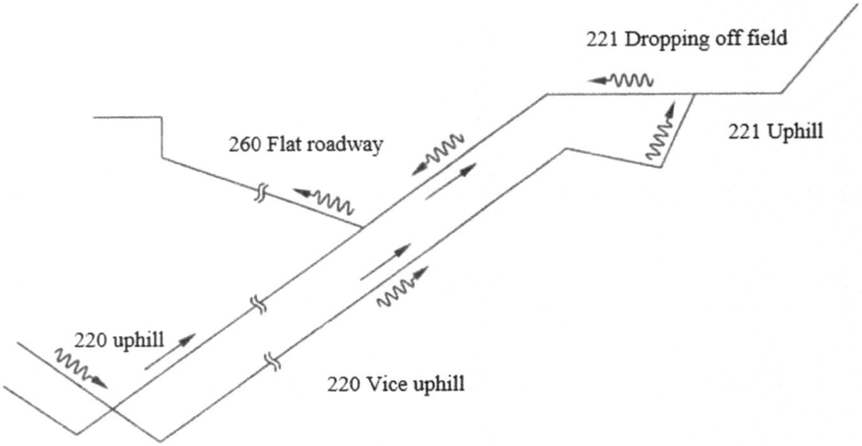

Fig. 2.24 Diagram of the third air flow reversal

shaft. Reversal air flow through the second dark slope shaft into the II level, and threaten all the mining from the II level into the wind. At around 6:30am on the next morning, the air flow reversal between the 220 main uphill (upper section) and the 101 lower yard (section) was ignited, with the conveyor belt burning from the point of fire origin to both ends. 220 secondary uphill had a regulating draught window 10 m from the lower end, which was destroyed by the burning conveyor belt, and the fire spread rapidly upwards. 220 sub uphill (upper section) and 221 positive uphill off part of the field air flow reversal, reversal route as shown in Fig. 2.24. Dense smoke through 220 positive uphill upper section reversal downward into 260 road, the third day of the fire area temperature is still up to 43 °C, 220 vice uphill fire heating air pressure seriously hindered the progress of relief work.

This incident was a typical descending air flow fire. A fire in a descending air flow is much more complex than a fire in an ascending air flow, but the damage could have been avoided or reduced if air flow control had been carried out effectively and in a timely manner. Firstly, the fire heating air pressure can be reduced by restricting the supply of air to the fire origin. Secondly, temporary containment walls or hanging air curtains can be constructed in the second dark inclined shaft to increase the fan air pressure on the first dark inclined shaft, thus preventing air flow reversal in the first dark inclined shaft. In addition, if the main shaft fan air can be reversing after a fire, its possible to overcome the fire heating air pressure generated by the fire and also to avoid smoke intrusion into the 233 track uphill [4].

[Case 5] A major fire accident in a coal mine lifting jack shaft

On 5 January 2010 at 12:20 pm, a major fire accident occurred in a coal mine, resulting in 25 deaths and 3 missing persons.

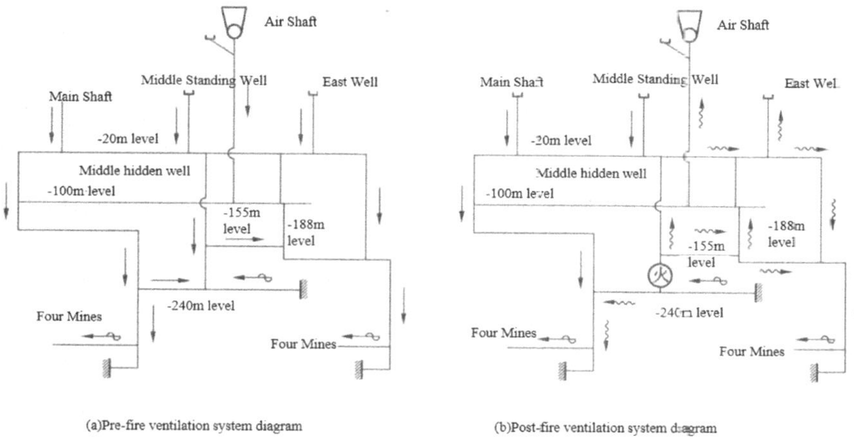

Fig. 2.25 Mine ventilation system before and after a disaster at a mine

The mine is fed by various shafts on the surface. The ventilation system in the production area of the east shaft and the production area of the west shaft are ventilated in tandem. In the vicinity of the mine, there have been several instances of confinement at the two penetrations. But these have been broken on several occasions. middle well production area air is exhausted via auxiliary fans into the east shaft production area into the incoming air flow. The rest of the mine is fed by a local fan. The mine ventilation system diagram is shown in Fig. 2.25.

Noon on the 5th, middle well −20 m winch's driver Tan received a call from Ma, saying that the power was out below, requesting to send power from the −20 m main switch. Tan immediately went to send power, but could not send power. 2 min later, Tan found that there was smoke and rubber smell in the roadway, the concentration was getting bigger and bigger, reported to the ground by phone and then ascended the well. West well −240 m level winch driver Ma called to inform middle well −155 m winch driver Tang that three vertical shafts were on fire and asked for a power outage. After the power outage, middle well −240 m horizontal transport worker Liu to middle well three vertical shafts bottom direction to understand the situation, on the way to smell the rubber smell, and see smoke, to three vertical shafts bottom to find the vertical shaft on fire, immediately let west well −240 m well bottom yard winch driver Ma report the ground, after ascending the well. West well −240 m, Ma quickly reported to the dispatch office and went to middle well three vertical shafts to inspect the fire, and successively found the cable, plastic pipe and wooden bracket burning under the crane in three vertical shafts.

After the fire in the cable of the three concealed shafts of the middle well, a large amount of toxic and hazardous gases were not discharged to the surface with the air flow, and spread to the surrounding roadway under the action of fire heating air pressure, causing the air flow of the three concealed shafts of the middle well to reversal and finally flow into the −240 m horizontal roadway with the air flow. The fire caused the air flow reversal in the middle well and finally flowed into the −240 m horizontal roadway with the air flow.

This incident was a descending air flow fire. The reason for the air flow reversal was that the main fan stopped running and the buoyant effect of the vertical jack shaft was in the opposite direction to the mechanical ventilation, the buoyant effect was significantly greater than the mechanical ventilation. When a fire occurs in a jack shaft and the ventilation is downwards, air flow reversal is a high probability event. It is therefore necessary to prevent this in advance by establishing a firewall to effectively direct the flow of toxic fumes through controlled ventilation facilities to avoid the whole mine being attacked by harmful gases [5].

References

1. Liu J, Li B. Anatomy of a typical accident case in coal mine. Beijing: Coal Industry Press; 2014.
2. Shi Z. Coal mine accident investigation techniques and cases. Beijing: Coal Industry Press; 2009.
3. Zhu H. Study on the law of gas explosion induced by fire in mine tunnel. Beijing: China University of Mining and Technology; 2008.
4. Wang H, Wang B. Thermodynamic hazards in mines. Beijing: Coal Industry Press; 2010.
5. Zhou X. Mine fire prevention and control. Xuzhou: China University of Mining and Technology Press; 2002.

Chapter 3
Airflow Control Technology in the Mine Fire Period

Bo Tan, Haiyan Wang, and Liyang Gao

In the event of a fire in a mine, it is important to take the correct measures to control airflow in order to ensure the safe evacuation of underground operators, prevent the spread of fire smoke everywhere and gas explosions, control the continued expansion of the fire and create favourable conditions for fire extinguishing.

In the event of a mine fire, airflow control is a complex technical issue that requires knowledge of disaster ventilation theory and some experience of accident management.

The basic requirements for ventilation in the event of a fire in a mine are as below [1].

(1) To put people first and protect workers in the affected and threatened areas by evacuating them quickly to safety or to the ground.
(2) To limit the uncontrolled spread of smoke flow in the well roadway, to prevent the expansion of the fire and to create conditions for direct fire suppression as close as possible to the source of the fire.
(3) Gas shall not be allowed to accumulate to an explosive concentration in the vicinity of the source of the fire, and gas airflow reaching an explosive concentration shall not be allowed to flow through the source of the fire, or to spread to areas of high gas concentration.
(4) To create favourable conditions for ambulance work.

B. Tan (✉) · L. Gao
School of Emergency Management and Safety Engineering, China University of Mining and Technology-Beijing, Beijing, China
e-mail: tanbo709@126.com

L. Gao
e-mail: gaoliyangcc@163.com

H. Wang
School of Civil and Resources Engineering, University of Science and Technology Beijing, Beijing, China
e-mail: b2146431@ustb.edu.cn

B. Tan et al. (eds.), *Mine Fire Prevention and Control*,
https://doi.org/10.1007/978-981-96-0466-1_3

(5) To prevent secondary sources of ignition.
(6) To prevent airflow reversal caused by fire and wind pressure.

Airflow control can be either regional or mine-wide. Airflow control can be achieved with the aid of main and local fans as well as ventilation devices; or by using only ventilation facilities such as ventilation doors, temporary confinement and regulation air windows, or a combination of several.

3.1 Main Fan Control Airflow Methods and Examples

The working condition of the mine's main fan is directly related to the ventilation status of the whole mine. During normal production periods, continuous operation of the main fan must be ensured. During mine fire periods, the working condition of the main fan should be decided according to the location of the fire, the development trend, and the general methods that can be adopted are: maintaining normal operation, stopping operation, air reversing and regulating fan conditions. Whatever the method, it must be considered carefully and comprehensively to ensure safe rescue. The following examples illustrate the management of major fans during a mine fire.

3.1.1 Keeping the Main Fan in Working Order

In general, it is important to keep the main fan in normal operation to ensure that the mine ventilation system is stable and the air volume is sufficient to enable smooth rescue work to be carried out in the event of a fire generating large quantities of toxic and harmful gases, resulting in casualties from poisoning or asphyxiation. There is much experience in this area, but also many lessons.

[Case 1] An example of the expansion of an accident caused by ceasing ventilation in the handling of a fire accident

A fire in a French coal mine was mishandled, resulting in a number of deaths by poisoning.

As shown in Fig. 3.1, fresh airflow enters the bottom of the shaft from the intake shaft and flows in two separate wings at the 320 m level. One stream exits the surface via the east limb mine area by air shaft 1 and the other stream exits the surface via the west limb mine area by air shaft 2. As part of the east limb mine area was closed by fire 14 months ago, the main fanf_1 on wind shaft 1 is no longer in operation, while the main fanf_2 on wind shaft 2 is still in operation to keep the west limb mine area supplied with air.

[Accident Overview]

On the morning of the incident, f_1 was activated to seal the old fire area. But unfortunately, the old fire area was not completely extinguished and reignited shortly

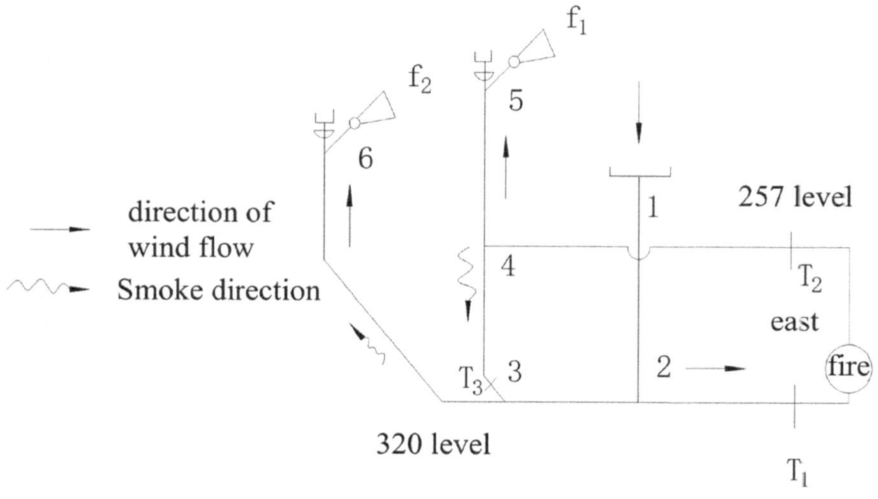

Fig. 3.1 Diagram of a fire in a French coal mine ventilation system

afterwards. The personnel on the surface noticed smoke being exhausted from f_1 and immediately decided to stop the operation of f_1 and open the explosion-proof cover of the No. 1 air-return shaft opening in order to control the air supply to the fire area and reduce the development of the fire. 4–5 min later, the airflow was considered to be back to normal after a paper test. However, shortly thereafter an alarm signal was received down-hole and a zone manager descended from the intake shaft to the 320 m level to find out what was going on. He met two poisoned workers at the bottom of the shaft in the depot and immediately assisted them to ascend the shaft.

It was not until the poisoned workers were raised that surface personnel realized there was a disturbance in the down-hole airflow and immediately activated the main fanf_1 and sent men down the shaft to check the situation at the workings. One worker was first found dead from poisoning near the bottom yard of the intake shaft at the 320 m level and a further six workers were found dead from poisoning in the large roadway of the air intake to the western mine area, 600 m from the inlet shaft.

The main fanf_1 stopped running for just ten minutes and already the fire smoke filled the roadway from the eastern mine area and invaded the western mine area and caused a serious accident with multiple poisonings, which was totally unforeseen by the mine leaders.

[Accident Analysis]

The incident was caused by the reversal of the airflow in the lower section of air-return shaft No. 1, sections 3–4. The reversed airflow caused smoke and fire to enter the western mine area through the poorly sealed T_3, resulting in the death of several people.

The reason for the reversal of airflow in zone 3–4 can be analyzed from two aspects. Firstly, the air pressure of the outer part of the system, which is divided by the boundary of branch 3–4 until f_1 stops running, is greatly reduced compared with the air pressure of the inner part of the system. The opening and closing of the fire area confinement (T_1, T_2) also makes the air resistance of the inner part of the system significantly reduced, which inevitably makes the airflow of branch 3–4 reversed. Secondly, the 3–4 branch is an angularly connected branch and its airflow direction is governed by the main fanf_1 and f_2 on the two air shafts. When f_1 stops running, its airflow direction must necessarily flow from node 4–3 due to the action of f_2, to the extent that the airflow hair of sections 3–4 is reversed.

Under no circumstances should the main fanf_1 be stopped if it is necessary from the point of view of keeping the original wind direction in sections 3–4 unchanged and avoiding this accident. And it is preferable to stop the main fanf_2 and open the blast cover of the air-return shaft No. 2.

[Experience Summary]

(1) When a fire occurs in a branch airflow, the original working condition of the main fan should be maintained. Especially during the fire-fighting phase, no measures should be taken to reduce the air or stop the operation of the main fan. In multi-fan ventilation mines, the fan that undertakes the task of smoke extraction should not be stopped, except in the case of fires in the intake shaft and its under-shaft yard.
(2) In the event of a fire accident in mine, active measures should be taken to strengthen the control of the ventilation system, analyse the location of the fire and wind pressure and the extent of the hazard, and evacuate people from the shaft in time to prevent the fire and wind pressure from expanding the disaster. At the same time, creating conditions to extinguish the source of the fire directly is an effective way to prevent the generation of fire and wind pressure and the hazards it causes. Therefore, a scientific grasp and application of the laws of fire and wind pressure activity is an important guarantee for good rescue and relief work in underground fire accidents.

3.1.2 Stopping the Main Fan

Consider stopping the main fan work in the following cases.

(1) Fires occur in the intake shaft barrel or at the bottom of the intake shaft, when air reversing is not possible due to conditions and the fire gases cannot be short-circuited into the return airflow.
(2) The fire has been occurring for a longer period of time at the single-headed digging face, and the gas concentration has exceeded the upper explosion limit, at which point no more air can be sent.
(3) When the main fan has become a resistance to ventilation.

When stopping the main fan, the blast door of the air-return shaft should be opened at the same time so that the airflow is automatically air reversing under fire wind pressure. Caution should be exercised when using this ventilation measure.

Ceasing ventilation measures can easily cause gas to build up to dangerous explosive concentrations. The method of stopping the main fan operation should never be used lightly, and only when there is certainty, otherwise it will expand the accident. For example, on 22 May 1962, a mine in the Huabei Bureau had a fire in the air intake large roadway. After stopping the main fan, it resulted in the death of 13 people, including two engineers and a mine manager. Another example is a coal mine in Guizhou Province on August 9, 1993 in the air intake slope shaft in the bottom of the car yard in the substation fire and ignited air intake slope shaft of the wooden bracket. Only two of the 27 people on the return airflow side of the shift were evacuated and 25 people were killed. In the course of the rescue, the mistaken stopping of the main fan caused the airflow to reverse, resulting in the deaths of a further 23 rescue personnel, including three firefighters, one paramedic and the mine's chief engineer and safety section chief. The mine had air reversing conditions and its also capable of short-circuiting the fire smoke (central parallel ventilation). However, the mine leaders wrongly gave the order to stop the main fan reversing, which caused the disaster to expand and resulted in a vicious accident with a cumulative loss of 48 lives.

As shown in Fig. 3.2, Xuzhou Bureau of a mine in dealing with the digging face of the gas explosion accident in 1958, because of the error in judging the disaster (thought to be open fire fire), mistakenly stop the main fan for 2 h, its so that a large number of explosive gases trapped in the roadway and work surface. This amplified the casualties, resulting in 43 deaths and 26 injuries. This shows that the method of regulating the wind to stop the operation of the main fan should be cautious and should not be decided lightly.

3.1.3 Mine Air Reversing

In the event of an underground fire, air reversing equipment and facilities are used to change the direction of the fire's smoke flow so that people downwind of the fire are in the fresh airflow of the "upwind" side of the fire. Air reversing and local air reversing. The main focus here is on air reversing by controlling the main fan and its ancillary facilities (full mine air reversing and area air reversing).

(1) The full mine air reversing is air reversing through the main fan and its appurtenances. Axial fans use the main fan reversing air reversing and centrifugal fans use the air reversing channel air reversing.
(2) Regional air reversing regulates the air reversing facilities of one or several main fans in the event of a fire in the air intake of a ventilation system in a large roadway in a multi-inlet, multi-return mine, in order to achieve an air reversing

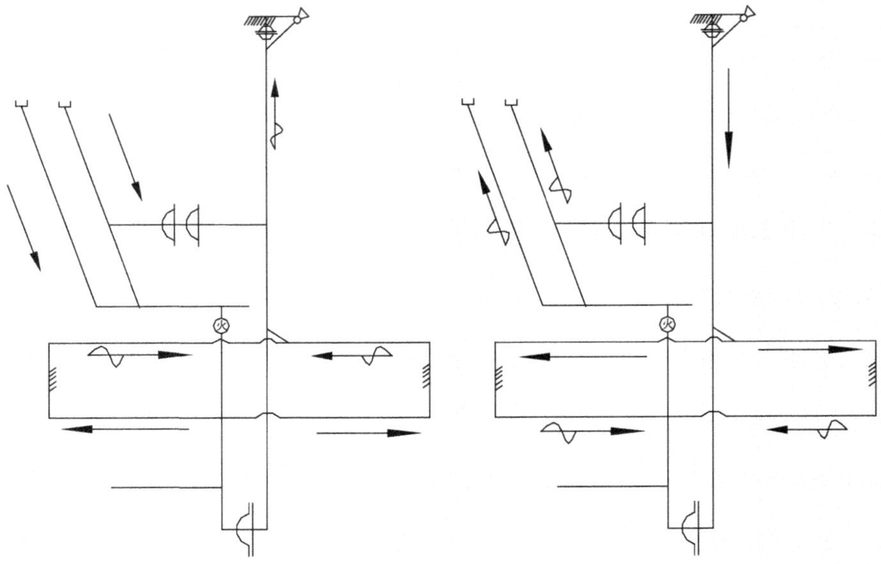

Fig. 3.2 Diagram of the ventilation system for the poisoning of people caused by a false stop of the main fan. **a** Normal ventilation system **b** ventilation system for air reversing

of the airflow in a part of the mine reversing method. This is known as zone air reversing.

Sometimes mine air reversing is required when dealing with mine disaster incidents.

Firstly, when the source of the fire occurs on the air intake side of the mine, after all personnel have been evacuated, full mine air reversing can be carried out and the fire smoke stream can be discharged in time. Secondly, in mine where multiple fans are working in combination, when the blast wave in one system has opened the blast door and changed the air direction and the disaster area gases may enter other systems. In this case, combined air reversing of the non-accident main fan should be carried out immediately. There have been many successful experiences with mine air reversing and some hard lessons learned. This is combined with examples of what to look for in mine air reversing.

[Case 1] Lessons from air reversing due to careless work

A fire broke out in the lower part of the air intake shaft in a mine, directly threatening the lives of several hundred people in the underground production, so it was decided that the whole mine should be air reversing. When the people on the upwind side of the fire were informed to evacuate, the people in a roadway connected to the air intake shaft were not evacuated in time. After the air reversing, several hundred people were saved, but some of those who had been in the air intake died of poisoning after the

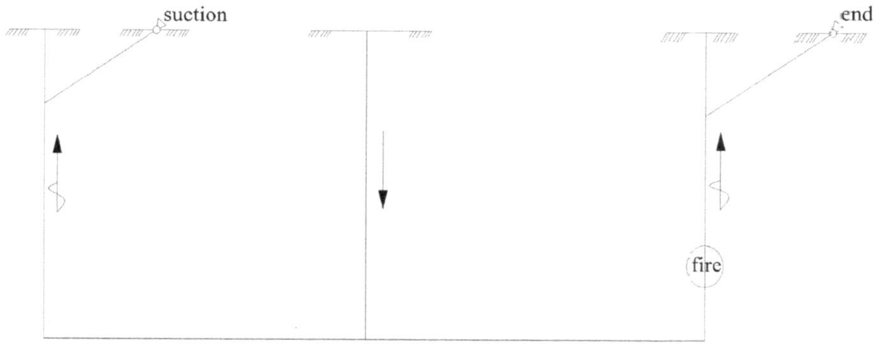

Fig. 3.3 Example of mine air reversing with multiple fans working in combination

air reversing. Afterwards, according to the dispatch office. Calls were made but not answered and it was mistakenly thought that the men had been evacuated. This is a profound lesson in casualties caused by careless work and we must remember it carefully.

[Case 2] Issues to bear in mind for mine air reversing with multiple fans working in combination

A coal dust explosion occurred in a mine west limb and high pressure shock wave broke open the blast door of the west wind shaft and stopped the main fan. This is shown in Fig. 3.3.

The mine was pumping air for two fans in combination and the fan at east limb was operating normally, which resulted in the pumping of gases from the west limb disaster area into the east limb mine area, poisoning personnel and amplifying the accident. After the discovery, it was requested to immediately restore the ventilation of west limb to prevent the harmful gas from continuing to enter east limb. But at this time, the fan of west limb could not restore the ventilation, so the air reversing of the fan of east limb was immediately notified before it caused more serious consequences.

This illustrates the mine of multiple fans working jointly. Once an explosion occurs in a certain zone, the fan stops or a fire occurs requiring air reversing, all must consider the air reversing of other non-accident zone fans at the same time. In theory, joint air reversing should be air reversing at the same time, but in practice this is not possible for the following reasons. Firstly, the operation of each fan cannot be carried out at the same time. Secondly, the mine grid cannot accommodate multiple large motors starting at the same time. Therefore, in order to achieve full mine air reversing, it is necessary to make careful arrangements and operate according to a plan. The non-accident area fan should be air reversing in sequence firstly. Finally, the accident area fan should be air reversing again. The process of joint air reversing should be shortened as much as possible when conditions permit.

[Case 3] Regional air reversing technique for localized fires

[Basic Information]

The ventilation system of Yanzishan coal mine is a multi-air shaft and multi-main fan combined operation system, the ventilation system is extraction type and the ventilation method is mixed type. Mine has 7 intake shafts and 3 air-return shafts, which are air-return shaft, west air-return shaft and east air-return shaft. The total intake volume of mine is 35,374 m^3/min and the total return air volume of mine is 36,309 m^3/min. The three air-return shafts are responsible for the ventilation of the whole mine, among which the return airflow is not responsible for the supply of air to the 309 (south wing) pan area of 14-3# layer, 302 pan area of 5# layer, 8# layer and the end of the main shaft in the East Zone. The east air-return shaft is responsible for supplying air to the 4# layer 302 pan area and the 3# layer 302 pan area in the east area, while the west air-return shaft is responsible for supplying air to the 14-3# layer 303 pan area and the 4# layer and 5# layer in the west area, and the non-ventilation system is reasonable and reliable. All the underground pans are arranged in three roadways, with a special return airflow roadway, and all the air-using locations are zoned for ventilation. The ventilation capacity is approved at 6,848,000t/a.

An external fire broke out on the main inlet airway in the 4# layer pan area in the east end, and smoke spread to all working faces in the 4# layer pan area with the airflow. If full mine air reversing is adopted, it may result in the existence of localized breeze zone and the airflow is not controllable to spread the smoke to other areas of mine, so regional air reversing is adopted. The main fan of the east air shaft was designated for separate air reversing, and the extract type was changed to a press-in type. The main fan, the main fan of the west wind shaft continue to pump-out type operation of the program, the program will be implemented, the mine will reach the east side of the airflow reversal, air intake into return airflow, smoke direction single and with the reverse airflow directly from the east side of the talk wind shaft discharge, without affecting other main fan area. But by the dry east wind non from pumping out type to press-in-type, the original ventilation network has been changed significantly. Therefore, the regional air reversing period of the east wind shaft will inevitably result in a breeze afterwards and the existence of a windless area. Technical means are needed to ensure that regional air reversing can control the fire zone and at the same time ensure that the ventilation system is stable and reliable.

[Technical Solutions]

In order to ensure that regional air reversing is carried out safely and smoothly, to avoid the emergence of breezy or windless areas which can cause gas overruns and to prevent secondary disasters from occurring. Therefore, it is necessary to base on the network settlement results. At this time, the positive and negative pressure airflow convergence point is the location of the track stone gate and belt stone gate on the 4# layer in the east area. It is necessary to construct ventilation facilities at the intersection of positive and negative pressure in time to isolate the mine and positive and negative pressure systems during regional air reversing, and to eliminate windless and breezy areas.

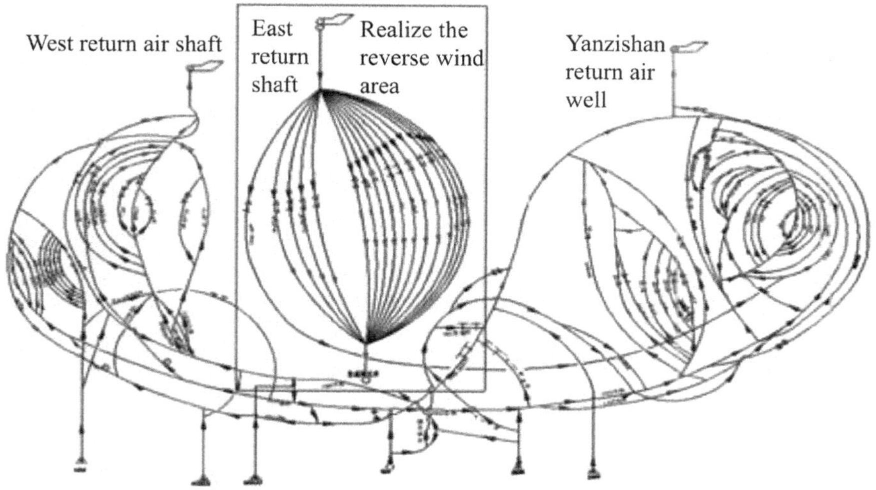

Fig. 3.4 Changes in the mine ventilation network when the main fan of the east wind shaft is press-in-type

(1) Constructing two temporary positive air reversing doors at the 4# level of the eastern part of the rail gate to separate the airflow at the rail gate and to prevent the interaction of positive and negative pressures throughout the mine;
(2) Constructing two temporary positive air reversing doors at the 4# layer belt stone door in the east to separate the airflow at the belt stone door and prevent the interaction of positive and negative pressure throughout the mine;
(3) Improving the unidirectional adjustment of the east end inlet and air-return shaft bottom link roadway to forward and reverse adjustment, preventing the airflow from reversing during air reversing in the east wind well and then short-circuiting here, causing the east end to be in a state of no air;
(4) The main fan of the east wind shaft was run in reverse with no load and the main fan air volume and external air leakage quantity were measured to ensure that the air reversing quantity during air reversing was not less than 40% as required by the Regulations.

The location of the temporary positive air reversing door construction for the track stone door and belt stone door on level 4# in the east area is shown in Figs. 3.4 and 3.5.

3.1.4 *Regulating the Working Conditions of the Main Fan*

Sometimes, depending on the actual situation of the relief work, when the three methods mentioned above cannot be satisfied, or are not economical or easy to

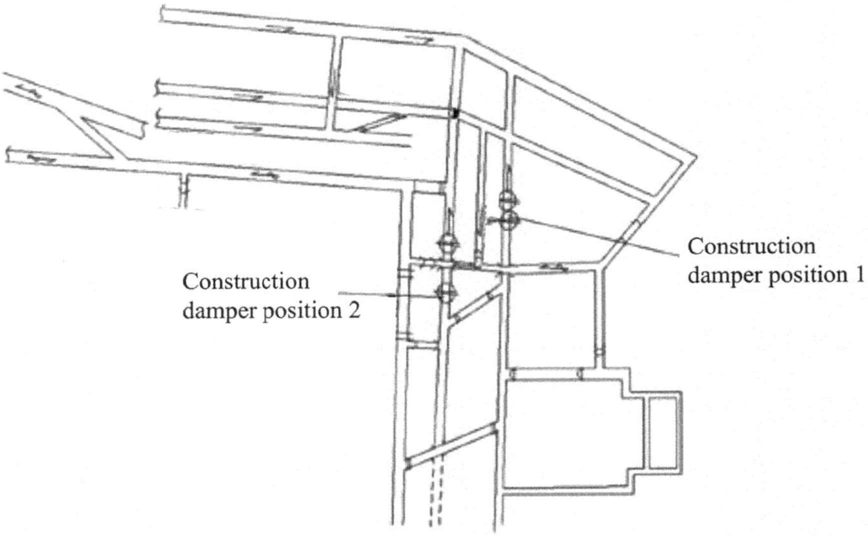

Fig. 3.5 Example diagram of the working conditions for regulating the main fan

achieve, the working conditions of the fan can be considered to be adjusted to achieve the purpose.

If there is a large area of fire area down-hole, while carrying out down-hole airflow air pressure regulation, consider reducing the total air pressure of mine to further reduce the pressure difference and air leakage quantity inside and outside the fire area. Furthermore, under specific conditions, due to the change of wind resistance value, it may make the axial fan work in unstable zone. The air volume of centrifugal fan increases too much, the power is too large and burns the motor. At this time, the fan working condition must be adjusted in time to make it safe to operate.

The ventilation system of a mine is shown in Fig. 3.6, which is a multi-fan joint extraction ventilation. Fan I, fan II, fan IV and fan V are four fans working together in four shafts, with air intake in the secondary shaft. Fan III was originally an air-return shaft, but now the fan is stopped and used as an intake shaft. The fan III used to be an air-return shaft, but now the fan is stopped and used as an intake shaft.

A fire broke out at 3 o'clock during the night. After receiving a report from the shaft, the dispatch office immediately informed the ambulance team to go down the shaft and organize the evacuation of the personnel, while finding the source of the fire and extinguishing it. The smoke was found to have entered the first and second levels from the III wind shaft, and the source of the fire was found against the smoke, with the wooden supports and coal already burning. There was no water source nearby to extinguish the fire directly. The smoke was a direct threat to the lives of hundreds of people down the shaft. At this point, the better solution was to use all mine air reversing, but it was more complicated to use four fans at the same time. In case there was a problem with one of them, it would still fail to achieve the purpose, so

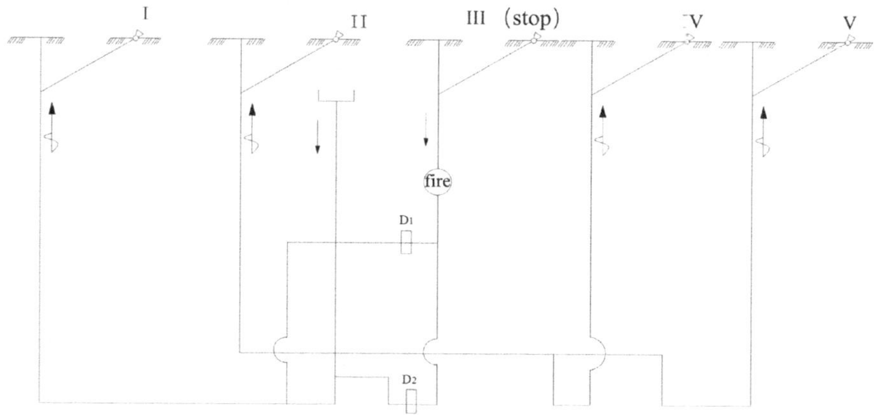

Fig. 3.6 Example diagram of the working conditions for regulating the main fan

the decision could not be made easily. After analysis and research, the D_1 and D_2 boards are built at the first and second places to temporarily control the smoke from entering the production mine area, and actively restore the fan operation of the III air shaft to discharge the smoke from the mine.

The temporary panel closures were built and fan III was in operation. However, the fire was still burning and it was feared that if fan III was out of service for a long period of time. The smoke and fire would still enter the mine area and endanger the safety of the personnel if something went wrong during operation. The decision was made to build a brick closure and then to connect a water pipe to extinguish the fire directly. However, during the operation of fan III, it was relatively normal at first. But as the underground confinement was established, the fan air pressure rose sharply and when it rose to a certain level the fan would enter an unstable zone and the fan would be destroyed. When the height of the water column meter was found to rise sharply in the engine room, a timely method was adopted to adjust the working conditions to ensure stable operation of the fan. This was achieved by adjusting the external air leakage quantity according to the change in height of the water column meter to ensure that the smoke from the well could be discharged effectively without making the air volume too large, thus effectively controlling the fire. The fire is extinguished as soon as the pipe is connected.

3.2 Down-Hole Airflow Control Methods and Examples

3.2.1 Down-Hole Airflow Control Methods

Down-hole airflow control during mine fires can be achieved by stabilizing airflow, reducing or increasing air volume, airflow short-circuiting, local or regional air reversing and other measures. The underground airflow control method should be selected according to the gas concentration, ventilation mode, fire range, mine depth, especially the location of the fire.

(1) **Stabilizing Airflow**

When the fire occurs in a more complex ventilation network, it is difficult for the rescuers to understand the specific situation of the fire area, or mine fire occurs in the return airflow big roadway, changing the ventilation method may cause airflow disorder, increasing the difficulty of retreating personnel. There may also be a gas accumulation and other consequences. At this time, it should be normal ventilation. This means maintaining the original ventilation system, maintaining the original wind direction and air volume and stabilizing the airflow, otherwise unexpected consequences can occur.

Maintaining normal ventilation and stabilizing the airflow, the conditions under which this measure applies are as follows.

(1) The source of the fire is located inside the mine area, the smoke flow has spread over a large area and the underground personnel are distributed over a large area.
(2) In high gas mines with complex ventilation networks, the use of alternative ventilation systems can increase the risk of gas and coal dust explosions, or extend the disaster.
(3) The source of the fire is located in the lone digging roadway and cannot stop the local fan operation.
(4) The source of the fire is located in the mine area or mine main return airflow roadway, maintaining the original wind direction is conducive to the rapid discharge of fire smoke.
(5) When the exact location and extent of the fire and the area threatened by the fire are not fully understood.

(2) **Reduction of Air Volume**

When dealing with fires, it is generally not advisable to reduce the volume of air supply to the fire area without sufficient reason, nor to stop supplying air to the fire area, as this is a basic principle to be followed when dealing with fire incidents. Because this can lead to the appearance of fuel-rich type fires, producing the phenomenon of the fire's smoke flow rolling back, increasing the risk of gas combustibles exploding and making it more difficult to rescue the fire. However, the ultimate aim of dealing with fires is to extinguish them, reduce casualties and resume production as soon as possible. The specific actions are as follows. First is to ensure the safety of relief personnel; second is to avoid as far as possible the appearance of fuel-rich combustion. In some special circumstances, when the use of other wind control methods will make the fire expand and the concentration of combustible gases at the scene increase, the method of reducing air volume can also be considered. But it should be used with caution and corresponding safety measures must be taken. For example, production should be suspended and evacuated within the affected area. Special personnel should be set up to closely monitor the changes in gas and other gases, especially gas. In the process of reducing air volume, if the gas concentration is found to be increasing, the use of air volume reduction should be stopped immediately and normal ventilation should be resumed. If necessary, air volume can be increased to dilute and discharge the gas.

(3) **Increasing in Air Volume**

Increasing air volume should be considered as carefully as reducing it. Otherwise, it will be counterproductive. However, in the following cases an increase in air volume should be considered first.

(1) Elevated gas concentration in the fire zone or its return airflow.
(2) Fire and wind pressure occurs within the fire zone, presenting the possibility of airflow reversal.
(3) When persons in danger in the affected area have not been evacuated after a gas explosion occurred during the handling of a fire.

(4) **Airflow Short Circuit**

The airflow short circuit is divided into a fresh air short circuit and a fire smoke short circuit. The main function of fresh air short circuit is to reduce the supply air volume of the fire source, so as to achieve the purpose of controlling the fire. Fire smoke short circuit is to direct the smoke and toxic gases into the return airflow or return airflow roadway, so as to protect the staff from poisoning. When using airflow short-circuiting, due consideration should be given to the following. The smoke entering the affected area is reduced, but the concentration of toxic gases is not reduced. There is a risk that the oxygen level in the affected area will be reduced and whether oxygen-poor asphyxiation will occur, resulting in an extended accident. The lessons learned in this regard are tragic.

For central parallel ventilated mines, the source of the fire is located at the intake shaft, intake shaft opening. If air reversing cannot be carried out in time or due to conditions, the ventilation door in the connection roadway between the inlet and air-return shaft can be opened or sealed to short-circuit most of the smoke flow and flow directly into the air-return shaft, thus reducing the smoke flow into the mine area. This will reduce the amount of smoke flowing into the mine area, thus facilitating the evacuation of personnel and the rescue of the rescue team.

(5) **Local Air Reversing**

Local air reversing: When a fire occurs in the mine area, the main fan maintains normal operation and adjusts the preset opening and closing state of the ventilation

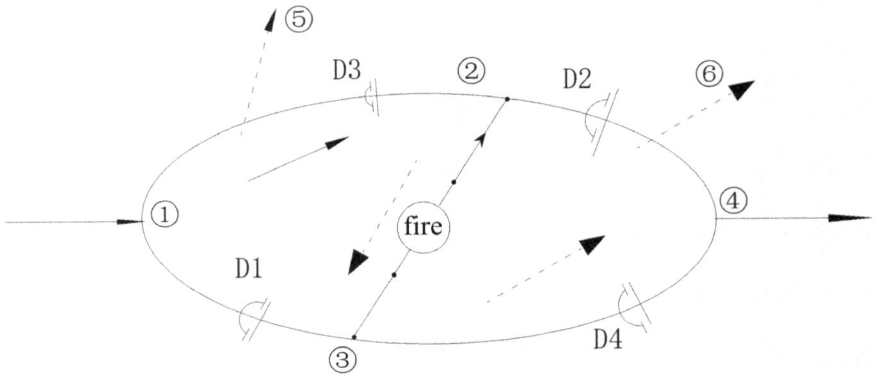

Fig. 3.7 Ventilation network diagram for a mine

door in the mine area to achieve local airflow reversal in the mine area. This air reversing method is called local air reversing.

The ventilation system of a mine area is simplified to a network as shown in Fig. 3.7, with the fire source in the diagonal roadway ②–③ branches. During normal ventilation, ventilation doors 1 and 2 are closed, ventilation doors 3 and 4 are open, and the direction of airflow is shown in the solid line. During air reversing, ventilation doors 1 and 2 are open and ventilation doors 3 and 4 are closed, the direction of the airflow is shown in the dotted line.

On 7 December 1979, a fire broke out in the lower part of the west limb belt at a mine of the Yima Bureau, and mine wide air reversing was used to rescue people in distress, but it was unsuccessful. On August 7, 1962, a fire broke out at the mouth of an intake shaft in a mine. Due to timely air reversing, none of the 1,300 underground personnel on duty were killed or injured. Air reversing enabled over 600 people to be safely evacuated.

(6) **Isolation Airflow**

The isolation of airflow is mainly a method of sealing the fire area if direct fire fighting is not effective. Depending on the concentration of gas and oxygen in the fire area, the containment measures may include increasing the air volume to dilute the gas or filling with inert gas to reduce the oxygen content, in order to prevent explosions during the containment process. This section is described in detail in Chap. 4.

3.2.2 *Example Analysis of Down-Hole Airflow Control*

[Case 1] Return airflow side angle connection airway's stabilizing airflow role

As shown in Fig. 3.8, the Figure (a) shows a complex crossover type ventilation network, consisting of airways 3-a and 2-b. This type of ventilation network is often

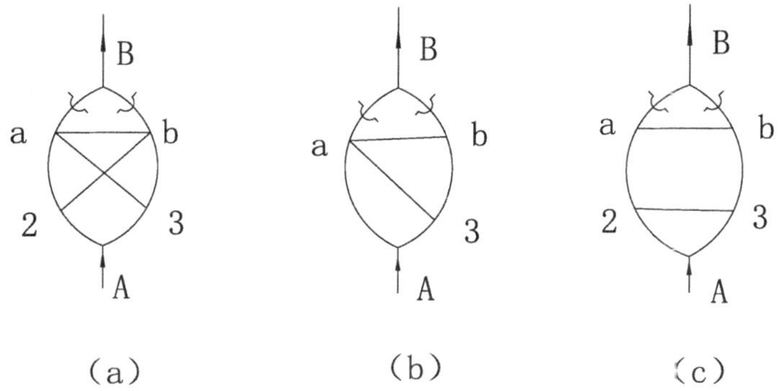

Fig. 3.8 "Stabilization airway" anti-airflow reversal example network diagram

found in the underground yard of the air intake and the main large roadway, as well as in the network across the mine area. When the ventilation door on the return airflow side is opened and the wind resistance $R_{a\text{-}b}$ or $R_{b\text{-}B}$ on the return airflow side of the roadway is increased or decreased, the airflow direction of the 3-a or 2-b angle connection airway will change and it is an unstable airway.

If the a-bangle connection airway is present, the direction of the airflow in the 3-a and 2-b cross-angle connection airway will not change, no matter how the return airflow side wind resistance $R_{a\text{-}B}$ or $R_{b\text{-}B}$ is changed.

For example, if $R_{a\text{-}B}$ is closed so that $Q_{a\text{-}B}$ is zero, $3 \rightarrow a \rightarrow b$. So the direction of the airflow in $3 \rightarrow a$ does not change. If there is no a-b airway, then the airflow in 3-a must be reversed. Therefore, it can be seen that returning the angle connection airway a-b on the airflow side keeps the airflow direction of the angle connection airway on the air intake side unchanged, as shown in (b) and (c) in the diagram. The presence of the angle connection airway a-b on the return airflow side increases the airflow stability and disaster resistance in the ventilation system.

The more angle connection airways, there are in a mine's ventilation system, the less stable the airflow is and the more likely it is that the airflow will reverse. If "stabilizing airways", such as a-b airways, can be identified, installed or restored in the ventilation network. This will reduce the possibility of airflow reversal in the angle connection airway on the air intake side, thus enhancing the stability and resilience of the ventilation system. This increases the stability and resilience of the ventilation system. This is also of great relevance to the control of airflow in the event of a fire in mine, in order to reduce the amount of harmful smoke that can spread.

Therefore, when the ventilation door on the return airflow side is opened or closed or when the wind resistance changes, it has no effect on the airflow direction of the other angle connection airways 3-a and 2-b, except for the airflow direction of the a-b airway. Therefore, we should have two different views on the understanding of the angle connection airway.

(1) An airway where the airflow can be reversed when the wind resistance of adjacent airways changes can be called a "true angle connection airway", such as an airway such as a–b.
(2) The position in the ventilation network is the angle connection airway But in practice, the possibility of the airflow reversal is small, or even the reversal cannot occur, the nature of this airway is similar to the parallel airway and can be called "quasi-angle connection airway", as in the Figs. 3.2b, 3.3a etc.

The airflow reversal of an angle connection airway depends on the proportional relationship between the wind resistance of the adjacent airways, which in essence is determined by the pressure difference between the two end points of the angle connection airway. Any angle connection airway that changes the ratio of its four adjacent air resistances will inevitably result in airflow reversal, otherwise it is not an angle connection airway. For example, the 2-b and 3-a angle connection airways in the diagram will also have airflow reversal if the air intake side air resistance $R_{a\text{-}2}$ and $R_{a\text{-}3}$ are increased.

Therefore, the presence of an air intake in the ventilation system, especially the angle connection airway on the return airflow side, is not necessarily a bad thing. It reduces the resistance to ventilation in the mine and regulates the fan pressure and the distribution of the return airflow resistance. At the same time, it also serves to stabilize the airflow in the air section of the mine area.

How to recognize the "true" and "accurate" angle connection airway of a complex ventilation network is of special significance for daily ventilation management, especially for disaster prevention and fire handling.

[Case 2] Example of CO overrun caused by improperly positioned ventilation door

[Accident Overview]

Figure 3.9 shows the ventilation system and ventilation network of a mine area. Mined-out area is closed by 4 closures 6, 7, 8 and 9. In order to control the air leakage quantity of the mined-out area and the intake volume of the working face, two regulating ventilation doors are set at the return airflow path A and one regulating ventilation door is set at B. When the coal mining face is finished and the coal mining face starts production less than 2 months later, the ventilation system in the mined-out area confinement appears to be incomplete. When the coal face was finished, the coal face started to produce less than 2 months later, CO appeared in the mined-out area confinement and became more serious. The CO content reached 1.5–2.0% in confinement No. 8, sometimes not in confinement No. 6, but not in confinement Nos. 7 and 9.

[Analysis of the Causes of CO Occurrence]

(1) From the ventilation network, it can be seen that the mined-out area controlled by the four containment 6, 7, 8 and 9 is exactly at the angle connection airway position, which caused air leakage due to poor containment.

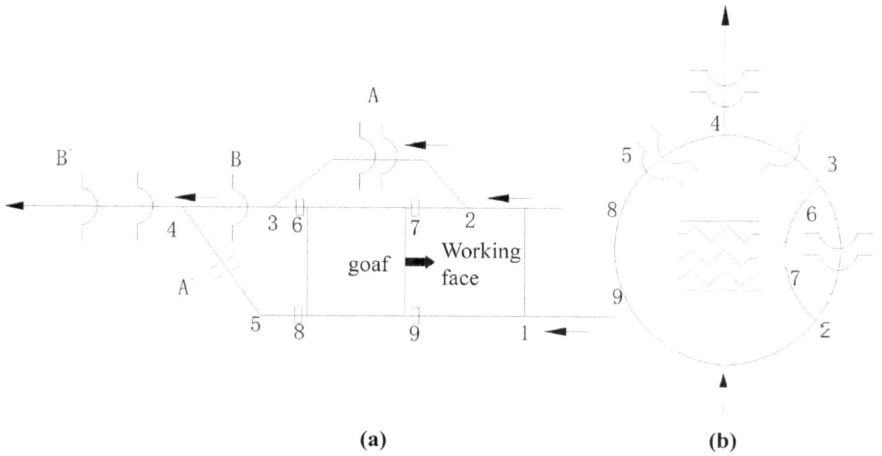

Fig. 3.9 CO overload due to improperly positioned ventilation doors. **a** Ventilation system diagram **b** ventilation network diagram

(2) $1 \rightarrow 2 \rightarrow A \rightarrow 3 \rightarrow B \rightarrow 4$ is a tandem airway. When the ventilation door at A is closed, the 2–3 parallel airway h_{2-3} air pressure increases and the 6–7 section pressure or differential pressure increases.
(3) When the B ventilation door is closed, the air pressure increases in parallel circuits 1–4 h_{2-3} and the pressure increases even more in sections 6–7.

As a result, the air in sections 6–7 must leak through the mined-out area to sections 8–9.

After testing: when the ventilation door is closed at A and B, the differential pressure of the No. 8 confinement can be 88.2–117.6 Pa. When the ventilation door is open at B and closed at A, the differential pressure of the Nos. 6–8 confinement is basically equal, both being 49–68.6 Pa.

The air leakage in the mined-out area is caused by the frequent opening and closing of the ventilation doors at A and B. The air leakage in the mined-out area is sometimes large and sometimes small. The airflow of the air leakage in the mined-out area goes in and out at times, just like a bellows, thus causing the "breathing phenomenon" in the mined-out area. This results in spontaneous combustion of the floating coal in the mined-out area.

The main reason for this was that the ventilation door was improperly positioned at A and B, which caused an imbalance in the differential pressure across the parallel circuit, sometimes large and sometimes small. At the same time, the mined-out area was located in the angle connection airway, and the combination of the two contributed to the spontaneous combustion of the coal.

[Measures to Deal with CO Overload]

Firstly, set up 2 ventilation doors at A, add 1 ventilation door at B to control airflow and air pressure balance, and remove 2 ventilation doors at A. Finally, set up 2 regulating ventilation doors at B, and then remove all the ventilation doors at B and A. At the same time, reinforce the 4 confinements (6, 7, 8 and 9) and inject yellow slurry into the mined-out area. The four containment doors (6, 7, 8 and 9) were reinforced and the mined-out area was filled with yellow slurry. Eventually, the signs of spontaneous combustion gradually disappeared.

[Case 3] Fighting fire by wind reduction method

[Accident Overview]

A high gas mine-20 m return airflow stone-gate was treated for a fire that formed when an electrical appliance caught fire (Fig. 3.10). The fire was located in the main return airflow path of the mine area and spread rapidly. Within 30 min, it had burnt the wooden shed supports, which put the main return airflow road and the main fan on the ground south of the mine in a very dangerous condition.

[Emergency Response]

Upon arrival of two teams from the rescue team, the workings of W_1, W_2 and W_3 were first stopped and the personnel evacuated. This reduced the gas concentration in the return airflow. In order to weaken the fire and allow access to the fire source, the ventilation doors D_1 and D_2 were opened by one third to reduce the air volume in the affected area. However, after opening the doors, the gas concentration at point A quickly increased to 2.2%. The ventilation door D_1 and D_2 had to be closed and normal ventilation resumed to reduce the gas volume to 0.7% within a short period

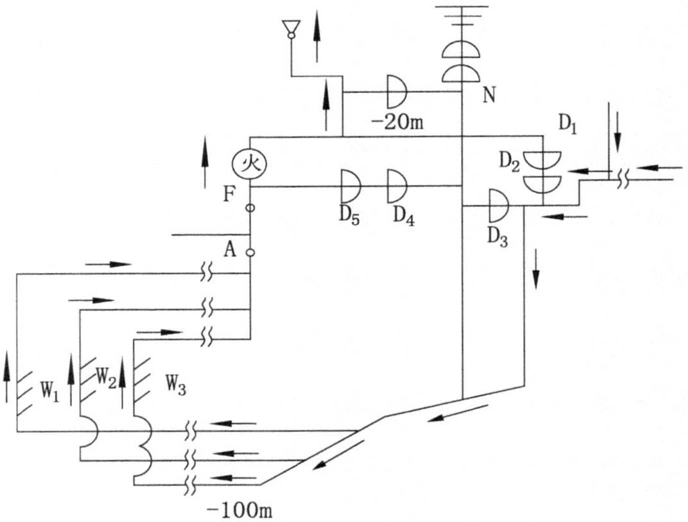

Fig. 3.10 Ventilation system diagram for an example of the wind reduction method of fire suppression

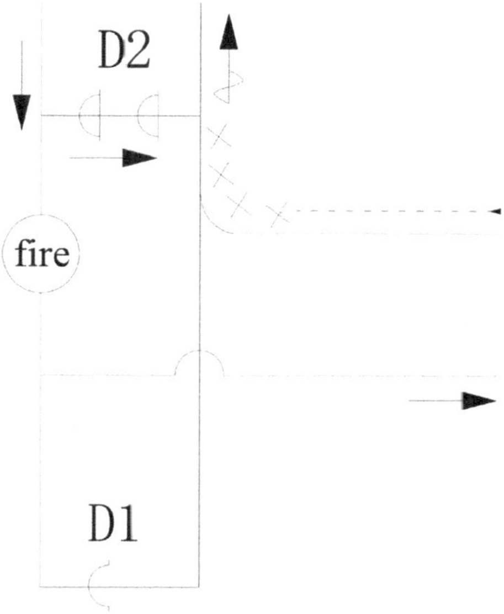

Fig. 3.11 Example of reducing the number of casualties caused by air volume

of time. The fire was then extinguished using the air reduction method and direct fire fighting.

[Case 4] Reducing air volume accidents causing injuries and fatalities

[Accident Overview]

A fire broke out in a mine tape downhill in Poland, as shown in Fig. 3.11.

[Emergency Response Process]

Nearly 100 people were working in the mining working face. In order to control the fire, the dispatch office brought the source of the fire under control without evacuating the personnel, which resulted in nearly 100 people dying of asphyxiation due to oxygen deprivation. It was later discovered that most of the personnel had gone up the hill near the return airflow and some were almost at the ventilation door. All were wearing filtered self-rescuers. All were analyzed as having died of asphyxiation due to oxygen deprivation. The fire may have been controlled by adjusting the ventilation door D_1 to control the source of the fire air intake, but it would have caused asphyxiation.

[Experience and Lessons Learned]

Filtered self-rescuers, used in an environment where the oxygen content cannot be less than 17%. If the oxygen is below 17%, it is not possible to ensure that people can save themselves. In fire fighting, premature control of the source of the fire air intake, and cause personnel to be oxygen-poor suffocation lessons.

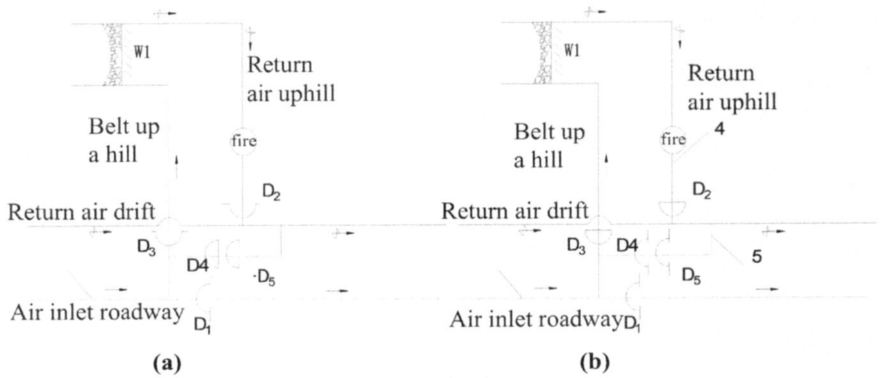

Fig. 3.12 Layout of the additional dedicated air reversing path in the upper mountain mine area. **a** Ventilation system in normal times **b** ventilation system in fire times

[Case 5] Specialized smoke exhaust roadway with ventilation door to achieve airflow short circuit

[Basic Information]

As shown in Fig. 3.12, the roadway arrangement system in the up hill mine area, its return airflow up hill is actually downward airflow. If a fire occurs in it, the airflow of the main airway reverses under the action of the reverse fire wind pressure as well as the smoke flow reverses and invades the work surface personnel.

[Disposal Method]

If a dedicated fume extraction roadway is excavated during the layout of the working face and a ventilation door D_4 is installed, it is normally closed (see Fig. 3.15a) and opened during the mine fire. In the upper hill of the air intake in the mine area and in the upper hill of the return airflow, there are normally open ventilation doors D_2 and D_3 respectively. In the large air intake roadway, there is a normally open ventilation door D_1. As shown in Fig. 3.10, in the event of a fire in the upper hill of the return airflow the mine area can be opened. When a fire occurs in the return airflow uphill, the airflow in the mine area can be short-circuited to discharge the smoke flow, and finally close the D_2, D_3 ventilation door to limit the development of the fire, as well as use the normally open ventilation door D_1 in the large roadway to adjust the air volume to dilute the smoke flow. This is done as follows.

As shown in Fig. 3.10, the roadway arrangement system of the downhill mine area, in the upper entrance of the air intake downhill, creates a special roadway for smoke extraction 5, and normally sets up a normally closed ventilation door D_4, D_5 to isolate it (Fig. 3.10a), in order to ensure normal ventilation of the downhill mine area. In normal times, the airway line is intake airway $\rightarrow$ D_3 $\rightarrow$ working face W_1 $\rightarrow$ D_2 $\rightarrow$ return airway.

If a fire occurs in the air intake up hill, the high temperature fire smoke with airflow downhill produces local fire wind pressure, the direction of its effect is opposite to the original wind direction, as in Fig. 3.10b. With the development of the fire, local fire wind pressure increasing, air intake downhill main airway may occur when the smoke flow backwards, immediately open the special air reversing roadway normally closed ventilation door D_4, D_5 so that the mine area air intake short circuit (through the airway line is the air intake roadway → D_4 → return airflow roadway). The back-flow will be discharged into the return airflow roadway via the dedicated air reversing roadway. At the same time, close the normally open ventilation doors D_1 and D_2 to restrict the development of the fire, and open the normally open ventilation door D_3 to regulate the air volume to dilute the smoke flow in the dedicated roadway, as shown in Fig. 3.10b. At this point, an ambulance team can be sent down the return airflow system to rescue people in distress and deal with the fire

[Case 6] Airflow short-circuit roadway on angle connection airway caused accident

[Accident History]

As shown in Fig. 3.13, the mine, which was low gas mine, caught fire in the −210 level intake airway 3–4 roadway due to a short circuit in the cable, causing the wooden support and cable to catch fire. After the incident, the work-face personnel were the first to be exposed to the smoke from the fire. The workers then retreated from return airway 6 → 7 → 8 to liaise with bypass 8–10 and through ventilation door D_1 to the safety of air intake large roadway 1 → 2 → 10. When the D_1 ventilation door was opened shortly, the fire fumes also followed the bypass from 8–10 and were strung upwards with the incoming airflow to 10 → 11 → 12 → 13 → 14 and the W_2 working face. At the same time, because the local fan at 11 o'clock did not stop, the toxic fumes were sent into the W3 digging face. At the location of the fire. the work-face personnel were safely evacuated though. However, the personnel of W_2 working face and digging face were instead greatly affected, and its enlarged the scope of the disaster and increased the casualties.

[Cause of Accident]

When the D_1 ventilation door is opened, the fresh airflow should have been discharged from 10 → 8 → 9 via return airflow up the hill. But because 10–8 bypass for angle connection airway. At the same time, 8–9 return airflow up the hill due to the old and dilapidated, roadway section reduced, wind resistance R_{8-9} increased, forming the discriminatory Equation.

$$\frac{R_{8-9}}{R_{2-8}} > \frac{R_{10-9}}{R_{2-10}}$$

In the 2–8 airway, the hot fire gases create a fire wind pressure which drives the airflow in reverse from 8 to 10, thus expanding the affected area and increasing the extent of damage.

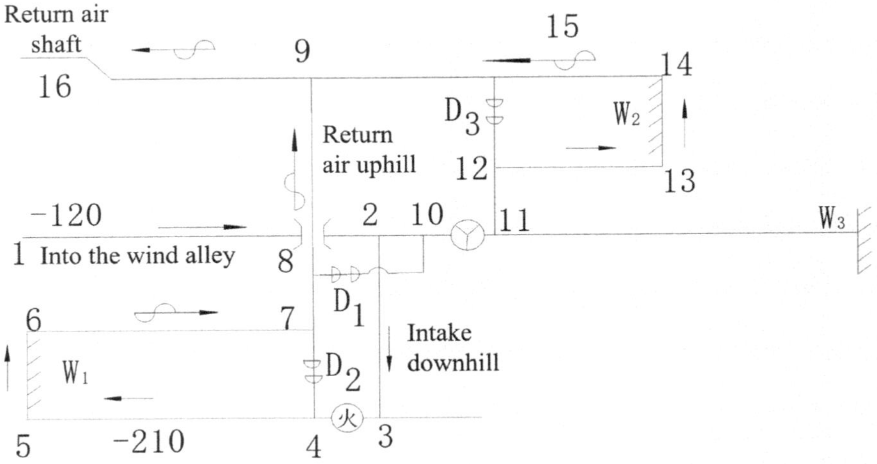

Fig. 3.13 Example of an accident caused by an airflow short-circuit roadway located on an angle connection airway

[Lessons from the Accident]

The lessons to be learned from this accident case are as bellow.

(1) The angle connection airway should be strictly managed at ordinary times, so that you know what you are doing. You should know that the direction of the airflow of the angle connection airway is affected when the ventilation door is opened or closed.
(2) When using the "short-circuit airflow" method to deal with fires, the direction of airflow flow after a short circuit must be correctly determined by applying ventilation theory.
(3) The sectional return airflow or the total return airway should not be too small, and the wind resistance R should be small to reduce the phenomenon of excessive local resistance. Otherwise, it is not only unfavourable to the usual ventilation, but also when a fire occurs it is easy to cause the airflow to reverse and reverse, which increases the difficulty of rescue and relief and causes casualties. Therefore, the management and maintenance of the total return airway should be strengthened in normal times.

[Case 7] Failure to take airflow short-circuit measures causing injury or death

A local coal mine—40 m air intake large roadway A where an open fire (shown in Fig. 3.14), fire smoke invasion route A $\rightarrow$ 2 $\rightarrow$ 3 $\rightarrow$ 4 $\rightarrow$ 5 $\rightarrow$ 6 $\rightarrow$ 7 $\rightarrow$ 8 $\rightarrow$ 9 $\rightarrow$ 10. Due to the fire is strong, W_1 work surface personnel upwind evacuation, and due to the smoke large vision, work surface personnel all in 3 $\rightarrow$ 4 downhill poisoning collapse. After the ambulance team arrived at the accident mine, the fire source A was burning badly and the personnel could not reach the 3 $\rightarrow$ 4 lower hill through

A. Afterwards, the D_2 ventilation door was opened to rescue people from the return airflow path 9 → 8. But due to the low return airflow path (0.5 m high), thick smoke and high temperature, it was difficult for the rescue team to enter, the D_3 and D_4 ventilation doors could not be opened and the airflow could not be short-circuited (the route of airflow movement after the short circuit: A → 2 → 8 → 9).

The main fan air reversing on the ground could not be carried out because there was no air reversing equipment. Later, the airflow was reversed by pressing the bureau fan downwards (the reverse airflow was 10 → 9 → 8 → 7 → 6 → 5 → 4 → 3 → 2 → 1). The rescuers opened the D_1 ventilation door and went into the 3 → 4 downhill, where they rescued the people in distress. The personnel were rescued, but all were poisoned and killed due to the length of time it took. When the mine fire accident first started, the ambulance crew could have quickly opened the D_3 ventilation door, short-circuited the smoke airflow, as well as concentrated on rescuing and guiding the personnel through the 9–8 return airflow channel, opened the D_2 ventilation door and entered the air intake ramp. In this way, the W1 workings could be quickly evacuated. However, the 9–8 return airflow path was too low for the ambulance crew to pass with their instruments, so the escapees suffered.

[Case 8] Example of airflow short circuit relief failure

[Accident Process]

A mine air intake inclined shaft shaft 150 m for the electrical substation due to electrical fire (Fig. 3.15, ignited the fire caused by the board, after the fire, W_1, W_2 working surface personnel affected by the fire gas (airflow route is 1 → 2 → 3 → W_1 → W_2 → 4).

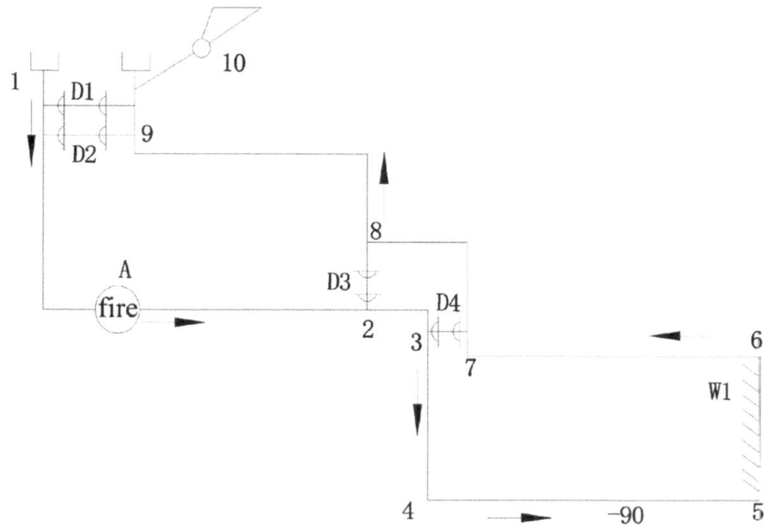

Fig. 3.14 Schematic diagram of a local coal mine—40 m intake airway large roadway firing

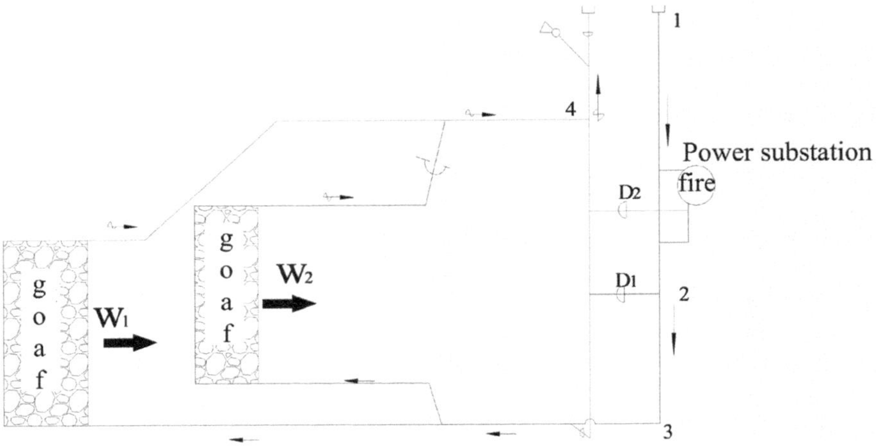

Fig. 3.15 Diagram of a mine air intake slope shaft 150 m into the shaft for the substation and pump room M due to electrical fire

[Emergency Relief Process]

When the rescue team arrived at the accident shaft, they decided to air reversing quickly according to the situation of the people in distress at the fire location. Air reversing allowed the rescue team to go from the return airflow level—40 m to the W_1 and W_2 working faces to rescue and guide the people in distress. But due to the failure of the main fan air reversing device on the ground, the airflow could not be reversed (after reversing the airflow route was 4 → W → W → 3 → 2 → 1). However, due to the failure of the main fan air reversing device on the ground, the airflow could not be reversed (after reversing the airflow route was 4 → W_1 → W_2 → 3 → 2 → 1) and the escapees could not be rescued. Later, a team was sent to open the two ventilation doors of D_1 to short-circuit the airflow (after short-circuiting the airflow path: 1 → 2 → 4), but the fire was so strong that the smoke was thick. The team approached the D_1 ventilation door several times, but failed to open it. The airflow short-circuiting failed to achieve its purpose. At this time, the personnel at the W_1 and W_2 working faces were already in a very dangerous and dense smoke environment, and the escapees were seriously poisoned, resulting in very tragic consequences.

[Experience and Lessons Learned]

Through the painful lesson of the fire accident in this mine, the mine air reversing device must be tested regularly. When airflow reversing cannot be carried out in time, the two D_1 ventilation doors should be opened quickly and promptly in order to short-circuit the airflow. Through this fire accident, in the event of a fire accident, no matter how bad the environment at the accident site is, the rescue task can be completed and the people in distress will be able to escape.

[Case 9] Airflow short-circuit disaster relief success story

[Basic Information]

As shown in Fig. 3.16, a low gas mine N_1, N_2 and N_3 are intake and return airflow downhill. The mine area substation is a wooden bracket chamber. A fire in an electrical appliance causes the brackets in the substation to burn, and smoke from the fire flows to the four working faces (W_1, W_2, W_3, W_4) and four digging heads (C_1, C_2, C_3, C_4), exposing 30 personnel to the smoke (as shown in Fig. 3.16). At this point, the personnel of the W_1 and W_2 faces retreated quickly when thin smoke was detected at the beginning of the fire, and evacuated to the N_1 and N_2 downhill air intake sides via ventilation door D_1. A few people from the W_1 working face were slow to retreat and were fumigated by CO at the W_1 working face return airflow roadway B. When the smoke was detected on the W_3 and W_4 faces and at the four boring heads, it was too late to retreat to the N_1, N_2 and N_3 lower hills, and due to the poor visibility of the smoke, they were all killed near the three lower hills.

[Rescue Process and Disposal Process]

After the ambulance team arrived at the scene, on the one hand, using direct methods to extinguish the fire. At the same time, the two ventilation doors of D_2 were opened, and the direction of the thick smoke airflow changed from the original flow to A → D → E → F route into the return airflow channel. The other areas were only thin smoke state. At this time, the ambulance team after one and a half hours of intense rescue. In addition to the lower part of the fire source (N_1 down the hill at a) three

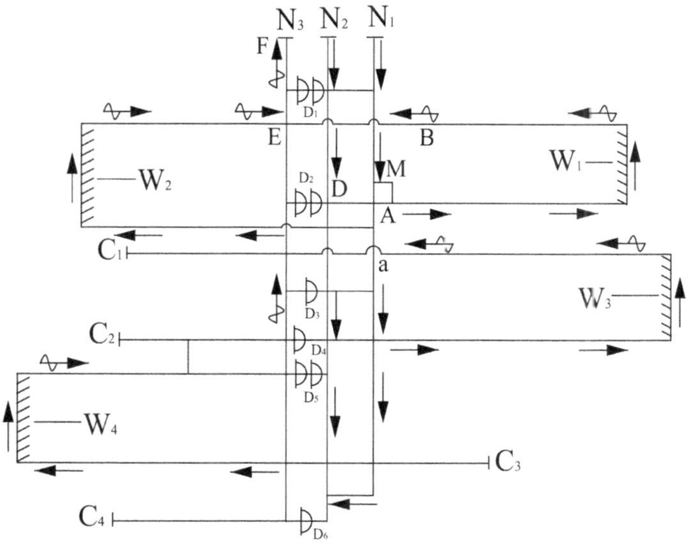

Fig. 3.16 Example of an electrical fire in a wooden braced chamber in a mine area substation

comrades were killed due to severe poisoning, the rest of the people in the disaster area were all rescued.

Because the mine is a low gas mine, airflow short circuit, there is no danger of gas accumulation. When open D_2 ventilation door, the area affected by dense smoke into a thin smoke state, the rescue squad in the rescue line of sight clear, fast action, the victims quickly out of danger. If D_2 ventilation door does not open, the rescue team in the smoke reconnaissance rescue, action is very slow, even if the victims will be rescued, the victims eventually died because of poisoning for too long. Therefore, according to the specific situation of the mine fire, the airflow short circuit control measures were very effective and correct.

In addition, the mine area substation should be set up in a non-combustible bracket roadway. Although an electrical fire occurred, it would not have caused such a serious disaster. A quantity of fire extinguishing equipment should be stored in case of emergency.

[Causes and Lessons Learned]

This accident could have been better if local airflow regulation had been applied in time. Perhaps no one would have been killed, and it is analyzed as follows.

If the wind barrier is hung or the board is closed at A and B, the D_2 ventilation door is half-opened to control the air volume to ensure the smooth discharge of fire smoke and not to short-circuit the airflow in large quantities, as well as to maintain the original airflow direction. The new air enters from the lower hill N_1, flows through the mining surface and the lower part of the lower hill, dilutes the smoke and then discharges from the lower hill III. This will quickly expel the smoke from the affected area, so that the trapped personnel will be in the new airflow and carry out self and mutual rescue, which are likely to be all out of danger.

[Case 10] Belt up hill fire accident working face local air reversing

Figure 3.17 shows a centralized belt going up a hill in the five pan area of the third level of a mine, where a fire was caused by the belt rubbing against the main drum, spreading the smoke stream and killing a number of workers.

From the ventilation system diagram, it is entirely possible to use local air reversing, the original ventilation system is $1 \rightarrow 2 \rightarrow 4 \rightarrow 5 \rightarrow 6 \rightarrow 7 \rightarrow 8$, as long as the ventilation door is opened. While closing the D_3 and 7–8 return airflow in the upwind side of the fire source D_2 ventilation door, you can achieve local air reversing by $1 \rightarrow 2 \rightarrow 7 \rightarrow 6 \rightarrow 5 \rightarrow 4 \rightarrow 9$ air reversing system, personnel along the $5 \rightarrow 6 \rightarrow 7 \rightarrow 2$ route to withdraw, fire smoke from the belt up the hill $4 \rightarrow 9$ outflow. But because there is no prior consideration and no construction of preparatory control ventilation door D_3 and D_2. This led to serious consequences.

With the usual 2 min of preparation a partial return of wind can be achieved and people can be rescued in time. However, delays due to the absence of a preparation ventilation door increase casualties.

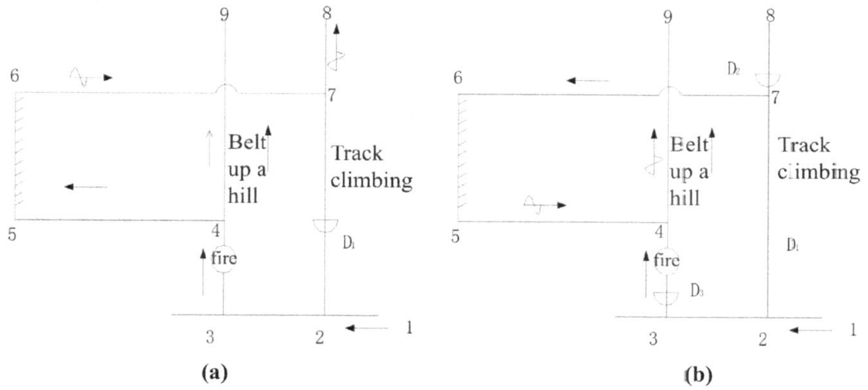

Fig. 3.17 Example of partial air reversing at an accidental working face on a belt. **a** Diagram of pre-ventilation system for air reversing **b** diagram of rear ventilation system for air reversing

[Case 11] Local air reversing using corner networking paths

[What Happened]

Figure 3.18 shows a schematic diagram of a low gas mine with an annual production capacity of 300,000 tonnes. The coal is prone to spontaneous combustion and the coal dust is explosive. Diagonal ventilation system, with air intake for the belt downhill and return airflow for the track downhill. the downhill slope length is 350 m. As the fifth belt in the belt downhill caught fire, 65 workers were trapped in the smoke down the shaft at the time and the fire spread quickly down the airflow.

[Handling Measures]

In order to rescue the personnel, it was decided to use air reversing measures according to the ventilation system diagram and the ventilation network diagram.

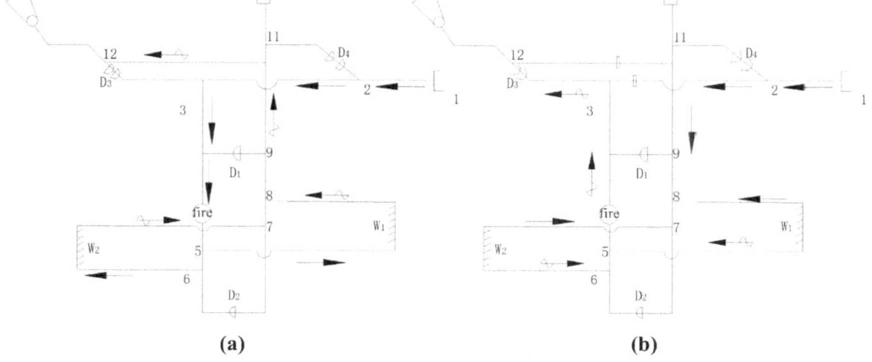

Fig. 3.18 Regional air reversing example ventilation system diagram. **a** Diagram of pre-ventilation system for air reversing **b** diagram of rear ventilation system for air reversing

The steps for adopting measures based on the characteristics of the corner network are as follows.

(1) Opening of the two ventilation doors A and B.
(2) At the same time in 2–3 and 10–12 airway to build E and F two confinement. The airflow is changed from the original system of 1 → 2 → 3 → 4 → 5 → 6 → 7 → 8 → 9 → 10 → 12 to an air reversing system of 1 → 2 → 11 → 10 → 9 → 8 → 7 → 6 → 5 → 4 → 3 → 12. Personnel can be safely evacuated by the 7 → 8 → 9 → 10 → 11 → 2 → 1 route.

It should be noted that the direction of action of the fire wind pressure generated by the fire is the same as the direction of the airflow after air reversing. The direction of the airflow after air reversing is therefore stable. If the direction of the airflow after air reversing is not the same as the direction of the fire wind pressure, it is important to prevent a sudden reversal of the airflow, to avoid bringing unpredictable hazards to the rescue and relief work.

[Case 12] Example II of belt conveyor downhill on fire local air reversing

In August 2000, a fire was caused by high temperatures in the hydraulic coupling at the head of the second part of the belt conveyor downhill of the belt conveyor at a mine in Datong west limb. Local air reversing was carried out in order to rescue the affected personnel, as shown in Fig. 3.19.

Pre-ventilation system for air reversing: fresh airflow from the west large roadway → car park → ventilation door C → belt conveyor downhill → mine area flat roadway → mining working face → track downhill → ventilation door D → total return airflow roadway.

Local air reversing of the rear ventilation system: first close ventilation doors C and D as well as open ventilation doors A and B. Fresh airflow from the west large

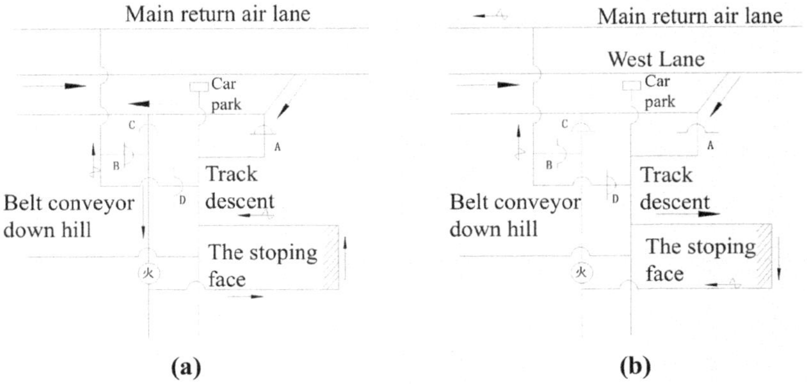

Fig. 3.19 Belt conveyor downhill fire local air reversing example 2 ventilation system diagram. **a** Diagram of pre-ventilation system for air reversing **b** diagram of rear ventilation system for air reversing

roadway → yard → ventilation door A → track downhill → mining working face → district flat roadway → belt conveyor downhill → ventilation door B → total return airflow roadway.

[Case 13] Example of local air reversing during a mining working face fire

A fire broke out in the third part of the chute head of the 305 mining working face transport chute in the upper mine area on the west side of a mine. Local air reversing was carried out to rescue the affected personnel, as shown in Fig. 3.20.

Pre-ventilation system for air reversing: fresh airflow from intake airway → track up hill → ventilation door 1 → transportation down trough → mining working face → return airflow flat roadway → return airflow up hill → ventilation door 2 → total return airflow roadway.

Local air reversing of the rear ventilation system: first close ventilation door 1,2 and open ventilation door 3,5. fresh airflow from intake airway → track up hill → ventilation door 3 → return airflow flat roadway → mining working face → transportation down the chute → ventilation door5 → return airflow roadway → total return airflow roadway.

[Case 14] Local air reversing in mixed ventilation situation

At 6am on 27 February, a fire broke out in the 780 transport roadway of a mine at a distance of 3700 m from the shaft entrance. At that time, a heavy coal train drove out to the entrance of the shaft and when it reached the U-shack roadway at 3700 m, the wire frame motor car fell off the roadway and hit the conductive bow on the U-shack, causing a short circuit, which caused an arc spark and then ignited the wooden backboard and hook wood in the roadway, resulting in a fire. Upon receipt of

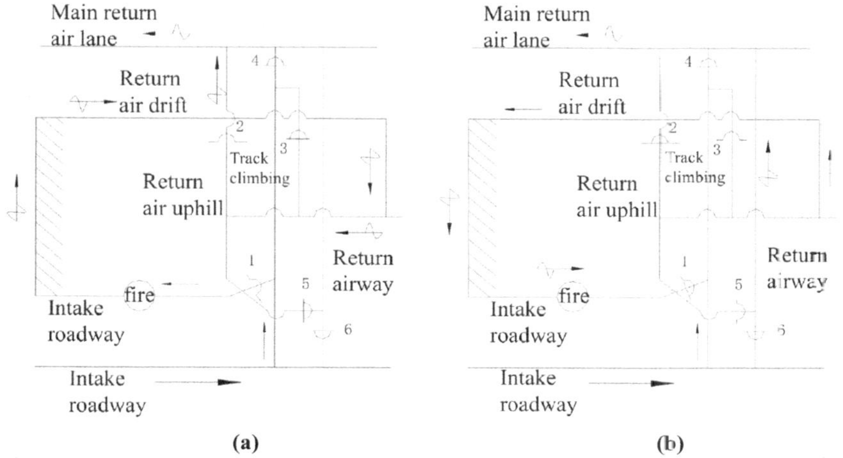

Fig. 3.20 Ventilation system diagram for example of local air reversing during a mining working face fire. **a** Diagram of pre-ventilation system for air reversing **b** diagram of rear ventilation system for air reversing

the report, an emergency evacuation order was issued and all personnel were safely in the shaft within 1 h.

The reason why air reversing was not carried out immediately at that time was that the fire in the 780 big roadway was very small at first and the smoke and harmful gases entering the 660 big roadway were still very small, the personnel from the four descents closest to the source of the fire could also smell a little smoke. There was still time to evacuate the underground personnel from the 660 big roadway.

The air reversing situation is shown in Fig. 3.21. A temporary wall of No. 1 plank was set up at the 780 large roadway air intake, the counter rotating fan was stuck to the wall. Then, the plank wall was sealed tightly with fast sealing material, all works were completed, as well as personnel within 20 m of the shaft were evacuated and alerted. The fan was activated and the air reversing was prepared.

[Case 15] Dedicated air reversing roadway air reversing

The air reversing is carried out using a dedicated air reversing roadway as shown in Fig. 3.22.

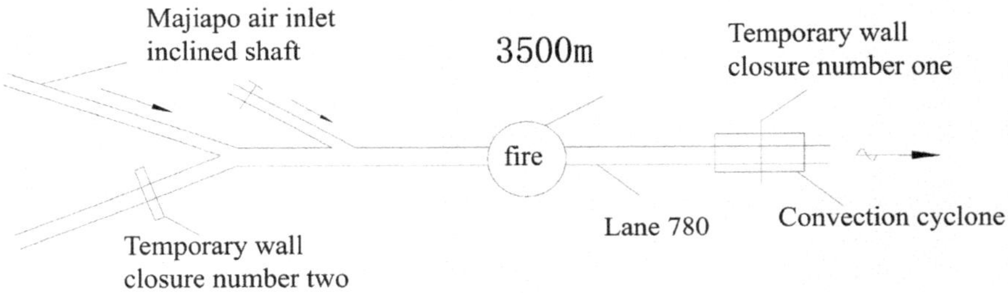

Fig. 3.21 Local air reversing example ventilation system diagram

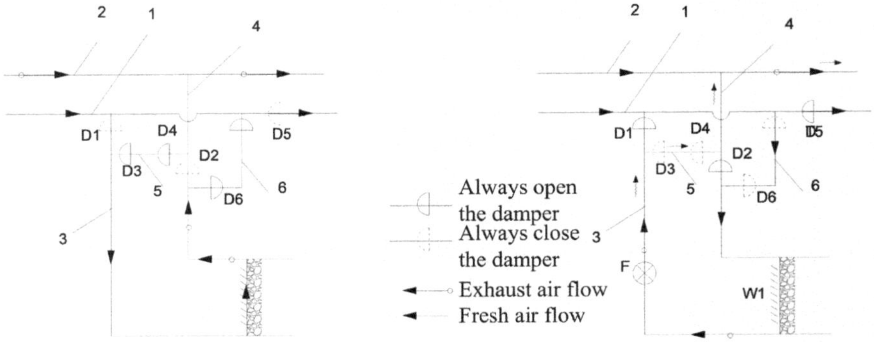

Fig. 3.22 Layout of connection roadway and dedicated air reversing roadway. 1—air intake large roadway; 2—return airflow large roadway; 3—air intake downhill; 4—return airflow downhill; 5—connection roadway; 6-dedicated air reversing roadway

Maintain a connection roadway 5 between the inlet and return airflow downhill, and open an additional short roadway6, with a normally open ventilation door D_1, D_2 in the inlet and return airflow downhill, and a normally open ventilation door D_5 ventilation door D_3 in the air intake large roadway. D_4, D_6 and D_7 are normally closed (Fig. 3.22) to ensure that new air is sent to the working face through the air intake downhill. Then, the lack of air is fed into the main return airflow downhill, thus forming the ventilation system in the downhill mine area. The airway line is 1 → 3 → 4 → 2.

If a fire occurs in the air intake downhill as shown in Fig. 3.22. Generally, the fire smoke will go down with the airflow and invade the working surface, and then discharged into the total return airflow roadway. But the high temperature fire smoke downstream produces local fire wind pressure, the direction of its effect is opposite to the original airflow direction, with the development of the fire. The local fire wind pressure is also getting bigger and bigger. At this point, the main airway smoke flow reversal will be difficult to avoid. If the reversal of the smoke flow occurs, the fire smoke will come up to the entrance of the lower hill and merge into the horizontal air intake of the large roadway, which will probably invade all the subsequent mine areas with the air intake, with unimaginable consequences. A reliable way to deal with this is to open the normally closed ventilation doors D_3, D_4, D_6, D_7 in the connection roadway and the special air reversing roadway as soon as you notice the first signs of smoke flow reversal in the main airway, and to close the inlet and return airflow downhill and the large air intake roadway. The normally open ventilation door D_1, D_2, D_5 enables air reversing in the mine area, creating favourable conditions for the ambulance crew to go down through the air intake after air reversing to rescue people in distress and deal with fires. The state of the ventilation door and the direction of airflow after air reversing is shown in Fig. 3.22. The airway line after air reversing is 1 → 6 → 3 → 5 → 4 → 2.

[Case 16] Partial air reversing playing confinement fire prevention

The high tensile strength tape in the upper part of a mine caught fire and ignited the connection roadway between the upper part of the tape and the upper part of the track. 80 m of the connection roadway, most of which was semi-coal rock, had more than 30 m of severe roofing and could only be crossed by wind and no pedestrians. The connection roadway was filled with debris, supports, various materials and cable skins and other combustible materials. The connection roadway was burning very intensely. In order to put out the connection roadway fire as soon as possible and to resume production in the mine area, we tried to put out the fire directly without leaving the fire area. Since it was not possible to pass through, the method of high-foaming was used, and high strength tape was built at the mouth of the connection roadway up the hill to launch high-foaming, but it was not effective (Fig. 3.23).

Later, the connection roadway outlet hit the board closed and loess wall, to be closed, its attempt to fire directly from the return airflow side. At this point, the fire had already burned to the connection roadway outlet, and the use of high-pressure water guns to extinguish the fire was actually equivalent to fighting the fire alone. Without a clear return airflow path, the water volume was too small to suppress the

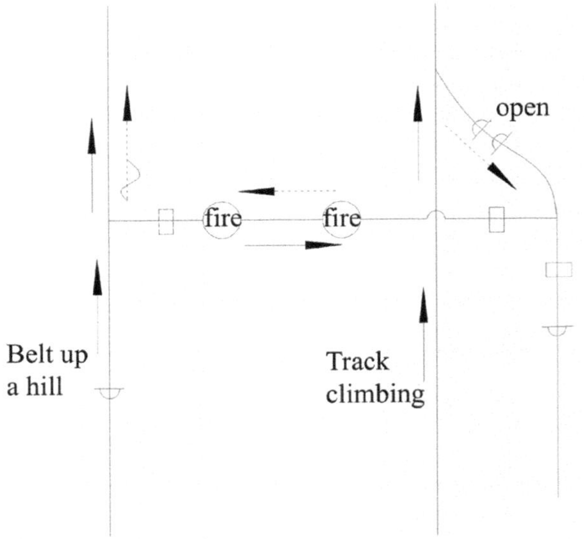

Fig. 3.23 Diagram of the ventilation system for controlling the airflow to beat the airtight example

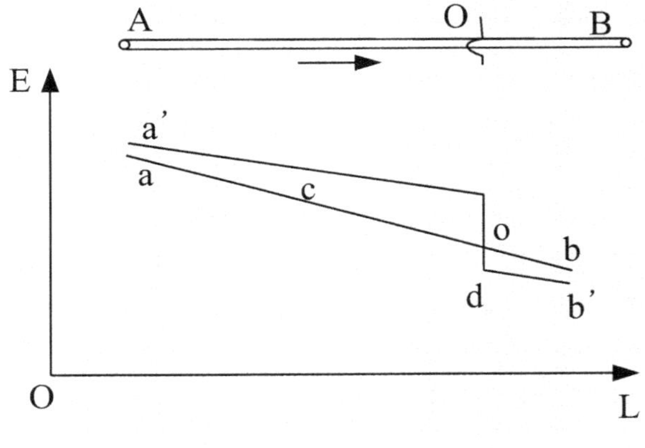

Fig. 3.24 Principle of regulating air window

fire, while the water volume was too large to discharge the water vapour produced. This intermittent extinguishing for a few days, but the effect is not obvious, and several small water and gas explosions occurred. Then in the return airflow mouth, build plate closed, send high-foaming, its effect is still not satisfactory. After several times of direct fire extinguishing failed, the direct fire extinguishing solution had to be abandoned and a complete closure was decided.

It is not feasible to work on the return airflow side, because the fire burns out immediately once the water stops. Therefore, the containment had to be carried out on the air intake side. The sequence of closure was decided upon by first closing the return airflow opening which is connected to the upper hill of the track, then removing the closure and the loess wall from the air intake opening of the high strength tape and constructing a high weight closure. Before closing, the ventilation system is adjusted so that the airflow of the connection roadway is reversed by the track air intake, the tape up the hill return airflow. After the air reversing of the connection roadway, the connection roadway through the high strength belt at, closed within the escape of flame up to several meters away, can be permanently sealed. The closures are connected by large diameter steel pipes and are temporarily unsealed. In the second step, the original system is restored and the airflow of the connection roadway is directed from the high-strength belt to the orbital entrance. Successively, the holes left were sealed tightly to achieve a complete closure. Finally, grouting was carried out to extinguish the fire completely.

This example illustrates the importance of using the correct ventilation method when dealing with fires. Without the use of local air reversing, it would not have worked on the downwind side of the fire, so close to the source of the fire, where the temperature reached hundreds of degrees.

3.3 Localized Ventilation Pressure Regulation Techniques and Examples

3.3.1 Ventilation Pressure Regulation Technology

(1) Principle of uniform pressure fire prevention technology

Ventilation pressure is the driving force behind the flow of airflow. If air leakage airflow exists, air leakage air pressure must be applied. In order to reduce the amount of air leakage, the air leakage pressure must be regulated. This method of trying to reduce the pressure difference between the two ends of the air leakage channel in the mined-out area in order to reduce air leakage, inhibit spontaneous combustion of coal, inert and extinguish the fire area is called uniform pressure fire prevention technology.

According to the law of resistance, the difference in air pressure acting on the two ends of the air leakage channel [2].

$$H = R_1 Q + R_1 Q^2 \tag{3.1}$$

In the Equation,

H—ventilation resistance of the wind tunnel, Pa.

Q—air volume of air leakage, m^3/s.

R_l—air leakage channel laminar wind resistance, Ns/m^5.

R_2—air leakage channel turbulence wind resistance, Ns^2/m^8.

From the Eq. (3.1), it can be seen that the air leakage quantity $Q = 0$ in the mined-out area if the air leakage wind resistance R is constant and measures are taken to make $H = 0$ [3].

But the actual use of equal pressure method of fire prevention, and not possible to make mined-out area channel at both ends of the air leakage pressure difference is absolutely equal to zero. The reasons are as bellow: (1) mined-out area air leakage channel at both ends of the pressure difference at any time by mine ventilation pressure, atmospheric pressure, fire area fire wind pressure and other factors; (2) There is a "flammable wind speed zone" for the spontaneous combustion of mined-out area coal, and neither high nor low wind speeds are likely to cause spontaneous combustion of mined-out area coal. Therefore, it is not necessary to make the differential pressure between the two ends of the air leakage channel in the mined-out area equal to zero. As long as the differential pressure between the two ends of the air leakage channel is reduced to a certain fire safety value that can prevent spontaneous combustion of coal in the mined-out area, it is sufficient. According to the relevant literature, its value is about 1–2 mm H_2O. In this paper, according to the formula for the maximum permissible air leakage quantity in uniform pressure fire prevention derived earlier, to determine the safety value of the differential pressure in the process of equalizing pressure, i.e. the redundancy of pressure regulation.

According to the mechanism of equal pressure and different conditions of use, uniform pressure fire prevention technical measures are broadly divided into open area equal pressure and closed area equal pressure. The open area uniform pressure is to establish the uniform pressure system in the production working face, to reduce the air leakage in the mined-out area, to inhibit the spontaneous combustion of coal relics, to prevent carbon monoxide and other toxic and harmful gases from gathering beyond the limit or gushing out to the working area, so as to ensure normal production. The closed area uniform pressure is to take uniform pressure measures to prevent the spontaneous combustion of coal in the closed area where spontaneous combustion of coal is likely to occur or has already occurred. The closed area equal pressure is a confined area where spontaneous combustion of coal may occur or has occurred.

(2) Mechanism of uniform pressure fire prevention technology

Uniform pressure fire prevention techniques in open areas include air window pressure regulation, fan pressure regulation and combined fan-air window pressure regulation measures.

(1) Adjusting the air window to regulate pressure

As shown in Fig. 3.24, the ventilation conditions change when the air window is adjusted in the airway. A comparison of the changes in pressure ramps before and after the installation of the air window shows that.

① The pressure energy of the airflow on the upwind side of the air window increases and the pressure energy of the airflow on the downwind side of the air window decreases, with the increase and decrease decreasing with increasing distance from the air window.
② Air window before and after the airway due to the reduction of air volume pressure slope line becomes slower. Air window regulating method if the ventilation door is properly set. In the mined-out area wind resistance is unchanged, the working

face diffusion air leakage is bound to be reduced. This is beneficial to suppress the spontaneous combustion of coal in the mined-out area caused by diffuse air leakage, and often the spontaneous combustion that has developed will be extinguished as a result. Even the spontaneous combustion of the coal left behind from the mined-out area (the airflow branch of the air leakage is equivalent to the diagonal branch of the diagonal air network) can be suppressed to a certain extent. However, it must be noted that the area of the pressure regulating ventilation door cannot be reduced indefinitely. On the one hand, the reduction of the ventilation door area will reduce the effective air volume at the working face, which is not conducive to safe production. On the other hand, an excessive reduction of the ventilation door area will reverse the airflow direction of the air leakage in the mined-out area.

The mechanism of air window pressure regulation is to increase the resistance and reduce the airflow, changing the pressure distribution on the airway in question in order to achieve the purpose of pressure regulation. Therefore, its application presupposes that the air volume of this airway can be reduced.

(3) Fan regulation

The air window is an energy consuming device and the fan is a power device. The two regulating mechanisms are opposite.

As shown in Fig. 3.25, the equalizing fan is installed in the airway so that the air volume is greater than the original air volume, then the pressure ramp at the working surface changes before and after the pressure adjustment.

A comparison of the two shows bellows.

(1) A sudden increase in the pressure energy of the airflow at the installation fan, the increase being equal to the full pressure of the fan.
(2) The upwind side of the fan decreases in airflow pressure energy and the downwind side increases in airflow pressure energy, with the magnitude of the change decreasing as the distance from the fan increases.
(3) Increase in air volume on the airway and steepening of the pressure slope.

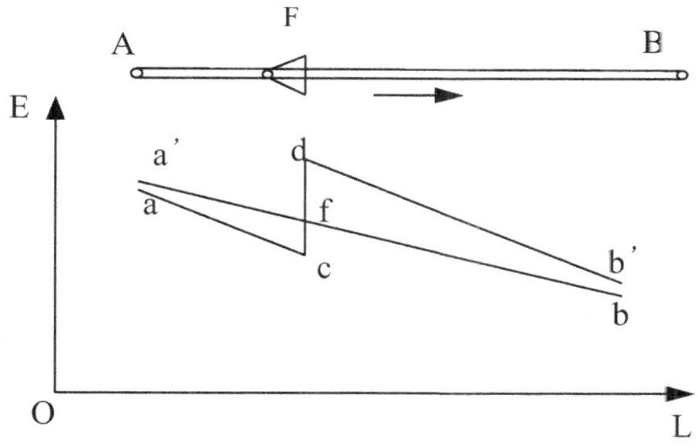

Fig. 3.25 Principle of fan regulation

If the fan is properly selected, the fan regulating method has a certain inhibiting effect on the spontaneous combustion of the mordant carbon caused by the external leakage of airflow from the mined-out area, which is beneficial to the normal ventilation of the working face. It should be noted that when using this method of pressure regulation, the fan capacity should not be increased indefinitely simply for the purpose of disappearing carbon monoxide or other signs of spontaneous combustion in the corner of the working face. If the local air pressure at the working face is too high, it can create a supply of air to the inside of the mined-out area, which is quite dangerous.

(4) Combined fan-air window pressure regulation

As mentioned above, the use of air window or fan alone will inevitably result in a reduction or increase in air volume on the airway. In the process of regulating pressure, it will have different degrees of impact on the adjacent airway. In the actual work of fire prevention, there are certain requirements for the increase or decrease of air pressure and the size of air volume, which cannot be achieved by using air window or fan alone. Fan and air window are divided into two types of pressure boosting and pressure reducing regulation depending on the installation position of the fan and air window.

(1) Combined air window-fan pressurization

By pressurization adjustment, it is meant that the pressure energy of the airflow on the airway between the two regulating devices is increased. The fan should therefore be installed on the upwind side of the air window, as shown in Fig. 3.26.

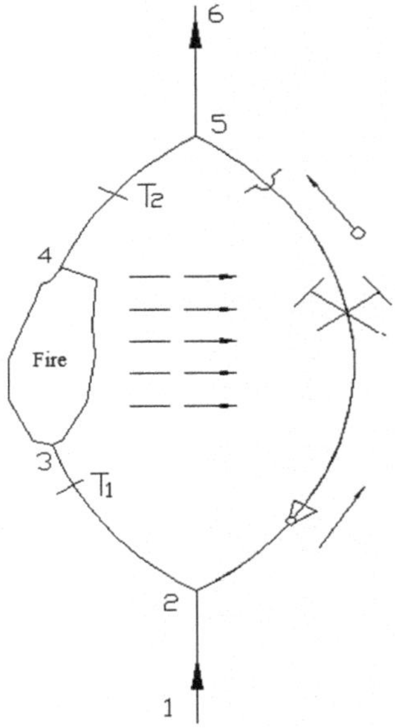

Fig. 3.26 Combined fan-air window booster set-up diagram

This setting method can improve the wind pressure of the working face. For example, for the spontaneous combustion fire that has occurred in the mined-out area of the coal mining working face, it may be due to the strong role of mine main fan, the air leakage of the mined-out area is larger, the fire is delayed to extinguish, the fire smoke flow and toxic and harmful gases will be constantly from the mined-out. At this time, this kind of fan can be installed in the air window before the equal pressure system, in order to increase the air pressure of the coal mining face to balance the gas pressure in the fire area, to achieve the purpose of gradually suffocating the inert fire area.

The booster adjustment can be divided into two types of adjustment: constant and reduced air volume for the airway.

In order to maintain a constant air volume in the regulating airway, the increased resistance of the air window must be equal to the pressure generated by the fan. In Fig. 3.27a, the pressure ramp before and after the air window is installed in the airway and the fan air volume remains unchanged. Because the air volume remains unchanged, the pressure energy on the outer airway of the two regulators is the same as before the regulation, i.e. the pressure slope line before and after the regulation is re-summed. The airflow pressure energy on the airway between the two regulators increases, which is parallel to the pressure slope line before the regulation.

If the air volume of the regulated airway is allowed to be reduced, the reduction in air volume allows the resistance of the air window to be greater than the pressure of the fan. As shown in Fig. 3.25b, the slope of the pressure slope line before and after the airway regulation is slower compared to that before the regulation. Between fan and air window, the increase in pressure energy is equal to the sum of the increase in pressure energy of fan and the increase in pressure energy of air window. The pressure energy of the airflow on the upwind side of the fan increases in the airway due to the reduction of air volume in the airway. The pressure energy of the airflow on the downwind side of the air window decreases because the resistance of the air window is greater than the pressure of the fan.

Combined fan-air window pressure regulation is suitable when the spontaneous combustion of coal in the mined-out area of the working face is due to the presence

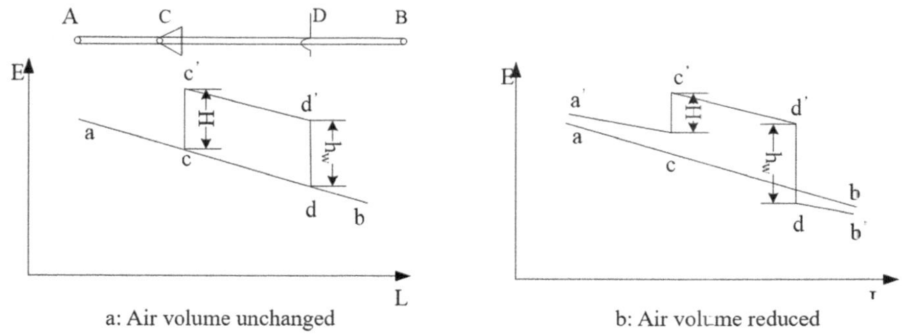

Fig. 3.27 Combined fan-air window boost regulation

of external leakage into the airflow. The pressure adjustment must not be too large, otherwise the airflow will be supplied to the inside of the mined-out area.

(2) Combined air window-fan pressure reduction

The air window is installed on the upwind side and the fan is installed on the downwind side when the pressure is reduced. The pressure energy of the airflow between the two regulating devices is reduced compared to the original, as shown in Fig. 3.28.

The main purpose of this setting method is to reduce the air pressure at the working face. When the pressure of the working face airflow is higher than the gas pressure of the fire area or mined-out area, the airflow of the working face will supply air to the fire area or mined-out area. At this time, the pressure regulating fan can be used to set up the pressure equalization system after the regulating air window to increase the negative pressure state of the coal mining face to balance the fire area or mined-out area.

The change in air volume before and after regulation can also be divided into two types of regulation: unchanged air volume and reduced air volume. Figure 3.29b shows the pressure ramp before and after the two regulating situations of constant air volume and reduced air volume during pressure reduction regulation.

(3) Relationship between wind pressure and air volume in the combined fan-air window equalization

Figure 3.30 is a graphical illustration of the relationship between air pressure and air volume of the regulating air window and the regulating fan. f is the equivalent characteristic curve of the total air pressure acting on the airway of mine major fan.

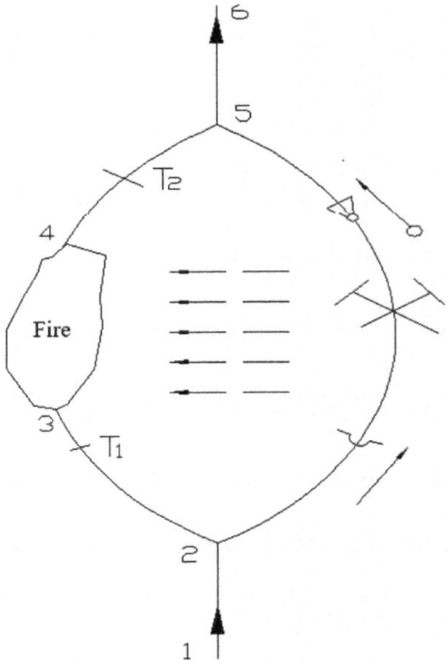

Fig. 3.28 Combined fan-air window buck setting diagram

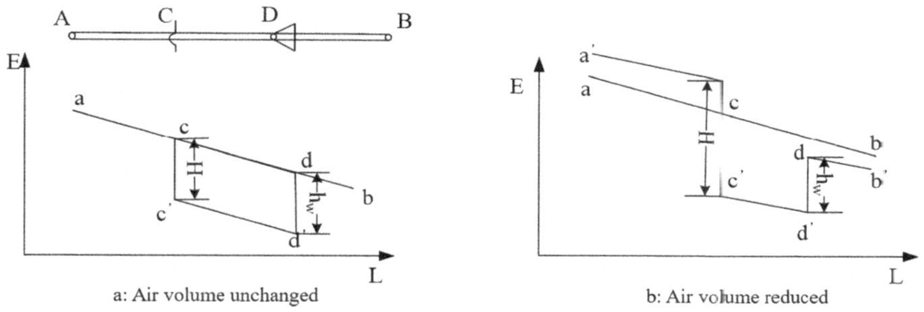

Fig. 3.29 Fan-air window combined buck regulation

R is the wind resistance of the airway before the regulating fan and the regulating air window are installed. The intersection of f and R is M. The vertical and horizontal coordinates of this point are the air pressure H_M and air volume Q_M before the regulating air window is installed. After the installation of the regulating fan and the regulating air window, the wind resistance of the airway is R_2, $R_2 = R + R_{window}$, and the characteristic curve of the regulating fan is f_1. According to the principle that air volume is equal and wind pressure is added, the joint curve f_2 can be synthesized with the equivalent curve f. The vertical and horizontal coordinates H_p and Q_p of the intersection p with R_2 are the wind pressure and air volume after equal pressure. It can be obtained that the air pressure $H_p > H_M$, but air volume $Q_p < Q_M$, it can not meet the work surface air volume requirements. If we want to keep the air volume of the working surface unchanged after equalization, we must increase the air volume supply, by adjusting the air window area and air window wind resistance, we can get $R_1 = R + R_{window}$, in order to make $Q_N = Q_M$, and get the air pressure as H_N, to achieve the purpose of equalization.

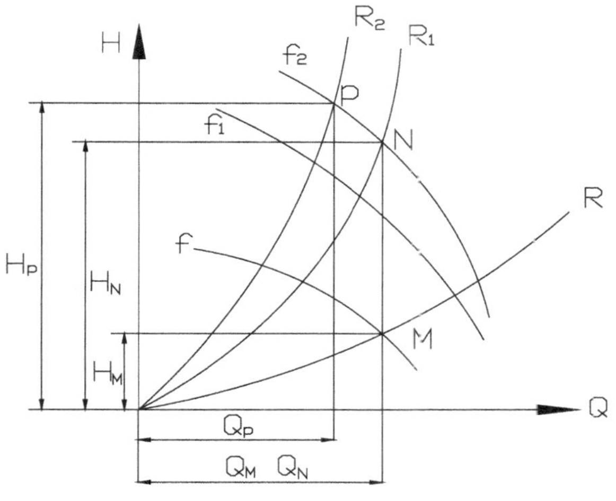

Fig. 3.30 Diagrammatic representation of the relationship between combined fan-air window mean pressure and air volume

(4) Mechanism of uniform pressure fire prevention techniques in closed areas

Depending on the pressure equalization device, the closed area uniform pressure fire prevention technology can be divided into fan-chamber pressure equalization and connecting duct-chamber pressure equalization. According to the number of chambers, they can be divided into single-sided and double-sided chambers. The characteristics of the fan-chamber pressure equalization are safe and reliable, simple management. However, the pressure equalization range is small, requires the construction of high quality confinement walls and is susceptible to atmospheric pressure. It is more costly in the case of longer fan-chambers.

Figure 3.31a and b are a unilateral, bilateral air chamber—connecting tube pressure equalization. The mechanism is connecting tube short-circuit diversion effect, which can not make the two ends of the fire zone air leakage pressure difference (unless the connecting tube wind resistance is zero) is zero, but the fire zone air leakage direction is constant. Figure 3.31c the placement of the connecting pipe so that the fire zone air leakage branching angular association, it is possible to make the fire zone air leakage pressure difference is zero, which is an ideal arrangement of the connecting pipe. But the shortcomings of the fire zone air leakage direction is variable.

Figure 3.32a and b are in the fire area return airflow side or air intake side and return airflow side of the construction of air chambers, in the air chamber wall directly installed or through a section of the air duct installed equal pressure fan, the use of fan to reduce or increase the pressure of the air chamber, to achieve the purpose of

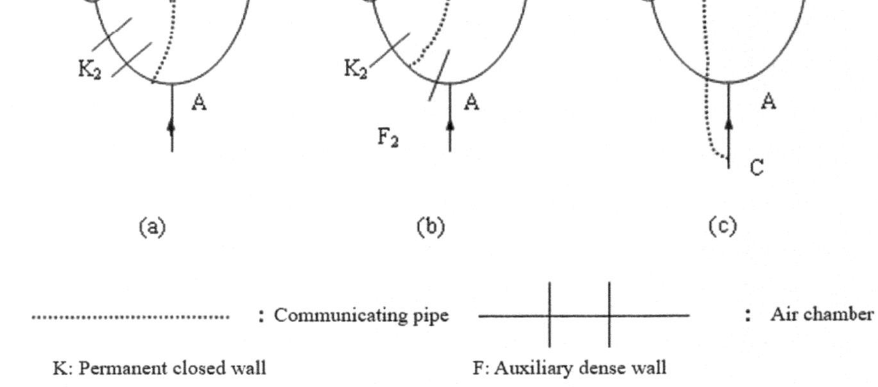

Fig. 3.31 Connecting tube—air chamber equalization method

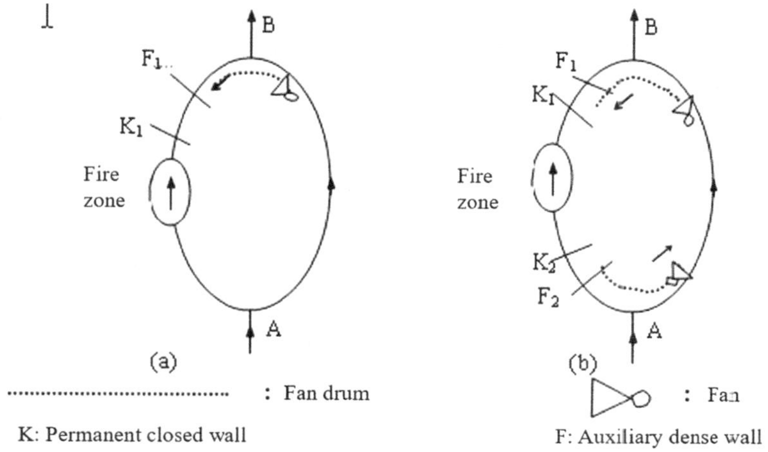

Fig. 3.32 Fan-air chamber closed area pressure equalization

equalizing the air pressure in the fire area. The characteristics of the air chamber pressure equalization: the pressure equalization range is large, the air leakage pressure difference in the fire area may be zero, the quality of the construction of the confinement wall is not high. However, there are strict explosion-proof requirements for the fan, the management is more complex.

(5) Other air changing pressure equalization

(1) The essence and characteristics of the air pressure regulation method

The essence of the air pressure regulation method is to add or remove fans, ducts and regulating facilities and to change the structure of the ventilation network to reduce and equalize the differential air leakage pressure and reduce the air leakage quantity.

The air pressure regulation method is divided into two cases: full mine-air changing and local air changing.

The air changing of a full mine involves a large scale and is generally combined with the renovation of the ventilation system. In order to meet the air requirements of mine, the main fan operating point is adjusted so that the main fan operates at reduced pressure to reduce the air leakage quantity in the fire area and to protect against fire.

Localized air changing is mainly used in the mine area and in the working face area. This can be achieved by changing the roadway layout of the working face, adding, removing or moving ventilation doors, adjusting air window facilities, etc.

Poor ventilation systems are one of the root causes of spontaneous combustion fires, which is why air pressure regulation is a fundamental fire prevention measure. The use of the air pressure regulation method allows for the organic combination of air and pressure regulation, ventilation management and fire prevention in spontaneous combustion mines.

(2) Methods and techniques for air pressure regulation

The following methods and techniques are commonly used in the local air pressure regulation method.

① Workings diverted to open parallel airway.

The common methods are W-ventilation and Y-ventilation.

The W-ventilation method is to add a waist roadway to the U-ventilation system, dividing the working face into two sections, usually using the upper and lower flat roadways as the intake airway of the upper and lower faces respectively, and the waist roadway as their common return airflow roadway, forming a W-shaped structure.

The Y-ventilation method is to leave a roadway along the return airway in the mined-out area as an intake airway, which is used to dilute the harmful gases in the return airway directly. Therefore, if the air volume of the mined-out area can no longer be increased by the U-ventilation method, the Y-ventilation method can solve the problem of gas accumulation in the corner of the working surface and reduce the air leakage in the mined-out area. The Y-ventilation method can not only solve the corner gas accumulation on the working face, but also reduce the air leakage in the mined-out area, which is beneficial to fire prevention.

② Opening of short-circuiting branch coupling branches or corner coupling branches.

The greater the pressure difference between the two endpoints of the short-circuit branch, the greater the pressure difference between the two endpoints of the short-circuit branch. Short-circuit branches can be in the network and the better. When the mined-out area is between the return airflow side and the air intake side, the better the short-circuit effect; the shorter the short-circuit branch, the better the short-circuit effect. If there is no existing roadway separated by a ventilation door, a short circuit can be created by using an existing air bridge, so that the return airflow side of the mined-out area is also on the air intake side.

③ Air leakage endpoint homolateralization.

By adjusting the position of the ventilation so that both ends of the air leakage path are on the air intake side of the ventilation system, the differential air leakage pressure can be significantly reduced. One of the techniques is to change the position of the ventilation door or air window in the ventilation system.

④ Air leakage channel angular linkage.

In a ventilation network, the angled branch is an unstable branch in the wind direction and has the conditions to achieve equal pressure. If the ventilation system can be adjusted so that the original non-cornered branch is transformed into a cornered branch. Then, adjusting the wind resistance of the adjacent branch can make the differential pressure between the two ends of the air leakage channel converge to zero, ultimately reducing the air leakage quantity in the mined-out area and achieving the purpose of fire prevention.

⑤ Positive pressure ventilation.

As a large amount of CO is stored in the mined-out area of the coal seam or in the adjacent mined-out area, under the negative pressure of the working face, the CO limit at the working face is often exceeded, posing a serious threat to workers' health or life safety, so it is necessary to adopt positive pressure ventilation technology. Positive pressure ventilation technology is divided into mine positive pressure ventilation technology and local positive pressure ventilation technology.

Mine positive pressure ventilation technology is a ventilation method that allows the ventilation pressure in the shaft to be greater than the local atmospheric pressure. Local positive pressure ventilation is the establishment of a relatively independent positive pressure air supply system at the working face, similar in structure to the "fan-air window" equal pressure ventilation system, in the case of negative pressure ventilation in the mine as a whole. Local positive pressure ventilation is the main technique used in mine to prevent and control CO overloads.

There have been a number of accidents in China where positive pressure ventilation has led to spontaneous combustion in the mined-out area and failed to provide timely warning. Positive pressure ventilation may lead to two problems. Firstly, it is not able to top the spontaneous combustion reverse oxygen absorption spread. Secondly, the mined-out area spontaneous combustion gas is pressed into the small mine or the surface, which is not easy to warn in time and cause sudden fire. For example, the Hegang Fuhua coal mine was a particularly serious fire accident.

3.3.2 *Example of Ventilation Pressure Regulation Technology*

[Case 1] Example of uniform pressure fire prevention techniques for mined-out area fires

A working face ventilation system in a mine is shown in Fig. 3.33b. W_1 mining working face has a fire which ignites the coal as well as direct fire suppression is ineffective and the face needs to be closed. The fire has ignited the coal and direct fire suppression is ineffective.

As the gas concentration has reached 1.5% during the containment process, the air volume is reduced and there is a risk of the gas reaching explosive concentrations. Therefore, it is necessary to use the method of increasing air volume to discharge the gas.

Firstly, the intake airway of the W_2 mining working face is controlled. The air volume of the W_1 mining face is increased by hanging wind barriers1 in the intake airway. When the gas content of the return airflow roadway at the working face drops to 0.1–0.2%, temporary confinement 2 and 3 can be quickly established in the intake airway and return airflow roadway to seal off the fire zone.

Secondly, open the side cutting eye ventilation door 4 and create a ventilation door at 5. Make the fire area in the upwind side of ventilation door 5. If ventilation

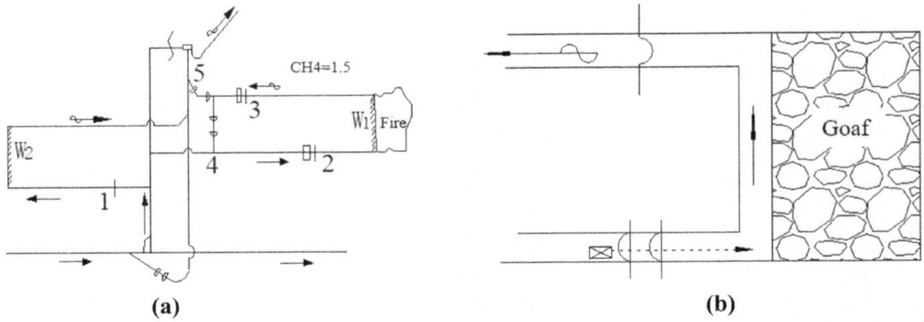

Fig. 3.33 Example of uniform pressure fire prevention techniques for mined-out area fires. **a** Ventilation system diagram **b** boost suppression CO diagram

door 5 is not air leakage, the pressure difference in the fire area is 0. Even if there is air leakage, it will mainly pass through the side cutting eye and the wind entering the fire area is very small.

Thirdly, once the fire area has been stabilized and there is no risk of explosion, a permanent confinement is built.

This closure ensures the stability of the ventilation system in the mine area and also serves the purpose of pressure equalization.

If the CO limit is exceeded in the coal mining face, the method of accelerating the advance is generally used to dump the fire area to the mined-out area. If there is a surge of CO to the mining face, the following method can be used to deal with it, as shown in Fig. 3.33b.

(1) Adding a regulating gate to the return airflow roadway to increase the pressure at the mining face and suppress the gushing of CO.
(2) If the purpose of suppressing CO is not achieved by only increasing the resistance at the wind roadway, the method of installing a fan at the machine roadway can be used to increase the pressure and increase the pressure at the mining face to a greater extent, so that CO will no longer gush out to the mining face.

After several days of continuous rapid advance, the fire area can be thrown into the unventilated area of the mined-out area as well as gradually suffocated and extinguished.

[Case 2] Example of uniform pressure fire prevention techniques for mined-out area fires

As shown in Fig. 3.34a, the 7203-1 mine area in the north wing of a mine is a double mining wing mine area with two strike long wall header workings. Both coal mining faces are U-shaped ventilation. The normal air volume is 700–800 m^3/min.

West limb7203-IW was mined to the stopping line and equipment and stand removal was completed. The east limb mine area 7203-IE is being retrieved. At the end of the 7203-IW retreat, a temporary (sand belt) firewall was installed at each of

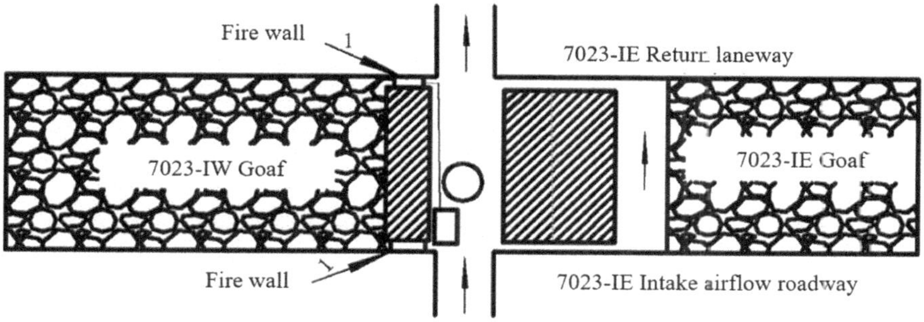

Fig. 3.34 Example of uniform pressure fire prevention techniques for mined-out area fires

the intake and return airflow roadway openings beyond the stopping line to prevent natural fires in the stopping line coal wall and 7203-IW mined-out area.

Shortly after the firewall 1 was set up, the CO concentration and temperature on the outside of firewall 1 on the 7203-IW return airflow side gradually increased due to the pressure difference between the incoming and return airflow sides of the air barrier. There was a slight smoke and other initial signs of spontaneous combustion and fire.

It was decided to adopt "equal pressure" fire prevention, and a local fan was installed on the wind retaining wall 2, and the air guide was configured to equalize the pressure directly to the firewall on the return airflow side. The pressure differential between the two sides of the controlled area was not eliminated, but caused the pressure differential between the two sides of the controlled area to be eliminated.

The fire gas reverses. Under the action of the local fan, a local closed loop airway is formed, as shown in Fig. 3.34c, where the local fan (fan) → guide duct → firewall of 7203-IW return airflow roadway → 7203-IW stopping line → 7203-IW intake airway firewall → local fan. Under the action of circulating ventilation, smoke and flame escaped from the firewall of 7203-IW intake airway and entered the local fan and the entrance of 7203-IE intake airway. Due to the timely detection of smoke and flame escape, the uniform pressure parameters were adjusted in time for control, which eventually achieved the purpose of uniform pressure fire prevention.

[Case 3] Uniform pressure fire prevention in closed areas using angle connection airway

As shown in Fig. 3.35a, the normal production ventilation system is a 1 → 2 → 3 → 4 → 5 → 6 → 7 ventilation system. To ensure this ventilation system, a D_2 ventilation door is required in the 2–10 airway and a D_1 regulating ventilation door is required in the 10–6 airway. When air reversing is required at the working face, the air intake l–2 and the return airflow 6–7 airway can be cut off. Open D_1 and D_3 ventilation door, change to 9–10 air intake, 2–8 return airflow, then constitute 9 → 10 → 6 → 5 → 4 → 3 → 2 → 8 air reversing system, as Fig. 3.35b.

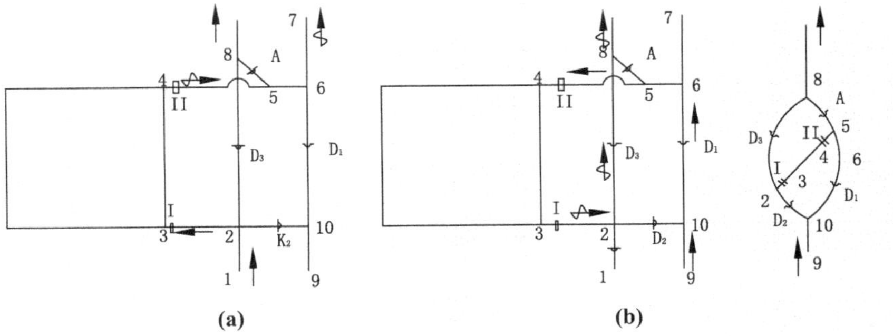

Fig. 3.35 Example of closed area uniform pressure fire prevention using an angle connection airway. **a** Pre-ventilation system diagram for air reversing **b** rear ventilation system diagram for air reversing **c** network diagram

When the workings reach stopping line 3–4, I and II containment can be used to isolate the airflow and prevent spontaneous combustion of the coal at the upper corner 4. However, due to the poor containment, air leakage is still serious. In order to prevent spontaneous combustion in the stopping line or mined-out area, the parallel airway 3–4 mined-out area (or stopping line) can be changed by changing the local ventilation system so that it can be located in the angle connection airway to facilitate the regulation and control of the mined-out area or the pressure in the fire area. Mined-out area or the differential pressure in the fire area to achieve the purpose of fire prevention.

In order to change the 3–4 stopping line into the angle connection airway 2–5, the upper yard material path 5–8 can be used and the ventilation door A can be adjusted by adjusting D_1, D_2, D_3. By changing the wind resistance of the corresponding airway, the following relationship can be easily maintained so that the air leakage quantity Q_{3-4} through the mine area or stopping line is at a minimum or close to zero, thus achieving the objective of reducing air leakage for fire protection.

As can be seen from the above discussion, the use of a 5–8 connection roadway of no more than 30–40 m and the addition of a few ventilation doors to the original roadway arrangement or ventilation system in the mine area can be a solution to an urgent need in times of emergency. Therefore, it is necessary to consider these important issues when designing and developing the mine disaster plan in order to facilitate the rescue and safe evacuation of people in times of emergency.

[Case 4] Uniform pressure fire prevention in closed areas using angle connection airway

[Basic Information]

During the recovery period of 8252 working face in Dongxingtai coal mine, the majority of the tail roadway along the roof of the working face can't be used because of gas pumping. The gas gushing out from the working face is large, and the gas concentration in the corner of the working face is high. In order to solve this problem, a trip to the gas pipeline was reserved before the return airflow stone gate was pushed through the working face. The mobile gas extraction pump is used to extract gas from the mined-out area of the working face. After the face is pushed through the return airflow gate, the mobile gas extraction pump starts to operate. When the workings were sampled, it was found that the CO volume fraction at the rear of the scraper conveyor reached 0.00677% and the CO volume fraction from the mobile gas extraction pump was 0.0157%, confirming that the mined-out area was showing signs of a natural fire. To this end, the following measures were taken: ① immediately stop the mobile gas extraction pump; ② re-seal the 31 roadway stone door sealing wall connecting the mined-out area and the return airflow roadway; ③ implement a small range of pressure regulation, so that the roadways leading to the mined-out area of the working face are in a negative pressure state; ④ accelerating the speed of the workings.

After the above measures, the CO gushing from the working face was suppressed. The CO volume fraction of the track roadway is stable at 0.0010–0 0015%. However, after the material transport roadway of the working face was penetrated, the air volume of the working face varied greatly, and the ventilation facilities of the 8252 track roadway were built on the slope, the ventilation system was not very stable, the CO concentration of the working face increased suddenly and showed a sharp upward trend, the CO volume fraction of the rear end of the machine was as high as 0.048%, and the working face was faced with the problem of shutting down production. The work-face faced shutdown. Through on-site investigation and analysis, the reasons for the natural fire on the lower 8252 working face are as follows. ① The top coal of the working face is not released thoroughly, and there is a large amount of floating coal in the mined-out area; ② The pushing speed of the working face is slow; ③ The two roadways of the working face, especially the section of the material transport roadway, are small. And the ventilation pressure difference between the two roadways of the working face is too large; ④ The airflow of the working face is long, and the air intake air is too large. The airflow course is long and the air intake is large, while the return airflow roadway connected with the mined-out area of the working face is short by dry airway, with small wind resistance and large air leakage quantity in the mined-out area. At the same time, a mobile extraction pump was switched on to extract the mined-out area gas, which further reduced the air leakage wind resistance to increase the air leakage in the mined-out area and continuously supplied oxygen to the mined-out area. This provided the oxygen supply for the spontaneous combustion of the mined-out area floating coal, which eventually led to the natural ignition of the workings.

[Technical Solutions]

After on-site investigation and analysis, the roadway of the lower 8252 working face and the mined-out area connected to it is a parallel ventilation system. The mined-out area of work-face 8252 is precisely the angle connection network of the two parallel systems. By taking advantage of the reversibility of the angle connection airway, pressure equalization measures were taken to suppress the natural firing of the mined-out area.

The roadway sealing wall (31 roadway gates and 8252 access road) connected to the mined-out area of the working face was sprayed to seal the air leakage and increase the air resistance as far as possible. In the 8252 working face return airflow roadway, a regulating air wall was built to increase the return airflow air resistance of the 8252 working face and reduce the working face air volume to 500 m^3/min. Increase the air volume to 400 m^3/min in the return airflow roadway to 31 sub-roadway to reduce the air resistance of the roadway return airflow connected to the mined-out area of the work-face. The ventilation door of the 8252 track roadway was fitted with a ventilation door interlocking device to ensure the stability and reliability of the work-face ventilation system and to speed up the work-face advance. According to the transport conditions of the working face, it was decided that the speed of the working face should not be less than 1 m/d. The mined-out area of the 8252 working face was grouted with yellow mud and locally re-injected with a dampening fluid (Fig. 3.36).

By taking advantage of the reversibility of the angle connection airway and adopting uniform pressure measures, the purpose of suppressing the natural fire in the mined-out area of the working face was achieved. By adopting the uniform pressure fire prevention technology of the angle connection airway, the air leakage resistance of the mined-out area CO gas flowing to the mined-out area of the working face has been increased. The total amount of CO gushing out has been gradually reduced from 0.0523 m^3/min to 0.0015 m^3/min, which has freed the coal resources forced to be confined due to spontaneous combustion of the working face by 54,600t. The total amount of CO gushing out was gradually reduced from 0.0523 m^3/min to 0.0015 m^3/min, which freed up 54,600t of coal resources that were forced to be confined due to spontaneous combustion of the working face. 130 m of excavation footage was saved.

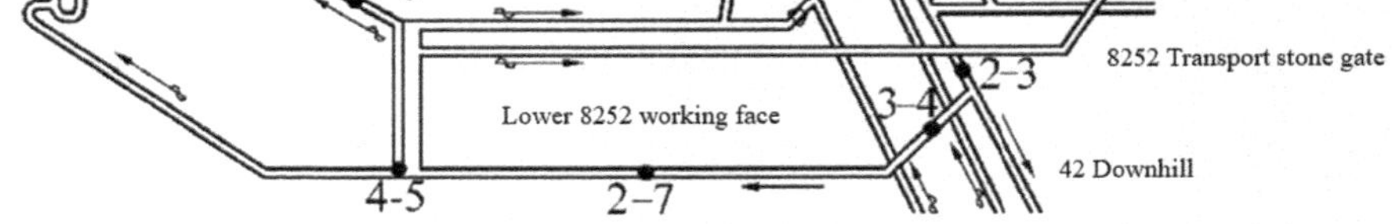

Fig. 3.36 Example of uniform pressure fire prevention in a closed area using an angle connection airway

[Case 5] Changing the ventilation system to increase the pressure at the working face

The 1127 working face of a mine is the first working face to be retrieved across the upper mountain and across the area. The retrieval period is 15 months. While the natural firing period of the 12S coal seam is 12 months, the retrieval period exceeds the natural firing period of this working face and has a natural firing potential. 1127 working face ventilation door D_1 found CO gas during the fire inspection check on 22 July, and the gas content in the confined area: CH_4 is 0.9%, CO is 0.5%, CO is 20%, and the air temperature is 25 °C. The ventilation system diagram is shown in Fig. 3.37a.

After investigation, it was found that the pressure difference between the front and rear of the ventilation door was large, as well as the fissure development at the ventilation door was serious. In order to solve the problem of CO overrun of the ventilation door D_1, the ventilation system was modified by opening the ventilation doorD_1, D_2, setting up the airtight or ventilation doors T_1–T_4, D_3, the layout of the facilities is shown in detail in Fig. 3.37b. Adjustment of pre-ventilation system: secondary shaft → a mining 12s uphill → 1127 transport roadway → 1127 mining face → 1127 airway → auxiliary mining 12 s → auxiliary mining five middle rail slope → auxiliary mining wind bridge → 1102 rail hill → −300 level return airflow roadway → east wind shaft → ground. Adjustment of rear ventilation system: When 1127 working face crosses the side eye of auxiliary mining 12 s, the ventilation mode of this working face is changed to downward airflow, and the direction of working face airflow is shown in Fig. 3.37b. That is, new air → ground → −450 big roadway a mining 12 s side eye air intake → 1127 air duct → working face lack of air 1127 canal → a mining 12 s up hill → a mining five in the rail slope → 1101 return airflow up hill → −300 return airflow roadway → east wind shaft → ground. The problem of CO overrun before and after the ventilation door is solved by this method.

[Case 6] Local positive pressure ventilation to solve the problem of CO overload

[Basic Information]

As shown in Fig. 3.38, the 81,201 header of the 12 # coal seam of a mine is located in the south central part of the shaft. The working face has a strike length of 950 m, a tendency length of 150 m and a mining height of 2.6 m. The distance between the working face and the upper 11 # coal seam mined-out area is 12.1 m. Five observation holes were drilled into the upper mined-out area prior to mining and no water was observed in the mined-out area. The harmful gas condition was 2.6% CH4, 1% CO2, 1.3×10^{-5} CO and 25 °C. According to the mining practice of a neighbouring mine, the mine decided to adopt local positive pressure ventilation for the comprehensive mining workings in order to prevent harmful gases in the upper mined-out area from escaping into the working face through the post-mining fissure.

[Technical Solutions]

Positive pressure ventilation is achieved by installing two auxiliary fans with ventilation doors on the incoming airflow roadway of the working face. At the same time,

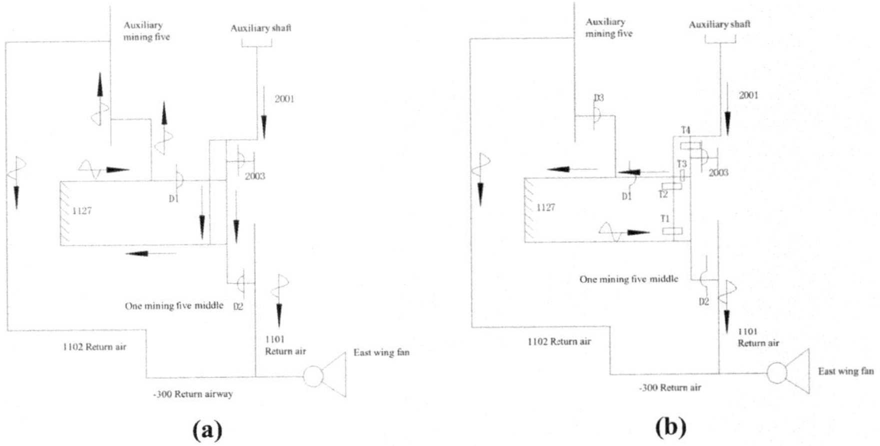

Fig. 3.37 Example of changing the ventilation system to increase pressure at the working face. **a** Air changing pre-ventilation system **b** air changing rear ventilation system

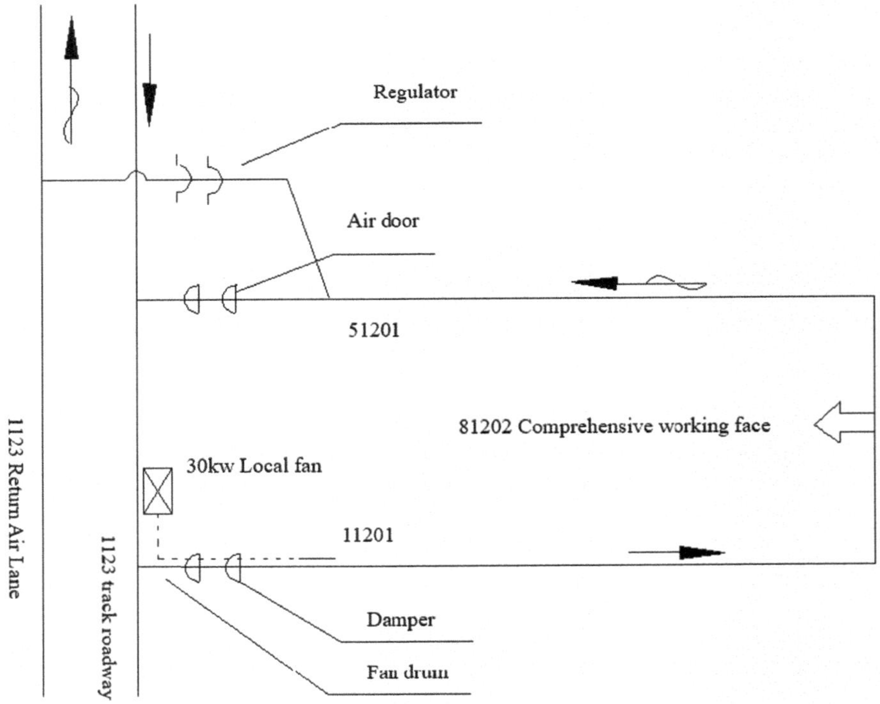

Fig. 3.38 Local positive pressure ventilation to solve CO overruns ventilation system diagram

2 regulating air windows are installed in the return airflow of the working face, using the pressure of the local fan, in order to create a certain positive pressure in the mining working face and to make the pressure in the mined-out area lower than the mining working face, thus forcing the harmful gases not to pour into the mining working face. Two 30 kW counter-rotating fans are installed on the large roadway of the intake airflow 1123 track at the 81,202 integrated mining face, one being used and one as a standby. Two regulating ventilation doors were installed in the 11,201 air intake chute on the face. At the same time, two regulating air windows are installed at the 51,201 return airflow of the working face, and the φ800 mm rigid duct enters the inlet airflow of the working face through the regulating ventilation door sealing wall. The air volume at the outlet of the duct is 650 m^3/min. The diagram of the ventilation system is shown in Fig. 3.38.

[Case 7] Local positive pressure ventilation technique to prevent CO gushing from mined-out area of upper coal seam

[Basic Information]

Longhua coal mine is located in Shenmu County, Shaanxi Province, with a production capacity of 4 million t/a. The main mining seams are all shallow and low gas mine, with generally undetectable CH_4 gas. It is found that the CO volume fraction in the upper 2–2 seam of the former small coal mine room column mined-out area was as high as 1–2% during the back mining of the 30,106 comprehensive mining face in the 3rd coal seam. 30,106 workings were mined using integrated mechanized mining with full height mining (2.8 m). All use the collapse method to manage the roof. Except for the coal pillar, the mined-out area has very little floating coal, the coal pillar is hard and the average advance speed is 15 m/d. Therefore, the risk of spontaneous combustion of coal in the mined-out area of the 3–1 seam is extremely small. The surface fissures caused by the shallow dry burial depth (90–100 m) can be seen 2 d after mining. 30,106 working face is about 1200 m long. 600 m has been mined and the adjacent 30,105 working face to the south has been mined for the mined-out area, leaving a coal pillar width of 20 m. The coal pillar is hard and intact, no obvious fissures. 30,107 working face is to the west. The roadway has been excavated, which is a preparation face.

As the spacing between 3–1 coal seam and 2–2 coal seam is 30–40 m, if no measures are taken to advance the coal seam mined-out area with the working face, it is bound to be connected, under the action of mine ventilation negative pressure, high concentration CO will enter the 3–1 coal seam mined-out working face, which has a very serious safety hazard. Therefore, through the fire area detection and CO source investigation and analysis, the possibility of coal spontaneous combustion high temperature points in the 3–1 coal seam mined-out area and the 2–2 coal seam mined-out area above the working face is very small. CO is mainly from the dry adjacent open pit coal mine spontaneous combustion fire area. According to the findings, a blocking wall was established by drilling down-hole to fill the 2–2 coal seam mined-out area with polymer foam, as shown in the figure. At the same time, the CO volume fraction in the mined-out area of the upper 2–2 coal seam of the

30,106 transport roadway was reduced to approximately 300 × 10^{-6} by injecting nitrogen into the blocked area and reducing the negative pressure of mine ventilation, and the CO concentration in the mined-out area of the upper 2–2 coal seam of the 30,106 return airflow roadway was reduced to zero. Although the CO control work was effective when the 3–1 coal seam and the 2–2 coal seam mined-out area were not connected, when the two mined-out areas were connected, the air leakage in the mined-out area must have a certain impact on the CO concentration in the 2–2 coal seam mined-out area. In order to ensure that the CO in the mined-out area of the 2–2 seam does not gush out from the working face in large quantities, the original negative pressure ventilation system of the working face was adjusted to a local positive pressure ventilation system in advance when the working face was pushed near the boundary of the 2–2 seam (Fig. 3.39).

[Technical Solutions]

The existing ventilation line on the 30,106 face was changed from full negative pressure ventilation to local fan pressure boosting. ① At the intersection of the 30,106

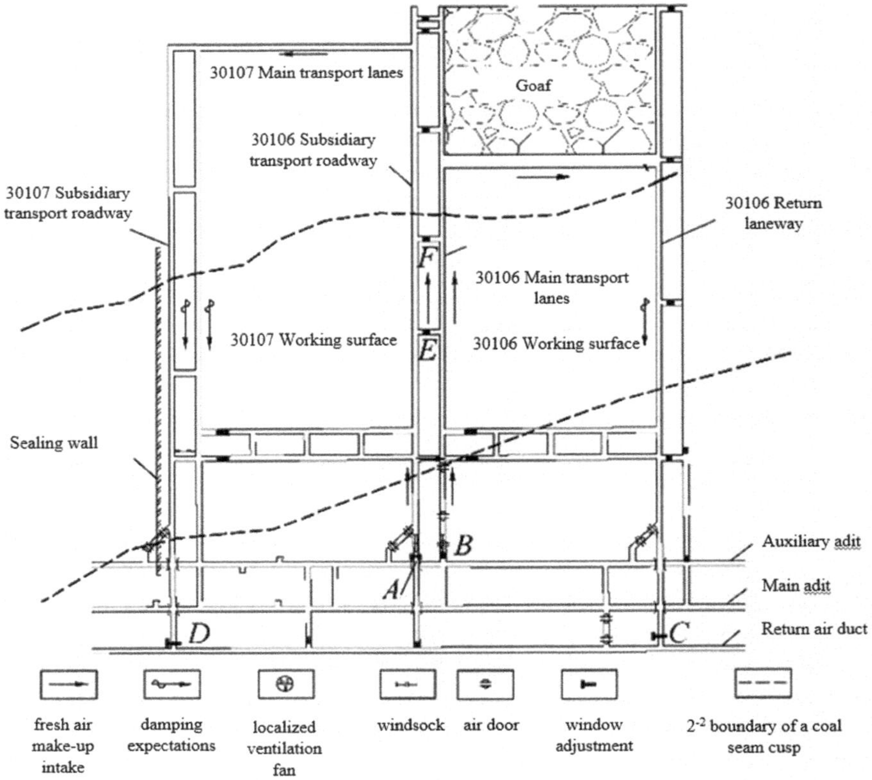

Fig. 3.39 Diagram of fan system adjustment

auxiliary roadway and the sub-level refuge bridge (point A in Fig. 3.39), four FBDN7, 1/2 × 45 local fans (two for use and two for backup) are installed by drilling holes in the protected coal pillars. The fan air intake is located in the secondary flat and is protected by a reflective iron fence. The local fan at point A supplies air to the two working faces. The air intake port is located at a point 25 m from the main roadway intersection with 30,106 (point B in Fig. 3.39), where three ventilation doors are installed (the spacing between the second and third ventilation doors is not less than the length of the mine equipment train). As the roadway is fitted with a belt conveyor ventilation door, it is difficult to close the door tightly, so 3 ventilation doors are installed. In addition, the fan supplies air between the first and second ventilation door via a duct. This creates a positive pressure chamber with high pressure energy to minimize air leakage. ③ The air windows of 30,106 return airflow roadway and 30,107 auxiliary transport roadway (points C and D in Fig. 3.39) are converted into movable adjustable air windows with scale markings to adjust the air pressure in the positive pressure area. ④ Closure of the 1st and 2nd connection roadway (points E and F in Fig. 3.39) between the main and auxiliary transport roadways of 30,106 to make the 2 working faces ventilation system relatively independent. ⑤ 30,106, 30,107 working face auxiliary transport roadway bypass outside the new or modified into 2 interlocking ventilation door, 3–1 coal seam and adjacent mined-out area confinement installed U-type differential pressure meter, monitoring the possible impact of positive pressure ventilation on the mined-out area. ⑦ The fan is fitted with an on/off sensor. Negative pressure sensors are installed at the 2 regulating air windows: CO sensors are added at the working face supports. In this way, the operation of the positive pressure ventilation system can be monitored in real time in the mine dispatch room. In addition, measures such as enhanced mine pressure monitoring are taken to prevent the mined-out area from collapsing after a large empty roof, resulting in CO gas being pressurized out.

[Case 8] Full mine positive pressure ventilation to solve the problem of CO

[Basic Information]

The mine uses three vertical shafts to develop the whole shaft field. The main shaft is responsible for lifting coal down and loading and unloading personnel as well as a safety exit, while the south and north air shafts are responsible for lifting gangue as well as a safety exit. The coal mining method is wall-mine, with the main shaft air intake and the south and north flanks extracting air.

[Technical Solutions]

The main fan is installed at the main shaft entrance, the main shaft frame is closed, two ventilation doors are installed in the shaft entrance yard, a pedestrian passage is set up, two ventilation doors are installed to prevent short-circuiting of the airflow at the shaft entrance, the wind shaft is dismantled and closed, and positive pressure ventilation is achieved. The direction of airway and airflow is the same as the original extraction ventilation, and the ventilation facilities under the shaft remain unchanged. Taking into account the high air leakage rate in the elevated shaft, the fan power is increased and two BK54-4-NO.12 counter-rotating axial fans are installed to ensure the air volume of mine. The total intake volume of mine is 1550 m^3/min, the total return air volume is 1450 m^3/min and the air pressure is 780 Pa. Repeated observation studies after the operation of this ventilation method have shown that the positive pressure ventilation method of the main shaft pressurization of the mine maintains all the advantages of the previous extraction ventilation, which completely avoids the problems faced by the late mine and suppresses the gushing of gas from the fire area and mined-out area. It also avoids the influence of the adjacent mine and, on the basis of enhanced containment and slurry injection, makes a significant difference to the condition of the fire area. The mine's positive pressure ventilation had no impact on the ventilation and safety of the adjacent mine.

[Discussion]

This method replaces negative pressure ventilation with positive pressure ventilation, which is generally suitable for small mines, where mine is relatively simple. Full mine positive pressure ventilation can solve the problem of CO over-limit in this mine, but improper equalization of pressure may aggravate air leakage, aggravate CO over-limit in the adjacent mine, and even fire.

[Case 9] Changing the position of the ventilation door air leakage endpoints homogenized

[Basic Information]

As shown in Fig. 3.40, a mine windrow is a sloping shaft with an inclination of 25°. The coal seam has a tendency to spontaneous combustion, and the firing period is 3–6 months. Wind shaft digging encountered a coal line, for the convenience of construction, coal gangue directly stacked in the wind shaft near the mouth, later built wind tunnel and fan room next to the gangue.

After the commissioning, 3 fires in 3 consecutive years in the wind tunnel against the gangue side. The temperature is 38–90 °C, CO concentration is 0.0024–0.05%, as well as the fire range is 20 m long and 3–8 m wide, as shown in Fig. 3.40.

After several ineffective fire-fighting attempts, the foundation of the fan room sank and the fan became inoperative. For this reason, a new wind tunnel was excavated and the engine room was rebuilt. In the old wind tunnel, a wooden and cemented confinement was constructed. Over the next three years, three spontaneous fires occurred in the vicinity of the wood and concrete enclosures and on the side of the inclined shaft adjacent to the old wind tunnel (i.e. to the left of the wind shaft). The latter fire was nearly 10 m in extent and 1.8 m deep, with a temperature of over 100 °C and a CO concentration of 0.05–0.10%. As the fire area was located near the exit of the wind shaft, it made it difficult to produce safely throughout the mine, so reasonable measures had to be taken to eradicate the source of the fire.

[Analysis of the Causes of Fire]

(1) Wind tunnel and the main fan room are located on the gangue pile, shallow from the surface, the gangue pile is loose, so there is a large amount of air leakage.

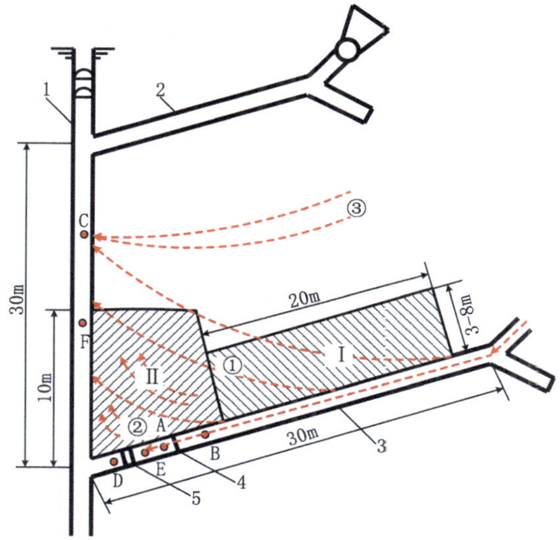

Fig. 3.40 Schematic diagram of wind tunnel firing. 1—inclined shaft; 2—new wind tunnel; 3—old wind tunnel; 4—wood panel confinement; 5—cement confinement; ①—air leakage from the ground to the inclined shaft via the old wind tunnel after the construction of the wood panel confinement; ②—air leakage from the ground to the inclined shaft via the old wind tunnel after the construction of the cement confinement; ③—air leakage from the ground directly to the inclined shaft leakage; I—the fire area in the old wind tunnel; II—the fire area in the new wind tunnel

Air leakage pressure difference is approximately equal to the reading of the fan room water column meter (2600 Pa).

(2) The anthracite coal mined at the mine has a sulphur content of 0.6–1.65%, which is a sulphur-bearing coal. The rainfall in the area is abundant, which provides favourable conditions for the oxidation of the coal to gather heat.

(3) After the construction of the new wind tunnel and the new fan room, the wind pressure at point A on the inside of the plank confinement and the starting fire point C in the inclined shaft are approximately equal, while the wind pressure at point B on the outside of the confinement is approximately equal to the surface atmospheric pressure. At this time, the pressure difference between points B and C is approximately 2400 Pa, resulting in the fire at point C developing in the direction of air leakage to the vicinity of the plank confinement. After the construction of the cement confinement, the wind pressure at point F in the inclined shaft is approximately equal to the surface atmospheric pressure. The wind pressure at point F in the inclined shaft is approximately equal to the wind pressure at point D on the inside of the cement confinement, while the wind pressure at point E on the outside of the cement confinement is approximately equal to the atmospheric pressure at the surface. So there is a pressure difference of about 2400 Pa between points E and F. This causes the fire in the inclined shaft

to spread to point E against the direction of air leakage after the construction of the cement confinement.

In summary, due to the loose accumulation of gangue in the wind tunnel fire area, easy air leakage and wet environment. So for a long time, sulphur gangue is easy to spontaneous combustion, the old wind tunnel and slant shaft in the fire to meet the four conditions for the occurrence of internal fire. So it is bound to cause a fire.

[Disposal Method]

The cracks in the ground are plugged and the external air leakage is treated anyway. The shaft opening of the old air shaft is sealed and the shaft seal is removed to make the E and Fair leakage endpoints co-lateralised.

[Case 10] Uniform pressure fire prevention at both ends of coal seam through ventilation door

In March 1986, the 13 permanent confinements and their enclosing rocks connected to the 503 roadway of the No. 14 seam produced fissures, resulting in a large amount of carbon monoxide gushing out from the 503 roadway, with the highest concentration reaching 300 ppm. The pressure energy in the western roadway of coal seam No. 14 was higher than the pressure energy in the eastern part, i.e. the pressure energy at point No. 2 was higher than the pressure energy at point No. 10 by 280 Pa. The pressure energy in the mined-out area and fire area of coal seam No. 12 was higher than the pressure energy at point No. 10 of the 503 large roadway by 400 Pa. The pressure-energy imbalance between the east and west and between the upper and lower seams resulted in fire gases from the upper fire zone gushing out from the eastern part of coal seam 14, 503roadway. In order to adjust its pressure-energy relationship, it was decided to construct two sets of ventilation doors, No.1 (shown in Fig. 3.41) and No.2 (shown in Fig. 3.41), on the downwind side of Point 10 and Point 2 in the east of 503roadway, No.14 coal seam, to increase the pressure energy of the 503 large roadway, so that it can be balanced with the pressure energy on the lean-out area of No.12 coal seam and the return airflow side of the fire zone. This inhibits the outflow of gas from the fire area into the production area and achieves safe production is achieved.

[Case 11] Main fan total air pressure and adjust air window to establish equal pressure system

One of the production pans in a mine is mining coal under the fire zone of the upper coal seam, one of the working faces is found to have CO overflow. The difference in air pressure between the working face and the fire zone of the upper coal seam is measured to be 160 Pa. Safe coal mining is achieved by mine the main fan total air pressure and regulating the air window to create a pressure equalization system. The plan and ventilation system diagram for this pan zone mining is shown in Fig. 3.42.

As can be seen from the mining plan, there are two production workings and three excavation workings in this pan area, with CO leaking from the mined-out area of the F workings, threatening the safe production of this workings.

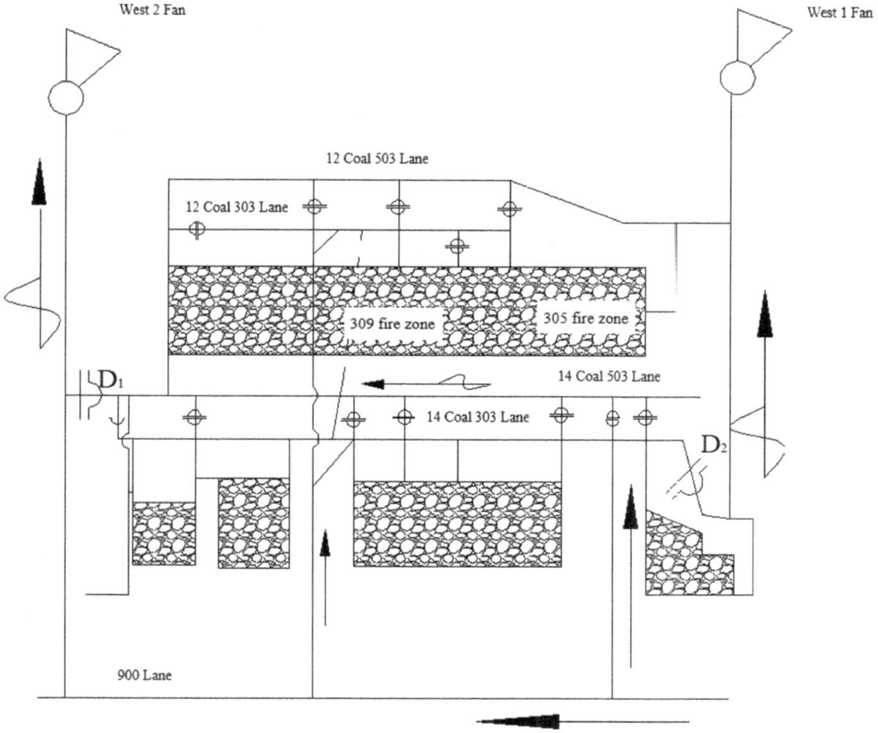

Fig. 3.41 Ventilation system diagram of an example of uniform pressure fire prevention through a ventilation door at both ends of a coal seam

The specific approach is without increasing the intake volume of pan area A. First adjust the air volume at B, C, D and E downward to the minimum air supply standard, then increase the air volume of F working surface, so that the air volume value of F working surface is 40–60% higher than the original air volume. Then in F working surface, set the air window at G behind the return airflow wind bridge, and adjust the air volume to the minimum air volume value required for equal pressure at the F working face, which can achieve the purpose.

Because after the air volume is adjusted at B, C, D and E. The air volume at the F working face increases accordingly, which is greater than the original air volume (the target is to increase the air volume by 40–60% compared to the original air volume). Then, the increased air volume pressure is adjusted to the minimum air volume value required by the equal pressure at the working face. The air volume at intake airway A is reduced accordingly, and the pressure at intake airway A is increased accordingly.

When a regulating air window is set up at the return airflow chute G in the F working face, the pressure energy in the intake airway A and the F working face will continue to rise, inhibiting the release of CO in the mined-out area of the F working face.

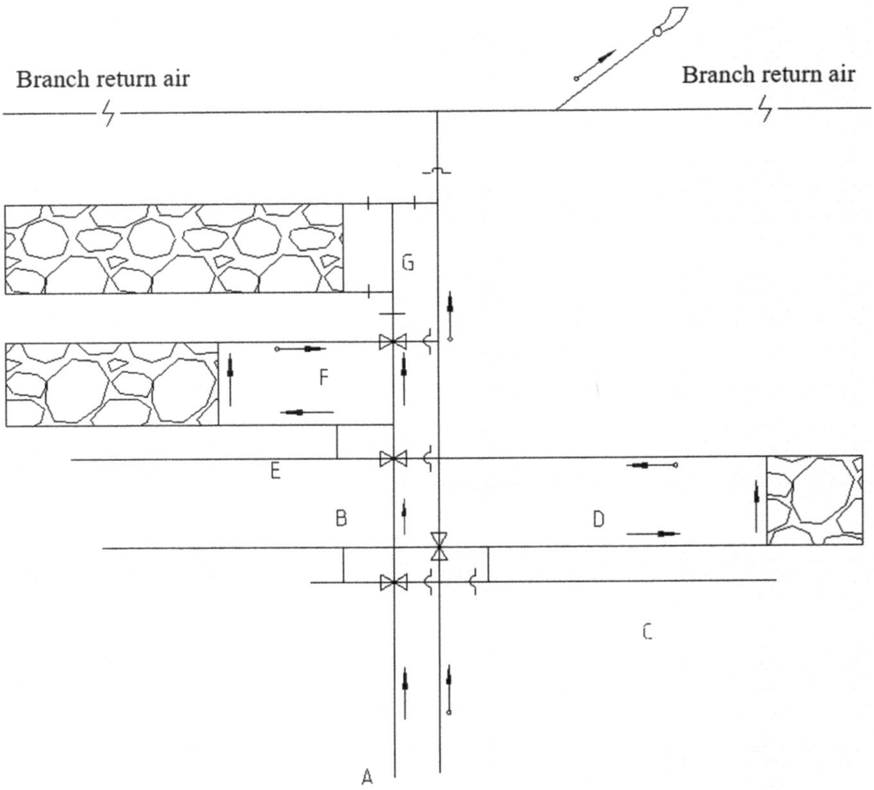

Fig. 3.42 Adjusting the air window even pressure system to regulate the wind example working surface ventilation system

With this method the air volume must be sufficient in the disc zone (branch) to be very effective. If there is a high resistance in the disc area, the air volume in the disc area must be increased by eliminating the high resistance section or by increasing the air volume in the disc area in order to achieve the objective.

[Case 12] Adjusting the ventilation door to control the pressure difference between the two ends of the working face

As shown in Fig. 3.43a, a fire broke out in the middle working face of the 5–6 roadway, closed with two confinements, while a confinement D_1 was built in the 3–6 airway. At this time, the air leakage quantity of the confinement of the fire A was as high as 4.8 m^3/s. After measurement and analysis, it was decided to set up a regulating ventilation door D_3 in the 1–5 airway to reduce the air pressure at point 5. The air pressure at point 5 was lowered and the containment in airways 3–6 was removed, as shown in (b). The pressure difference between points 5 and 6 was reduced from 248.2 Pa to 1.1 Pa, and the pressure difference between the two sides

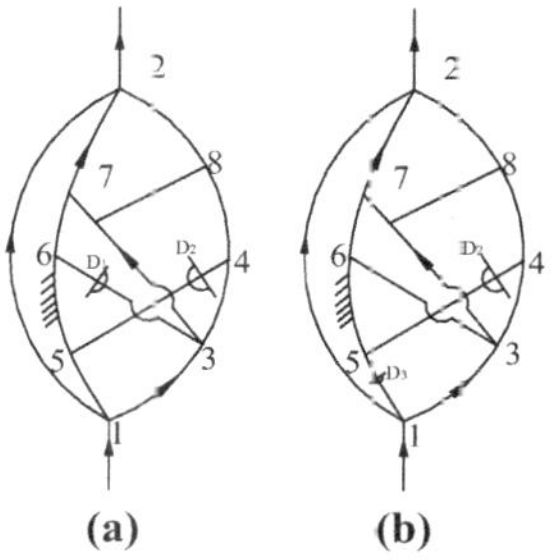

Fig. 3.43 Example of adjusting the ventilation door to control differential pressure at both ends of the working face to prevent fire suppression. **a** Ventilation network diagram before regulation **b** ventilation network diagram after regulation

of the containment tended to be close to zero. The air leakage quantity was close to zero, thus speeding up the process of fire fighting.

This example illustrates the importance of the position of the containment and regulation of the ventilation door in the fire zone network, which is sometimes as tight as, or even more important than, the containment wall.

[Case 13] Comprehensive example of closed area pressure regulation and fire prevention

In the upper and lower horizontal roadway, shown in Fig. 3.44, spontaneous combustion occurred in the fracture zone between 3 and 5, which formed the fire zone. Two confinements, A and B, were built at points 3 and 5 respectively.

[Pressure Energy Analysis]

The pressure energy at point 3 is higher than the pressure energy at point 5, and the pressure difference between the two points is 200 Pa, so the direction of air leakage between the fire zones is 3 $\rightarrow$ 5.

[Pressure Equalization (Regulation) and Control Measures]

(1) Adjusting the wind resistance of the airway with the adjustable air window

The installation of an air window in the roadway increases the wind pressure on the upper airflow side of the air window and decreases the wind pressure on the lower airflow side. Therefore, T_1 and T_2 can be installed on the upwind side (left side) of point 3 and the downwind side (right side) of point 5 respectively to regulate the air window, so as to reduce the pressure at point 3. The static pressure at point 5 is increased, thus reducing the pressure difference between points 3 and 5 and reducing the airflow and air leakage.

(2) Use of local fan regulation

A local fan is placed at point 3, with a negative pressure near the air intake side of the fan. The static pressure at point 3 can be reduced and the position of the local fan adjusted so that the pressure at point 3 is balanced with that at point 5.

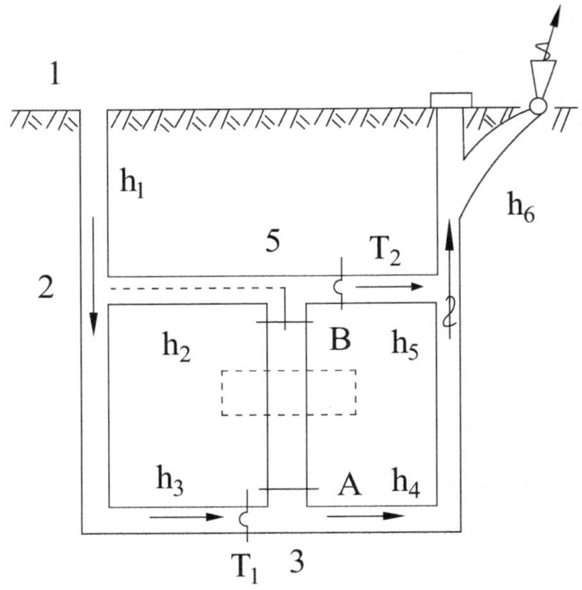

Fig. 3.44 Diagram of a closed area pressure regulation and fire suppression complex ventilation system

(3) Use of pressure regulating lines

As shown in Fig. 3.44, since the pressure at point 5 is low, a duct (iron or plastic) can be installed from point 3 to the point 5 containment, as shown in the dotted line in the Figure to transfer the pressure from point 3 to the point 5 containment, so that the pressure at point 5 is balanced with that at point 3.

[Case 14] Example of pressure regulating airbag with other pressure equalization methods

As shown in Fig. 3.45, a mine simultaneously mines a close coal seam with an inclination of 8° and a thickness of 1.2 m in the upper seam. It is mined by the long wall collapse method and is closed by permanent confinement T1 and T2 after mining. The lower coal seam is 1.5–3.0 m away from the seam, which is 2.0 m thick. As the two coal seams are close to each other, they share a common air intake and return airflow on hill VI. The return airflow returns to the main return airflow path through the VI, No. 1 uphill to the I air-return shaft and is discharged to the surface. The main fan installed in the I air-return shaft is 1 unit, with an air volume of 4000 m^3/min and an air pressure of 2000 Pa. In order to prevent the natural firing of the upper coal seam, mine total air pressure and adjust air window equalization system are adopted. The S1 regulating air window is installed on the return airflow uphill 10–11 airway.

The lower coal seam was mined in September 1965. Although measures were taken to equalize the pressure between the mine total air pressure and the regulated air window, the results of the analysis of the T_1 and T_2 confinement samples from

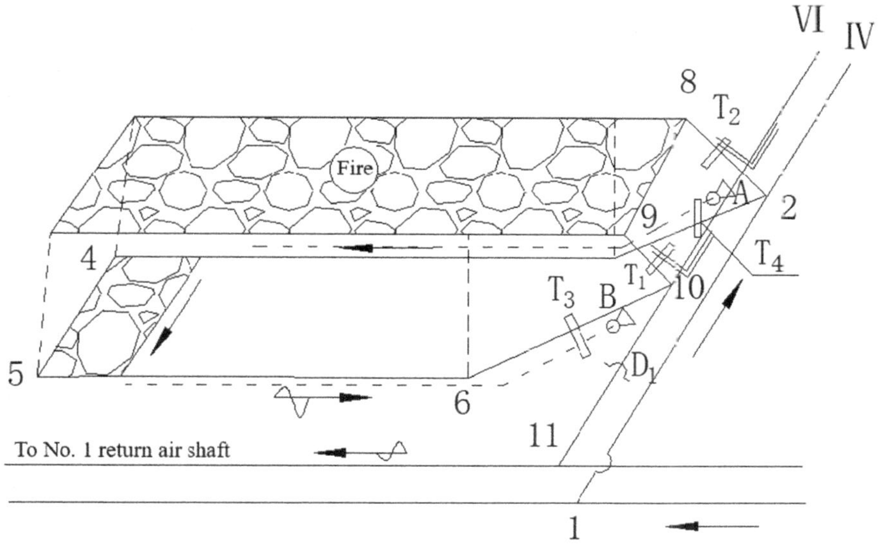

Fig. 3.45 Interrelationship of the upper seam fire zone lower seam workings

the upper coal seam showed that the oxygen content of the return airflow side of the confinement T_1 was much lower than that of the air intake side of the confinement T_2. The oxygen content of the air intake side of the confinement T_2 was basically the same as that of the air intake side of the confinement T. The oxygen content of the air in the lower coal seam roadway air was essentially the same.

In June 1966, nine months after mining the lower seam face, spontaneous combustion occurred in the mined-out area of the upper seam and the fire smoke flow penetrated through cracks in the roof and into the lower seam roadway and face. In order to extinguish the fire, the personnel had to be evacuated and the coal mining was stopped, and the fireproofing confinements T_4 and T_3 were built at the intake and return airflow roadway pillars to seal the lower coal seam face and the roadway so that no fresh airflow would flow to the upper coal seam fire area. Gas samples were taken daily from these confinements (T_1, T_2, T_3, T_4) to analyse the trend of the fire area. After one and a half months of closure, the gas samples were analyzed and proved that the fire was basically extinguished. After 2 months of closure, T_4 and T_3 were opened to prepare for the resumption of production, and attempts were made to smear the cracks in the roof of the upper and lower chutes with yellow mud to prevent fresh airflow from leaking into the fire zone of the upper coal seam. But it was difficult to achieve the objective because of the large number of cracks in the roof and the amount of work involved, which was also very time-consuming. In order to be able to both safely mining the fire area under the working face, but it also can extinguish the upper coal seam fire area. It must be in the lower coal seam working face throughout the mining process so that the upper coal seam fire area oxygen content of no more than 3%. So it must eliminate the role of mine major fan to eliminate want to fire area air leakage, eliminate the effect of changes in ground atmospheric pressure to eliminate the fire area gas exchange.

It was decided to adopt: (i) the use of a pressure-regulating fan and guide duct to supply and exhaust air to the working face; (ii) the use of pressure-regulating airbags to eliminate gas exchange in the fire area caused by changes in atmospheric pressure at ground level.

In order to reduce the air pressure difference in the lower coal seam to a minimum, a pressure regulating fan and air guide tube are installed in the inlet and return airflow chutes of the working face to supply and exhaust air to the working face. The pressure regulating fan is installed on the airtight wall with double ventilation doors at A and B of the inlet and return airflow chutes respectively. The air guide tube sends the air to the working face and uses the air guide tube L_2 and the pressure regulating fan to pump out the exhaust air from the working face to the uphill of the return airflow VI'.

In order to prevent air leakage into the fire zone of the upper coal seam, a low air volume supply is taken to the lower coal seam working face. The air pressure difference in this branch airway is always kept within the range of $\Delta H = 1–2$ Pa by adjusting the valves of the pressure regulating fan, which basically reduces the oxygen content in the fire zone. However, under the influence of atmospheric pressure, there is still an air exchange between the fire area and the nearby ventilation roadway. For this air exchange, when the atmospheric pressure increases the fire area into the fresh airflow, the fire area oxygen content increased. Atmospheric pressure decreases when the fire area smoke flow into the nearby roadway, threatening the lives of industry personnel safety. In order to make the upper coal seam fire area gas oxygen content does not exceed 3%, and does not make the fire gas leakage must eliminate the atmospheric pressure changes caused by the fire area air leakage and gas exchange, so decided to use the regulating airbag equal pressure.

With the two measures mentioned above, the air leakage in the fire zone caused by the total fan pressure and the gas exchange in the fire zone caused by the change of atmospheric pressure are basically eliminated, so that the gas in the fire zone is almost in a static state. The oxygen content in the fire zone starts to drop and in a relatively short time the harmful gases such as CO, CH_4 etc. basically disappear.

[Case 15] Equal pressure to open and seal the fire zone

There is a large amount of gas and fire gases accumulated in the mined-out area, the conventional ventilation and air locking methods cannot effectively control the outflow of harmful gases within a short period of time, thus not creating conditions for personnel to enter the working face to carry out cleaning and repair work. This is not conducive to re-ignition. The mined-out area can be sealed by equalizing the pressure, i.e. setting up a regulating ventilation door and local fan in the return airway (Fig. 3.46) to increase the ventilation pressure in the face and avoid the gushing of fire gas and gas in the mined-out area. In the process of implementation, to increase the air volume, two local fans were used to supply air in parallel, with an air volume of 800 m^3/min. By equalizing the pressure, the concentration of harmful gases in the mined-out area decreased significantly and the oxygen concentration increased significantly, creating safe working conditions for the workforce at the working face.

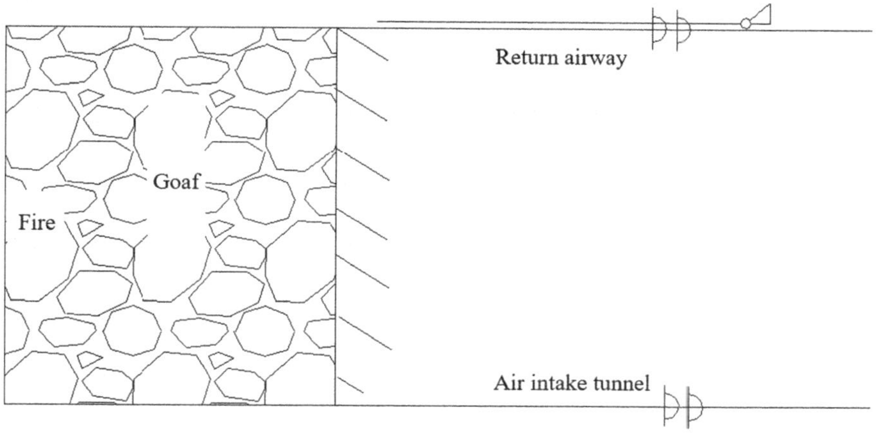

Fig. 3.46 Method of partial equal pressure ventilation to open and close the fire zone

References

1. Gang W, Weimin C. Practical measures for mine fire prevention and control. Beijing: Coal Industry Press; 2013.
2. Yongjin S. Coal mine equal pressure fire prevention. Beijing: Coal Industry Press; 2000.
3. Jiahua C, Yongde S. Argumentation of fire zone opening and closing parameters. Coal Technol. 2007;12(26):12–3.

Chapter 4
Mine Fire Extinguishing Method and Fire Zone Management Unsealing Method

Bo Tan, Hongqing Zhu, and Liyang Gao

Mine fire extinguishing method can be divided into direct fire extinguishing method, isolation fire extinguishing method and comprehensive fire extinguishing method. The direct fire extinguishing method is a method of extinguishing a fire directly or digging out the source of the fire using materials, equipment and facilities on site (e.g. water, sand, yellow clay, rock dust, dry powder extinguishers, etc.). isolation fire extinguishing method is in direct fire extinguishing ineffective or inaccessible to the source of fire when the method of fire extinguishing, that is, the construction of a seal wall to cut off the air to the fire zone, so that the oxygen content of the fire zone gradually decline, carbon dioxide and carbon monoxide content increased, so that the fire extinguished a method of its own. With the use of the isolation fire extinguishing method, the fire cannot be extinguished quickly after the fire zone is closed and there are still hidden dangers. The so-called comprehensive fire extinguishing method refers to the process of fire extinguishing in the field, direct fire extinguishing ineffective when using isolation fire extinguishing, but isolation closed fire zone can not achieve the purpose of timely fire extinguishing, and then the comprehensive use of direct fire extinguishing method (yellow mud grouting, pressure injection resist, N_2 injection, etc.) and isolation fire extinguishing method. The comprehensive fire extinguishing method can be applied not only to the extinguishing of mine fires, but also to the targeted prevention of natural fire hazards and fire zone threatened areas such as mining areas. After a mine fire zone is closed, in addition to strengthening

B. Tan (✉) · H. Zhu · L. Gao
School of Emergency Management and Safety Engineering, China University of Mining and Technology-Beijing, Beijing, China
e-mail: tanbo709@126.com

H. Zhu
e-mail: zhq@cumtb.edu.cn

L. Gao
e-mail: gaoliyangcc@163.com

B. Tan et al. (eds.), *Mine Fire Prevention and Control*,
https://doi.org/10.1007/978-981-96-0466-1_4

the management of the fire zone to prevent the spread of fire or explosion, it is important to know when and how to open and close the fire zone to resume production safely and early. To open and close the fire zone must be very careful to safely open and close the fire zone, and must first understand the fire zone closed, the air flow within the confinement area, gas changes in air pressure changes and some other circumstances [1].

4.1 Direct Fire Extinguishing Method Techniques and Examples

4.1.1 Removal of Combustible Materials

The combustible material that has been heated or is burning in the fire zone as well as its vicinity should be dug out and transported out of the shaft; the holes should be filled with non-combustible material after digging out; the means of digging out can be both human and mechanical. After digging out the burning coal, it should be promptly loaded into the mine car in the flat roadway and watered down, it is forbidden to pile the coal in the roadway; when digging out the source of the fire, rock powder should be spread liberally in the flat roadway near the source of the fire and sprayed with water. Direct extinguishing is the most economical method, but the conditions for using this method are that the fire is in its infancy and does not involve a large area; there is no danger of gas or coal dust explosion in the fire zone; and the source of the fire is located in a location directly accessible to personnel. This method should be used in such a way that personnel cannot stand directly underneath the fire source to prevent burns or injuries from hot objects; the roof of the fire area is in good condition as well as the gas concentration and temperature should be checked at all times when extinguishing the fire.

4.1.2 Use of Water to Extinguish Fires

(1) **Water-based fire-fighting techniques**

The direct water extinguishing method is a method of extinguishing fires by directly using water as extinguishing equipment to inject pressure water into the hot coal body to cool and extinguish the fire or to extinguish the fire directly near the exposed fire source. It is a practical, active and effective method of extinguishing fires with water, simple and easy to use, economical and effective. Its roles are as bellow. Strong water flow to the flame of the burning material pressed out, so that the burning material fully wet and prevent it from continuing to burn; water has a great ability to absorb heat, can make the burning material cool down; water evaporates into a large amount of water vapour in the fire, which can dilute the concentration of oxygen in the air in the affected area, as well as make the surface of the burning material and air isolation. Therefore, water has a strong fire extinguishing effect.

When using water to extinguish fires care should be taken to the following.

(1) Firefighters should stand on the air intake side of the fire source, not on the return air side. Because the return air side is hot and vulnerable to fire smoke, and the firefighters are easily burned by the hot water vapour.
(2) When extinguishing a fire, you have to spray uninterruptedly. So there should be enough water. When spraying do not spray the water jet directly to the centre of the fire source, but should be gradually sprayed from the periphery of the fire source to the centre of the fire source. When the water is not enough, the water jet directly to the centre of the fire source, water vapour in the role of high temperature, to produce hydrogen and carbon monoxide. There is a risk of explosive gas mixtures and explosions.
(3) Water conducts electricity and should not be used to directly extinguish electrical fires.
(4) If the oil fire is extinguished with water, only a mist of fine water can be used, so as to produce a layer of water vapour enveloped in the surface of the burning material, so that the burning material and air isolation. If you use water jets to extinguish the fire will make the burning liquid splash, and because the oil is lighter than water can float on the water surface, easy to expand the area of the fire.
(5) Ensure normal airflow so that fire smoke and water vapour can be discharged smoothly into the return airflow. At the same time, a tile inspector should be present to check the gas concentration and carbon monoxide concentration at all times.
(6) The roof of the shaft is easily damaged by the high temperature, which is prone to collapse and rise after being cooled by cold water.
(7) In order to provide an adequate water supply, pools should be provided at ground level, water supply lines should be laid and water valves installed at regular intervals.

Conditions that should be present when using water to extinguish a fire.

(1) The underground fire source is clearly located and accessible.
(2) The fire is small in size and scope.
(3) Adequate water and fire-fighting equipment is available down-hole.
(4) The gas concentration at the location of the fire is less than 2%, and the air flow is smooth enough to bring out the water vapour in time.
(5) The roof of the fire fighting site is strong and covered by supports.
(6) Sufficient manpower and material resources are available to extinguish the fire continuously.

(2) **Examples of water-based fire fighting**

Example of the "10.30" spontaneous combustion fire at the mining face of the Hongda coal mine.

At 08:15 on 30 October 2011, a natural fire broke out on the 21,021 coal mining face of the Hongda Coal Mine of Zheng Coal Group. The coal seam was thick, the old roadways were staggered, the coal body was broken and there were air leaks everywhere, making it difficult to extinguish the fire. The rescue team put in a total of 3 rescue squadrons, 56 combat squads, 418 combatants, through direct water extinguishing, drilling and water injection, injection of rockxiu, closed fire zone and other comprehensive fire extinguishing techniques, after 15 days of fire treatment, and finally put out the fire.

According to the situation of the fire at the 21,021 working face and the concentration of various gases in the return air stream, the following technical plan for fire-fighting was implemented by the rescue command.

(1) Water direct fire extinguishing

The ambulance crew installed a water pipe connected to the water source from the edge of the flame gradually to the centre of the fire source spray water direct fire extinguishing. At 13:15, the fire site area can no longer see the open fire, after careful observation. It is found that the fire point in the upper coal body there is a crack outside wide inside narrow, the depth can be seen more than 2 m, the crack in the coal body is still burning. The crews inserted a 3 m long steel pipe into the crack and injected water to extinguish the fire. However, the coal seam was thicker at the natural fire location, the working face was shut down for a long time, the working face was depressurized by a single hydraulic column, the coal seam sank and the roof was separated from the seam, a fissure was formed at the top of the coal seam, a small fire was burning inside the coal body along the fissure, and there was a development trend towards the upper part of the working face. At 21:15, the ambulance crew detected a CO concentration of 28 ppm in the return air stream of 21,021 and a local CO concentration of 554 ppm on the side of the mining area 40 m up from the fire point.

(2) Drilling and injection of water and rockxiu to extinguish fires

The fire was not extinguished by direct fire extinguishing, so the command requested the ambulance team to continue to inject water and closely observe the gas changes, while arranging for the mine's technical staff to cooperate with the ambulance team to set up boreholes around the fire point at the 21,021 working face and at locations where the coal seam temperature and CO concentration were abnormal, and to drill boreholes to inject water to extinguish the fire. At 05:15 on October 31, the rescue team detected that the CO concentration in the return flow of 21,021 working face reached 130 ppm. At 17:00 on October 31, the rescue team detected that the CO concentration in the return flow of 21,021 working face rose to 200 ppm.

According to the arrangement of the command department, at 00:00 on November 1, the shift started to inject rockxiu into the drill holes. After the injection of water and rockxiu in each drill hole, the CO concentration in the return flow of 21,021 gradually disappeared from the highest level of 200 ppm and showed a stable trend.

The fire was finally extinguished by using the technique of closing the fire zone and asphyxiating the fire.

4.1.3 Extinguishing Fires with Sand or Rock Dust

Sand or rock dust is low cost, easy to store for long periods of time, simple to use when extinguishing fires and is particularly suitable for extinguishing electrical fires. This is why it is important to have sand boxes and spades and shovels available in important locations such as electromechanical chambers, material stores, explosives stores, winch rooms and ventilator rooms. When electrical equipment is on fire, sand or rock dust can be spread directly onto the surface of the burning object to isolate the air and extinguish the fire.

4.1.4 Dry Powder Fire Extinguisher

Dry powder fire extinguishers are lightweight, easy to carry, simple to operate and can extinguish fires quickly. Commonly used dry powder extinguishing agents are sodium bicarbonate, ammonium sulphate, ammonium bromide, ammonium chloride, ammonium phosphate and so on. At present, the main agent for dry powder fire extinguishers in mines is ammonium phosphate powder. Under the action of high temperature, ammonium phosphate powder undergoes a series of decomposition and heat absorption reactions to extinguish the fire.

(i) **Principle of dry powder fire extinguishers**

(1) Ammonium phosphate powder is flown through the air in mist form, cutting off the flame chain reaction and preventing the development of combustion.
(2) A large amount of heat is absorbed during a chemical reaction, lowering the temperature of the burning material.
(3) The decomposition of ammonia and water vapour dilutes the concentration of oxygen in the air in the vicinity of the combustion products, causing them to be extinguished by lack of oxygen.
(4) The reaction eventually produces a paste-like substance, phosphorus pentoxide, which covers the surface of the burning material and penetrates into the burning material, isolating it from the air and extinguishing it.

(ii) **Use of dry powder fire extinguishers**

The main dry powder extinguishers suitable for underground fire fighting are fire grenades and powder sprayers.

(1) Fire extinguishing grenades contain 1 kg of ammonium phosphate powder, with a total mass of about 1.5 kg. The effective range of fire extinguishing is 2.5 m, and

a person with ordinary physical strength can throw 10 m. when used, unscrew the cover, pull out the pulling wire and immediately throw it into the fire zone, while paying attention to concealment to prevent shrapnel injuries.

(2) Portable powder extinguishers, according to the storage method of CO_2 are divided into two types of storage cylinder type and storage pressure type. The storage cylinder type powder extinguisher contains 6 kg of powder, using the liquid CO_2 in the cylinder as power, through the nozzle to spray the powder to form a powder mist. The effective range is about 5 m and the spraying time is 16–20 s. The volume of the high-pressure cylinder is 2L and the mass of the liquid CO_2 is not less than 240 g. When using, special attention should be paid to prevent blocking of the pipe, the powder extinguisher should be turned up and down several times to loosen the powder. Then, slowly open the pressure bottle. If the powder comes out, you can open the big switch. Otherwise, you should deal with the blocked pipe first before using. Nozzle distance from the fire source, according to the nature of the burning material. Oil, electrical equipment fires, the nozzle from the fire source distance can be larger. Because when it is too short, the powder flow rate is too fast, which may make the fuel splash, but accelerate the burning, or the powder can not adhere to the surface of the equipment and affect the effect. For coal and wood fires, the nozzle can be closer to the source of the fire, so that the high-speed dry powder flow into the internal combustion, to improve the effect of fire.

The use of fire extinguishers in underground mines is restricted because of the small space in the shafts and few straight alleys. In addition the use of fire extinguishers is difficult because of the unbreakable flight of the shells, the low hit rate and the high accuracy required for pulling the fire detonator, making it difficult for fire extinguishers to be used widely in underground coal mines. Fire extinguishing grenades and powder extinguishers are generally used in conjunction with each other. The fire extinguishing grenade is used first to extinguish larger fires and then the powder sprayer is used to extinguish residual fires. In some cases, a powder sprinkler can also be used alone. Both the grenade and the powder sprayer should be pre-assembled and in a state of operational readiness. Once assembled, the fillets and other leaks should be sealed with wax. The nozzles of the powder spray fire extinguishers should be tightly wrapped in plastic to prevent air leakage and the powder from absorbing moisture and caking. Normally, they should be placed in a dry and ventilated storage room, with a person to keep them, and checked every six months. If the powder is found to be lumpy, it should be poured out in time and dried up, crushed and then used again. The extension time of the detonator for fire extinguishing grenades must be above 2.5 s. It is strictly forbidden to use detonators that do not meet the requirements and should normally be stored in a dryer in the dangerous goods store.

4.1.5 Foam Fire Fighting

Foam fire extinguishing devices mainly include foam extinguishers, mechanical high-capacity foam fire extinguishing devices and the currently used three-phase foam fire extinguishing devices.

(i) **Foam fire extinguishers**

(1) Method of use

When extinguishing a fire with a foam fire extinguisher, at a distance of about 10 m from the point of fire, hold the lifting ring with one hand; hold the bottom ring of the body with the other hand and turn the extinguisher upside down so that the acidic liquid in the glass bottle flows out and mixes with the alkaline liquid and a chemical reaction takes place, forming a large amount of foam filled with CO_2 and spraying it out, covering the burning material and isolating the air to extinguish the fire. The CO_2 emitted from the foam also helps to extinguish the fire. As the spraying time increases, the effective spraying distance decreases and the user should gradually move closer to the burning area and keep the foam sprayed on the burner until it is extinguished. When spraying foam, always keep the body upside down, otherwise the spraying will be interrupted.

(2) Inspection and maintenance

① Foam fire extinguishers are a kind of fire extinguishing equipment that can be recycled. After each use, the lid should be opened, the cylinder and glass bottle cleaned and filled with a new fire extinguishing liquid.
② When storing, do not go near places with high temperatures to prevent the sodium bicarbonate from decomposing into carbon dioxide and becoming ineffective; take warming measures in the severe winter season to prevent freezing. And the nozzles should be unclogged frequently to keep them clear.
③ If the fire extinguisher is used for more than 2 years, it should be sent to the relevant department for a hydrostatic test every year. The test pressure shall be 1.5 times the design pressure and shall only be used after passing the test, and the test date shall be marked on the fire extinguisher.
④ After 1 year of loading, it must be sent to the inspection department to check the quality of the liquid.

(ii) **High-powered foam fire extinguishing devices**

High multiplier foam was used for mine fire extinguishing as early as 1956 and this method can be used for larger mine fire extinguishing jobs. It is widely used in China and has good results.

(1) Principle

The foam used to extinguish fires is usually liquid foam, which is a gas- and liquid-phase substance composed of a liquid film wrapped around a gas. A continuous

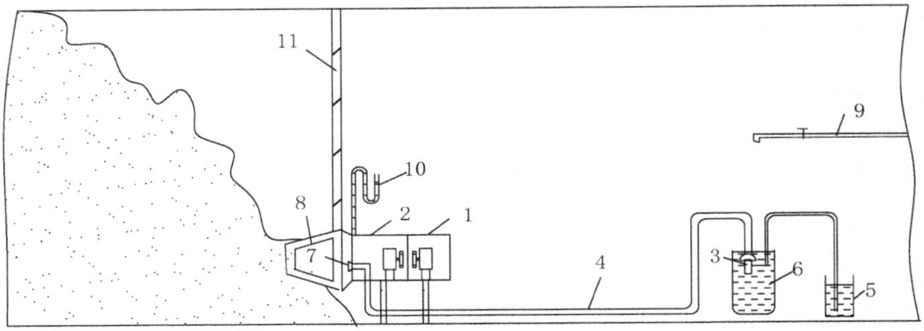

Fig. 4.1 High-powered foam extinguishing device

collection of foam, the liquid film of which coalesces with each other and closes separately. The volume of the foam is much larger than the volume of the liquid used to generate it, and the number of times the volume of foam that can be formed per unit volume of foam liquid is called the expansion factor of the foam. Liquid foam that is mechanically generated and wrapped in air is called air-mechanical foam. The expansion multiplier of this foam can be hundreds or even thousands of times. High multiples of foam (20 times the amount of foam generated for the low multiples of foam, 21 times to 200 times between the medium multiples of foam, 201 times to 1000 times between called high multiples of foam) is the expansion times in the hundreds of times more than the air mechanical foam short. Figure 4.1 shows the high multiplier mechanical foam fire extinguishing device, using the device can produce 500 times to 1000 times the air mechanical foam.

High multiplier foam fire extinguishing is the use of high multiplier foaming agent and pressure water mixture, in the ventilation fan under the wind flow to promote, produce a large number of air mechanical foam, when the foam filled the alleyway into the fire zone, the foam liquid film on the evaporation of water can absorb a lot of heat, play a cooling effect. A large amount of water vapour makes the air near the fire source relatively lower in oxygen content, in the oxygen content drops below 16%, the water vapour content rises above 35%, the fire will be extinguished. The foam itself is a very good insulation material, which has a high stability. It can isolate the burning material and air contact, play a role in closing the fire zone and suffocating the burning.

(2) Advantages

It is powerful and fast, and can be used to extinguish fires in safe locations away from the source of the fire, with little water loss, easy recovery work after extinguishing, low cost, non-toxic and non-corrosive. It has a wide range of applications and can extinguish general solid fires and oil fires, which can also extinguish electromechanical equipment fires when the power supply is cut off. It can send foam over a limited distance in horizontal alleyways inclined alleyways and from top to bottom or bottom to top in inclined alleyways.

(3) Precautions for use

① In the event of a hidden fire in a mine fire, after the foam has extinguished the open fire, other measures should be taken in a timely manner to eliminate the hidden fire and avoid re-ignition.
② Master the performance of the foaming device, operate it skillfully and keep the equipment in good combat readiness on a regular basis.
③ Equipment is installed correctly and reliably, when starting, supply the foam liquid first and then supply the air; when shutting down, stop the air first and then stop supplying the foam liquid to seal the fan inlet to prevent air leakage to the fire zone.
④ Note changes in the driving air pressure of the foaming machine to determine how the foam is advancing in the roadway.
⑤ To ensure that good foaming results are obtained, when powdered foam is used, the agent should be prepared as required before the fire extinguishing programme is determined so that it is fully dissolved.
⑥ The use of this fire fighting technique in gas mines should be used with particular caution, carefully accounting for the rate of gas build-up and considering an adequate safety margin. Before sending foam, start the ventilator and vent the fire zone gas before sending foam. When sending foam, gas observation should still be strengthened and all possible safety measures should be taken to prevent gas explosions.

(4) Scope of application

The high-capacity foam extinguishing method is suitable for fires in roadways far from coal mining faces or unsealed mining areas.

(iii) **Three-phase foam fire extinguishing devices**

(1) Principle of fire prevention and suppression with three-phase foam

Three-phase foam is produced in large quantities and stable for a long time, and can quickly fill the space to be filled to exclude the oxygen therein. At the same time, nitrogen is effectively sealed in the three-phase foam and drops to the bottom plate and is released as the foam breaks, giving full play to the inerting and explosion suppressing effects of nitrogen; the water in the foam plays a heat-absorbing and cooling role on the self-heating coal; the three-phase foam contains solid substances such as fly ash or loess, which form part of the three-phase foam mask and can maintain the stability of the foam for a longer period of time. After the foam is broken, the fly ash or yellow clay with certain viscosity can still be evenly covered on the floating coal and filled in the gaps, which effectively hinders the adsorption of oxygen by the coal and prevents the oxidation of the coal, thus inhibiting the process of spontaneous combustion of coal and achieving the purpose of fire prevention.

(2) Composition of three-phase foam

On the basis of water-based foam, solids such as yellow clay and fly ash are added to the blowing agent solution and mixed to obtain a homogeneous slurry, which is prepared by introducing nitrogen (or air) into the slurry using a foaming apparatus to obtain a three-phase foam with a gas–liquid-solid phase.

After adding foaming agent and injecting nitrogen into the fly ash slurry or yellow mud slurry to form a three-phase foam, the volume increases dramatically, which can be piled up to high places in the mining area and can effectively cover the floating coal at both low and high places, avoiding the loss of slurry and water along the channel of low resistance in the ordinary water injection or slurry injection process.

(3) Characteristics of three-phase foam for fire prevention and suppression

① High foaming multiplier and high production, the amount of foam produced is determined by the nitrogen source and water source. It can be stacked to high places, covers a wide area and can effectively extinguish fires in hidden locations.
② Good foaming performance, the water slurry becomes foam, which can effectively avoid the loss of slurry, and does not affect the working surface environment.
③ Foam stabilization time of up to 8 h or more, good performance in wrapping nitrogen and effective inerting of the filling position.
④ Make the distribution of fly ash more even, can effectively isolate oxygen and extinguish the fire thoroughly. Existing grouting system can be used, low cost, light equipment, simple operation, to achieve rapid fire extinguishing.

(iv) **Examples of foam fire extinguishing**

(1) Example of fire extinguishing by the second team of comprehensive mining in the north block of the downstream area of the first zone of the south 17 floors of the second level of Junde Mine

① Overview of the mining area

The mining area is 480 m in strike, 126 m in inclination, 2.4 m in mining height, 6.8 m in coal release height, 0.66Mt in recoverable volume and 56 days in natural fire period of the coal seam.

② How the accident happened

Despite nitrogen injection into the mining area and drilling and water injection and blocking agents into the mining area in the roof roadway, the floating coal in the mining area remained within the oxidation zone for a long period of time, which eventually led to a natural fire in the mining area, which was closed on 27 September.

After the closure continue to nitrogen injection to the mining area, the follow-up began to prepare for the use of the machine channel nitrogen injection tube injection three-phase foam. It was taken over on 24 November. On 25 November, three shifts began to inject three-phase foam, a total of 16,800 m^3. But due to the conditions after the closure was not injected to the correct location at 9:30 on 28 November after the first group of shelves appeared light smoke CO 5000PPm, mining area re-ignition.

③ Determination of firefighting plan and firefighting history

After the resignation of the mining area to the mining area firing point drilling and water injection to prevent the emergence of open fire, to buy time to extinguish the fire. After analyzing the location of the fire source, which was about 10 m from the working face, it was decided that the fire would be extinguished by opening the area and injecting three-phase foam. When drilling, every time a drill pipe was added, the change of harmful gas in the hole was checked, and the fire point was gradually judged according to the location of the maximum CO concentration in the drill hole, and the location of the fire point was finally determined. During the fire fighting process, the maximum concentration of harmful gases was CO 25000PPm, C_2H_2 61PPm, C_2H_4 990PPm. There was dense smoke, water injection was carried out to cool down and inject three-phase foam into the fire spot.

During the fire fighting and rescue process, a total of 47 boreholes were drilled from 16:00 on 28 November to 16:00 on 5 December, 23,000 m^3 of foam was injected and 200 m^3 of ash was added. The smoke soon disappeared, the harmful gases dropped rapidly and the temperature dropped from 50 to 26 °C. C_2H_2 and C_2H_4 were 0, CO 2PPm and the temperature dropped from 50 to 26 °C. The zone was successfully extinguished!

(2) Example of extinguishing a fire in a large, high-rising fire zone

① Overview of the working face

Henan Yima Group Gengcun Coal Mine 12,190 working face is located on the east side of the 2–3 coal track downhill in the west mining area of the mine, with 2–3 solid coal on the north and south sides; the working face has a strike length of 1060 m, an inclination length of 170 m, a coal seam thickness of 11–15 m and an inclination angle of 9–13°.

On 30 October 2005, the ventilation area discovered smoke coming out of the lower roadway of the working face at 892 m and immediately used the conventional method of drilling holes and injecting yellow slurry to deal with the fire zone, while reporting to the mine leadership.

From 8:00 to 6:00 on 5 November, the mine's second excavation team constructed deep boreholes from 880 m down the lower roadway inwards and outwards at the same time to accurately determine the extent of the fire zone, which was found to be between 840 and 950 m from the lower roadway.

② Preparation of three-phase foams

According to the characteristics of the Gengcun mine's fire zone, a new method of fire extinguishing is used to inject three-phase foam into the borehole. Firstly, the yellow soil is flushed with a high-pressure water gun to form a preliminary slurry, which passes through two filters and is made into a slurry with a water to soil ratio of 4:1 by stirring in a mixing pool, the slurry then flows into the down-hole pipe through a filter. At the entrance of the down-hole pipeline, the foaming agent is added to the slurry by the quantitative adding pump of the foaming agent, and the slurry and the foaming agent enter the down-hole grouting pipeline under the action of gravity, and enter the foaming device after being fully mixed in the pipeline. The nitrogen gas is connected to the nitrogen generator on the ground, and the nitrogen gas interacts with the yellow mud slurry containing the foaming agent to produce a three-phase foam, which is injected into the high bubble area under pressure.

③ Three-phase foam fire extinguishing

From November 6 to 7, a total of 16 h and 40 min of three-phase foam was injected into the fire zone. A large amount of three-phase foam quickly accumulated around the high rise zone in a three-dimensional shape, wrapping and covering the loose coal body in the high rise zone, effectively controlling the spread of the fire zone and rapidly reducing the concentration of CO and other index gases. In addition, the three-phase foam encapsulates a large amount of nitrogen and is injected into the high risk fire zone, releasing nitrogen after the foam bursts, which quickly plays its role in inerting and suppressing explosions. After the injection of the three-phase foam, the original intermittent gas deflagration did not occur again, further ensuring the safety of firefighters.

4.1.6 Inert Gas Fire Extinguishing

China's prevalent furnace smoke fire can be described as the prototype of inert fire extinguishing, later some mines use dry ice fire, liquid nitrogen fire, wet inert fire extinguishing are inert fire zone to extinguish the fire source as the basic principle of fire extinguishing methods.

Injecting liquid inert gas will quickly vaporize and absorb heat, taking away heat, can quickly reduce the fire temperature, while inert gas can dilute the concentration of explosive gases such as gas, play a role in inhibiting the explosion, the addition of a large number of inert gas can reduce the concentration of oxygen to inhibit the oxidation of coal.

There are two forms of liquid nitrogen fire extinguishing, one is to establish a liquid nitrogen vaporization system on the ground, a large liquid nitrogen tanker is transported from the nitrogen plant to the liquid nitrogen vaporization, with the help of vaporization pressure or compression pump through the water sand filling pipeline or special pipeline to the down-hole fire zone. The other form is to transport the liquid nitrogen to the down-hole with a small tanker, directly sprayed into the fire zone to extinguish the fire.

Wet inert gas fire extinguishing is through the jet turbine boat gas engine burning gasoline (diesel) to produce N_2, CO_2, water vapour as the main body of the wet inert gas, and then pressed into the fire zone, inerting the fire zone air, to achieve the dual purpose of preventing gas explosions and fire extinguishing.

Safety precautions to be taken when injecting nitrogen.

(1) Stable and reliable nitrogen source.
(2) The concentration of the injected nitrogen is not less than 98%.

(3) At least one dedicated nitrogen delivery piping system and its ancillary safety facilities.
(4) A monitoring system capable of continuously monitoring changes in the composition of the gas in the extraction area is available.
(5) Fixed or mobile temperature observation stations (points) and monitoring means are available.
(6) There are rules and regulations for regular testing, analysis and collation of relevant records by dedicated personnel, and timely reporting and handling of problems found.
(7) The nitrogen injection port should be strictly checked when injecting nitrogen into the extraction area, requiring tight pipelines and firm screws to prevent air from being exhaled into the pipelines.
(8) Personnel at the nitrogen injection port should pay attention to individual protection to prevent nitrogen asphyxiation or freezing to death accidents.

When injecting CO_2, in addition to meeting the technical requirements for nitrogen injection safety, the following issues should be noted.

Firstly, freezing occurs when CO_2 flows from the cylinder.

Secondly, in order to avoid damage to the eyes from CO_2, the operator of CO_2 must wear protective glasses.

Thirdly, if a tap is used to deliver CO_2, wet the tap beforehand.

4.1.7 Grouting (Gluing) to Extinguish Fires

(i) **Gel fire prevention and suppression technology**

(1) Composition of the gel

The gel is a material that has been more widely used in recent years to prevent fires in underground coal mines and consists of a base material (silicate, i.e. water glass) + a coagulant (salts such as ammonium bicarbonate) + water (around 90%).

(2) Gel flame retardant mechanism

The base material and the coagulant both have the effect of blocking, plus contains a lot of water, under a certain pressure, when injected into the coal body around the high temperature point. Because the gel is easy to flow before becoming gel, can penetrate into the interior of the broken coal body, so it can play a role in preventing oxidation, but also can seal the air leakage (fissure) channel, to prevent the leakage of air infiltration; its internal storage of a large amount of water, when high temperature evaporated by heat, can also play a heat absorption and cooling effect. Therefore, the gel is more effective in dealing with high temperature points and spontaneous combustion fire sources.

(ii) **Example of grouting and gluing to extinguish a fire**

(1) Overview of the mine fire zone

On May 26, 1996, a natural fire occurred in the coal seam of the west track flat of the 14,308 working face in a mine's 14 mining area. 14 mining area mines two coal seams, 3 up and 3 down, with a spacing of 0.4–1.1 m between the two coal seams.

(2) Fire extinguishing process

After 3 days of injecting 450 m^3 of colloidal mud, the fire zone was reconnoitered and most of the fire zone was extinguished. However, there was still a high-temperature negative combustion point in the center of the original fire zone, so we continued to inject colloidal mud for about 400 m^3. On January 16, the second time to open the fire zone reconnaissance, found that the original fire zone is still open fire, drilling did not hit the fire source area, the roadway basically no leakage of rubber.

As the second reconnaissance did not find any glue leakage, it was decided to drill a hole with a pitch angle of 3° and a depth of 8 m. After 3 days of injection of glue, the volume of injected glue was about 500 m^3, the fire zone was opened for the third time on 20 January for reconnaissance and a forced slurry jet was planned through the high temperature zone.

After the third fire zone opening, it was realized that the original formulation of colloidal mud was less permeable when exposed to high temperatures. After analysis of the location of the borehole, it was decided to drill a 12° downward sloping hole, 10 m deep, with the end of the hole about 1.7 m from the roadway, and to slow down the rate of formation. After 6 days of continuous injection of nearly 900 mm^3 of colloidal mud, the CO concentration in the borehole dropped to below 0.002% for a few days before the fire zone was completely extinguished.

4.2 Insulation and Fire Prevention Techniques and Examples

The essence of fire suppression in a closed fire zone is to directly or indirectly reduce or even cut off the supply of oxygen to maintain the three elements of combustion, and to control or extinguish the combustion of coal in a closed fire zone by combining other fire suppression measures such as inerting, grouting and gluing. In the event of a mine fire, fire containment measures mainly refer to the construction of seal walls to cut off the flow of air inside and outside the fire zone. On the one hand, seal walls can create a closed space and cut off the oxygen supply conditions for the spontaneous combustion of coal in the closed fire zone to control the fire. On the other hand, it can, to a certain extent, resist the effect of gas explosion shock wave and cut off the circulation line of high-temperature smoke flow to protect the lives of underground rescuers, so that rescue work can be carried out smoothly. At the same time, it can also protect the machinery and equipment outside the danger area [2].

4.2.1 Types of Seal Wall

The seal walls used to enclose the fire zone are divided into four categories according to their role: temporary seal walls, semi-permanent seal walls, permanent seal walls and explosion resistant seal walls.

(i) **Temporary seal wall**

The purpose of the temporary seal wall is to temporarily block the wind flow and control the development of the fire in order to prepare direct fire extinguishing equipment under its cover, to protect the rescuers and to ensure that the workers are protected from fire smoke and toxic gases while the permanent seal wall is being built.

The requirements for temporary seal walls are simple construction, local availability, rapid construction and not necessarily high tightness. The following types of seal wall are available.

(1) Brattice

Generally made of 4 m × 4 m or 6 m × 6 m canvas. When hanging the brattice, choose a location where the stand is complete, support 2–3 posts first, nail the brattice tightly to the top beam and posts of the stand with iron nails, and nail small boards around it. The bottom is pressed tightly onto the base with sand, stone and fluorine. Brattice is hung and the canvas is wetted with water. To hang brattice in a masonry tunnel, set up a wooden frame first and then nail the wind barrier.

(2) Wooden panel seal wall

This is shown in Fig. 4.2. This type of seal wall is more widely used. The first board is nailed to the shed legs and posts from the top, with the top edge of the next board pressing against the bottom edge of the previous board to form a step. Small boards are set around the perimeter of the aisle and the seams and boards are plastered with yellow clay. To facilitate access to the fire zone for observation and fire fighting, a 1 m × 0.8 m door hole can be left in the middle of the plank seal wall to install the dampers.

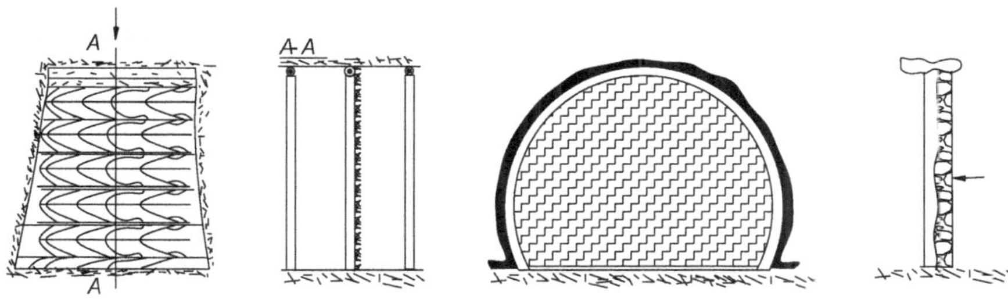

Fig. 4.2 Temporary seal wall with wooden panels

In addition, a number of quick temporary seal walls are used to enclose the fire zone, such as umbrella closures, inflatable closures and foam closures. Umbrella closures, also known as umbrella brattice, are made of latex glass fibre cloth. It is hung at the required location and quickly opens up in the tunnel with the help of wind flow pressure to block the wind flow. The brattice is easy to carry and does not require other ancillary materials, which can be hung by 1–2 people in 10–15 s. The inflatable containment (airbag fast temporary seal wall) is a flexible container made of flexible material (plastic, nylon, etc.) and filled with pressurized air (CO_2, N_2). By setting it in the tunnel, it can have the same blocking effect as other confinements. As the installation and removal of the inflatable confinement only requires inflation and deflation, it is simple, fast and reusable. Our metallurgical system has developed ball-type inflatable confinement, which has also achieved a good confinement effect. Foam containment is based on polyether resin and polyisocyanate, plus several auxiliaries, divided into two groups of A and B according to a certain ratio of combination, by strong mixing, sprayed by a gun on the seal wall backing (with grass curtain, linen and other breathable fabrics as the backing), within seconds that foam forming, forming a good airtight seal wall.

(ii) **Semi-permanent seal wall**

This type of seal wall lasts longer than a temporary seal wall. It has the effect of isolating the wind flow and eliminating the source of fire. They require good isolation and are easy to open and close.

(1) Wood segment seal wall

Scrap pit timbers are generally sawn into 0.8 m long sections, stacked up with a layer of wood sections and a layer of loess, then fastened with wooden wedges and plastered with yellow clay. As shown in Fig. 4.3, It is suitable for situations where the pressure of the surrounding rock is high, the handling of materials is difficult, the conditions of the workplace are poor, and rapid closure of the fire zone is required.

(2) Loess seal wall

It is usually built on the basis of a plank-seal wall. Before construction, a groove is hollowed out, and then two rows of pillars are made, 3–5 in each row. The pillars are nailed with wooden boards on the inside, filled with loess in the middle and pounded with a wooden hammer. This kind of seal wall has good isolation performance and can be used in the roadway with high pressure.

(iii) **Permanent seal wall**

The function of a permanent seal wall is to provide a tight, long-term insulation of the fire zone and to prevent the entry of air. For this reason, it must be strong and dense. This type of seal wall uses a lot of material, which is complex and takes a long time to build. Masonry seal walls are generally used.

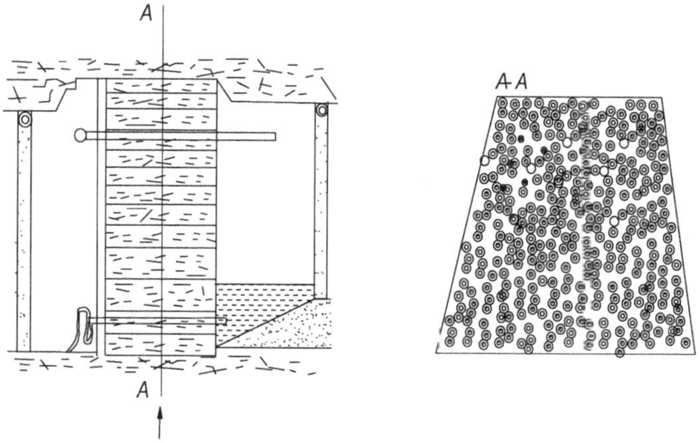

Fig. 4.3 Schematic diagram of a wooden section seal wall

(1) Masonry seal wall

The seal wall can be constructed of materials such as material stone, brick or concrete blocks with cement mortar, plastered inside and out, as shown in Fig. 4.4. Before masonry, a 0.5–1.0 m slot should be dug in the wall of the roadway around the wall to create a good foundation. To increase tightness, a layer of clay, mortar or water glass, rubber emulsion, etc. needs to be applied around the outside of the wall and the slot. Seal wall should be strongly supported within 5–6 m inside and outside to prevent air leakage due to risers or crevices. Each seal wall should have a monitoring sampling tube to monitor the gas composition and pressure inside the seal wall, as well as a drainage line should be provided at the bottom.

For locations with high ground pressure, to ensure that the seal wall can withstand a certain amount of pressure, the addition of timber panels in the seal wall can make the wall evenly stressed (Fig. 4.5) and cushion the pressure to which it is subjected.

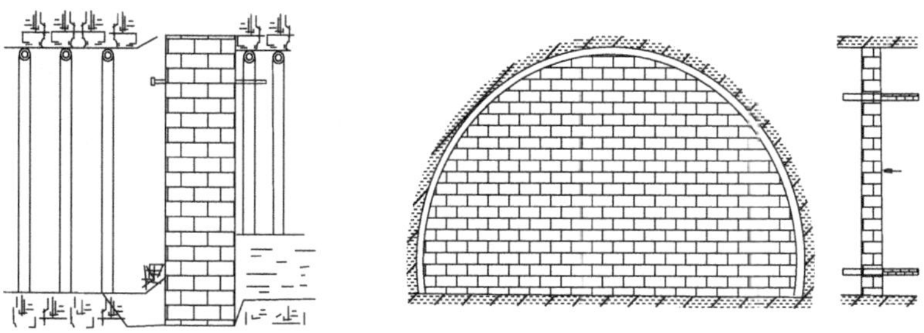

Fig. 4.4 Diagram of a brick wall with a permanent seal wall

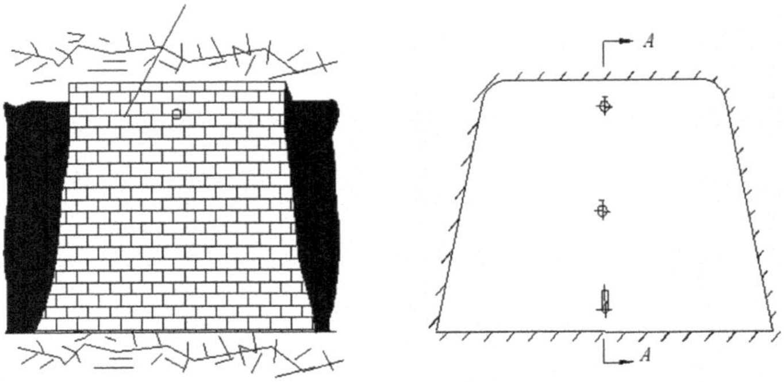

Fig. 4.5 Brick wall with timber slab in the middle

(2) Concrete seal wall

When the sealing, temperature and pressure resistance of the seal wall are high, concrete or reinforced concrete seal walls should be built. Concrete seal walls have good pressure resistance and reinforced concrete seal walls have not only good pressure resistance, but also high tensile strength.

The construction begins with the hollowing out of the trench, which is identical to that of a brick seal wall. The form-work is then erected and the concrete poured, which, when solidified, becomes a concrete seal wall with high compressive strength and good compactness.

The concrete seal wall is shown in Fig. 4.6. Its most important feature is that it is highly cushioned and can resist 1 MPa shock wave impacts. In order to prevent the concrete from generating cracks, bubbles or holes due to solidification and shrinkage, it should be filled in time after the form-work is removed, or the gaps should be filled with grout using a high-pressure pump, or the gaps and holes should be blocked by spraying concrete to prevent air leakage and oxygen supply.

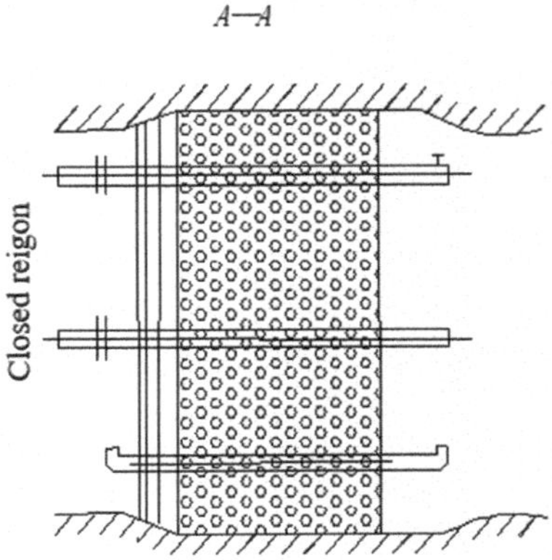

Fig. 4.6 Concrete seal wall

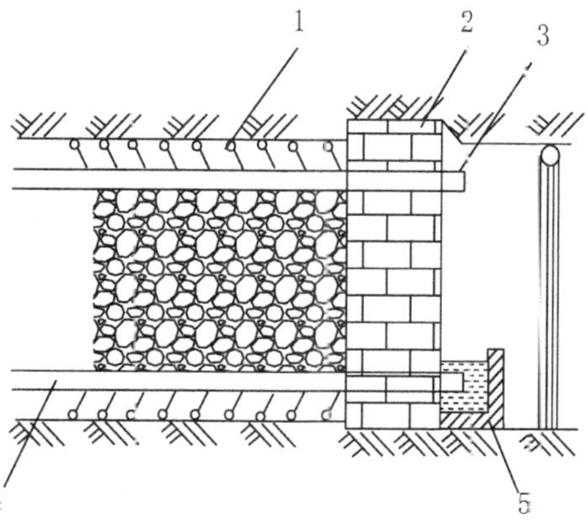

Fig. 4.7 Multi-layered mixed seal wall

(3) Multi-level hybrid seal wall

It is also possible to use one or more of these seal walls in combination, as shown in Fig. 4.7. This type of seal wall is much stronger, but is more complex to construct.

(iv) **Blast resistant seal wall**

Explosion-resistant seal wall is in the gas larger area closed fire zone, in order to prevent the fire zone gas or fire gas explosion on the closed fire zone outside the personnel caused by the construction of the explosion-proof function of the seal wall. Explosion-resistant seal wall must withstand a certain pressure, the requirements of different countries are not consistent, for example, the United States provides not less than 0.14 MPa, India's regulations range from 0.15 to 0.4 MPa.

(1) Commonly used explosion resistant seal wall

The blast-resistant seal wall must be able to withstand the effects of the blast wave for a period of time less than the shock wave pressure on the seal wall, i.e. it must have better pressure resistance. Therefore, the blast-resistant seal wall must have its own special requirements in terms of material and production. This seal wall is generally made of sandbags (Fig. 4.8), but can also be composed of sand sections.

① Gypsum blast resistant seal wall. currently, foreign countries also use gypsum fast filling blast resistant seal wall (Fig. 4.9). Gypsum seal wall thickness is usually 2–3.5 m. Construction of the first hollow conical groove, and then build two with reinforced support plank wall, with a pipe to the water and gypsum powder sent to the gypsum pump mixed, and then sent between the panels with a hose filling, hardening molding, after curing the compressive strength of gypsum up to 6 MPa. In 1985, a spontaneous combustion fire occurred in No. 406 pan area of coal seam L1 of Yongdingzhuang coal mine of Datong Mining Bureau, as well as the problems of air leakage and fire gas leakage were not solved by masonry and other forms of confinement.

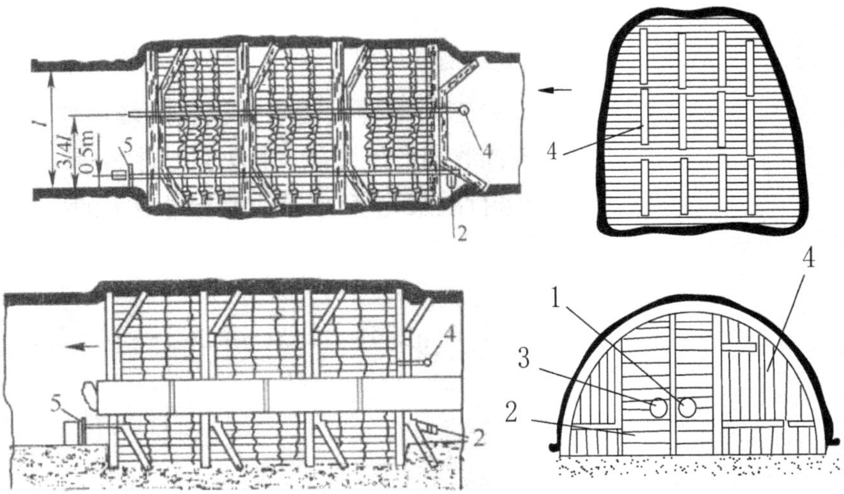

Fig. 4.8 Sandbag blast resistant seal wall

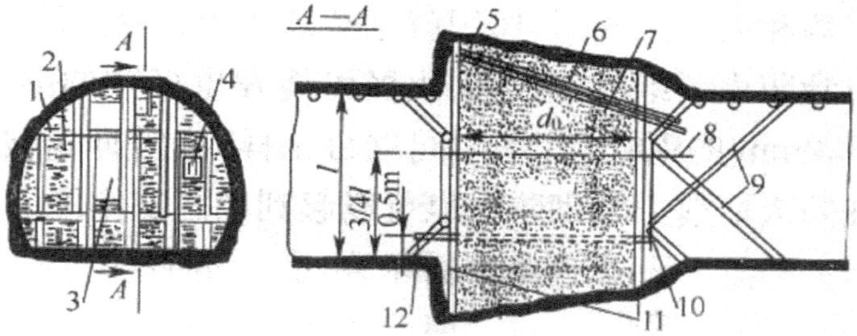

Fig. 4.9 Gypsum blast resistant seal wall. 1—Sampling tube; 2—temperature measuring hole; 3—testing hole; 4—U-shaped differential pressure gauge; 5—gypsum filling material outlet; 6—filling tube; 7—supply tube; 8—air sampling tube; 9—support frame; 10—drainage tube; 11—wooden board support; 12—mesh cover

In addition to the two traditional explosion resistant seal walls often used above, water sealing can also be used where appropriate and available. When water sealing is used, it can be operated remotely, safely and securely, which has the advantages of simple operation, good tightness, full explosion isolation and easy opening and sealing. Figure 4.10 shows a diagram of the water seal used in the high gas mega-fire zone of the Ningxia Bainiangou Mine. If there is a low-lying area in a part of the roadway, the roadway can also be partially water sealed, as shown in Fig. 4.11. The water sealing technique is different from flooding in that it only seals the low level of the fire zone air intake side or return air side of the roadway, which requires less water and is easier to implement.

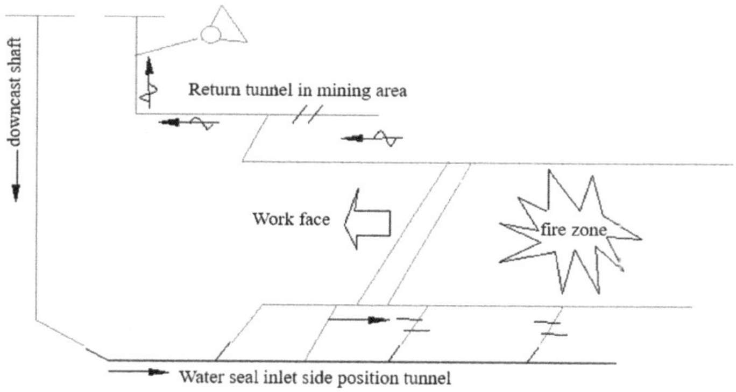

Fig. 4.10 Diagram of the water seal in the high gas mega-fire zone of the Bainiangou Mine

Fig. 4.11 Schematic diagram of the partial water seal fire zone

(2) Improved blast resistant seal wall

① Shape improvements. Of course, there are certain improvements in the shape of the seal wall to increase the blast resistance of the method. Figure 4.12 shows the blast-resistant seal wall, seal wall convex to the side of the fire zone, so that it can withstand greater impact pressure than the flat shape of the seal wall

② Add a buffer facility. In order to increase the blast strength of the seal wall, in addition to improvements in materials and shape, but also in the seal wall and fire zone between the auxiliary to some kind of pressure buffering facilities, as shown in Fig. 4.13. The blast-resistant seal wall with a wave-absorbing steel curtain and a row or rows of I-beams suspended from the roof beam of the roadway, which

Fig. 4.12 Convex blast resistant brick wall

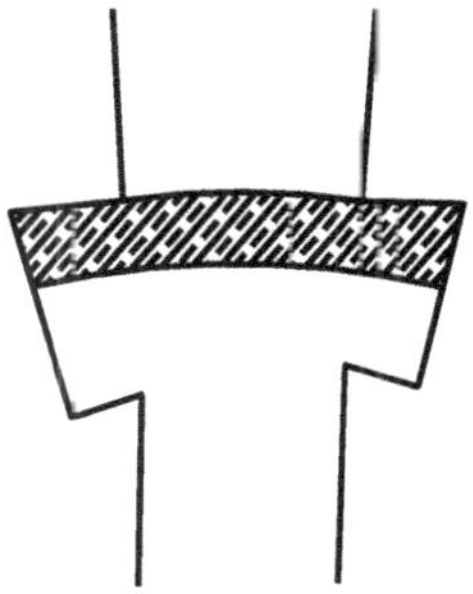

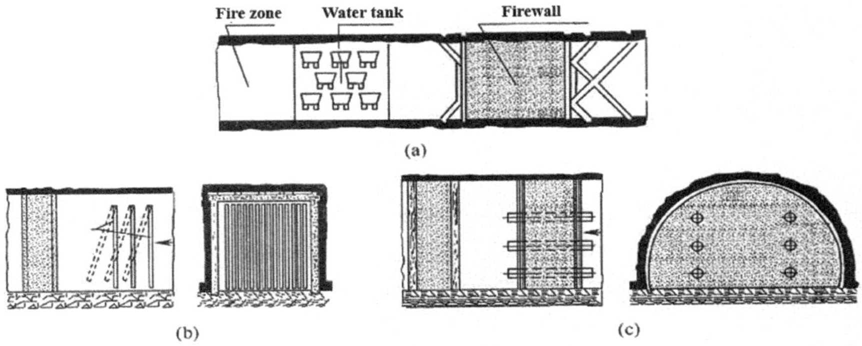

Fig. 4.13 Explosion-resistant seal wall with buffering facility. **a** Blast-resistant seal wall with sink installation; **b** blast-resistant seal wall with wave-absorbing steel curtain; **c** blast-resistant gypsum double wall confinement with wave elimination holes

can swing under the action of the blast wave. Steel curtain is located between the seal wall and the fire zone, its role is to slow down the pressure acting on the seal wall; with wave elimination hole blast resistant seal wall consists of two plaster seal wall, in the first seal wall facing the fire zone may occur in the explosion set a few diameter of 100–300 mm steel pipe. Its role with the same steel curtain, to slow down the shock wave pressure, to protect the second seal wall, the same, in front of the seal wall stacked to prevent the same effect.

4.2.2 Seal Wall Location

When constructing a seal wall, it is important to consider not only where the wall should be constructed, but also its feasibility. Due to the constraints of underground construction conditions and the complexity of the site, a focus on the "need" and ignoring the "possibility" can often lead to failure of the fire zone closure. The following is a brief overview of the basic principles for selecting a seal wall.

(1) Under the condition of ensuring the isolation of the fire zone, the confinement area is minimized and the number of seal walls constructed is reduced. The original seal walls and dampers of good quality can be fully utilized. When considering containment, care should be taken that the volume of the containment area is sufficient to withstand the expansion of the hot gases without endangering the miners constructing the seal wall or damaging the seal wall.
(2) Within 5 m from the front and rear of the seal wall construction site the surrounding rock must be intact and stable, free of faults and fracture zones, with a small roadway section and at a distance of about 10 m from the roadway intersection. Where temporary confinement is constructed, there should be room (6 m or more) for the construction of a permanent seal wall, and the rock at the bottom of the tunnel should be solid, preferably above the level of possible

flooding. The accumulation of water within the permanent seal wall will help to prevent air leakage, but its substrate should not be refractory clay. For temporary confinements, the accumulation of water may break the tightness of the seal wall, scouring the bottom of the seal wall and increasing air leakage pathways.

(3) To facilitate the rapid evacuation of construction seal wall personnel to a safe area, to facilitate the transport of materials and to facilitate the supply of fresh air to construction seal wall personnel.

Due to the complexity of the site environment, it is not always possible to choose the location of the seal wall to meet all of the above requirements and only relatively good conditions can be selected according to the specific environment of the site. When sealing the fire zone of a coal mining face, the location of the seal wall generally depends on the location of the fire source. The details are as follows.

① The fire occurs in the working face air intake roadway, the seal wall of the air intake side should be as close as possible to the fire source, while the seal wall of the return air side should be at a certain distance from the fire source, so that the temperature of the smoke stream is reduced. And it is easy to construct the seal wall operation.

② Fire occurs in the working face return air alley, from the principle of reducing the fire zone, air intake side of the seal wall should be as close as possible to the working face, and return air side seal wall should be a certain distance from the working face. In high gas mines, the air intake side seal wall close to the working face。Although the fire source air intake side air oxygen concentration and the possibility of gas explosion, but due to the large amount of gas gushing out of the working face, if the air intake side seal wall is not tight, there is still the possibility of explosion. For this reason, the air intake side should be injected with inert gas when the fire zone is closed, so that the fire zone inert.

If the fire occurs at the working face, the air intake and return seal walls should be built in opposite positions according to the development of the fire and the smoke and temperature conditions. If the fire does not reach the opening conditions after a long period of closure, the fire zone can be re-digged outside the seal wall to get rid of the fire zone; if conditions permit, the seal wall on the air intake side can be built closer to the working face to reduce the possibility of explosion during closure.

4.2.3 Sealing Wall Closure Sequence

Fire zone closure should only be carried out when it is ensured that no one is left inside. When constructing a seal wall in a multi-vented fire zone, the order in which the fire zone is closed should be determined by the extent of the fire zone, the size of the fire and the amount of gas gushing out. Generally, the fire zone should be closed first for the secondary wind-way, which has little effect on the fire zone, and then for the main intake and return wind-way of the fire zone.

The order of closure of the fire zone inlet and return air vents is very important, it not only affects the speed of fire control, but more importantly the safety of the rescuers. The following closure sequences are commonly used.

(1) Close the air inlet first, then the air return

Generally speaking, it is much easier to build a seal wall on the air intake side of the fire zone than on the return air side. As soon as the seal wall on the air intake side is closed, the amount of air entering the fire zone is greatly reduced, thus weakening the fire and reducing the amount of smoke gushing out, which facilitates the establishment of the return air side seal wall. Therefore, in non-gas mines, the air intake is usually closed first and the return air is closed later.

(2) Close the return air inlet first, then the inlet air inlet

A method generally used to quickly cut off the spread of a fire when it is not too large, not too hot and no gas has accumulated.

After the seal wall is established, the pressure in front of the wall rises locally and the pressure behind the wall falls locally. In gas mines, if the previous seal wall is built on the air intake side and there is an old void between this wall and the fire source, the amount of air flowing to the fire source will gradually decrease during the process of building the seal wall. At the same time, under the effect of local negative pressure, the amount of gas coming out of the old void area will increase, which will easily make the gas concentration in the air flow reach the explosion limit and cause an explosion. Therefore, it is extremely dangerous to close the air intake side of the seal wall when there is a gas source between the seal wall and the fire source. It is safer to close the return air side seal wall first. Because it creates a positive pressure in the fire zone, which more or less suppresses the outflow of gas from the old void.

(3) Inlet and return air openings closed at the same time

In the process of masonry seal wall, leave a certain area of the vent, to ensure that the supply of wind so that the gas in the fire zone does not exceed the limit of accumulation, when the wall work is completed, in the agreed time at the same time will be into return air side seal wall on the vent quickly closed and immediately evacuated personnel. As this method can quickly close the fire zone, cut off the oxygen supply, fire zone gas is not easy to reach the explosion limit, and can ensure the safety of personnel, so it is a common sequence of closure of gas mines to close the fire zone.

4.2.4 Method of Closing the Fire Zone

(i) **Classification**

There are three types of methods of closing the fire zone depending on the build-up of gas in the fire zone.

(1) Wind-break closure fire zone method

This method is suitable for situations where the air in the fire zone is confirmed to be oxygen poor and the oxygen concentration is below the gas failure threshold during the rescue process.

(2) Ventilated closed fire zone method

This method is to keep the fire zone ventilation conditions closed fire zone, in the fire zone into the return air on both sides at the same time to build a seal wall with ventilation holes. In the process of closure, the fire zone still passes a certain amount of wind, and the oxygen concentration in the fire zone decreases slowly. This air flow can also transport part of the gas, which makes the gas concentration in the fire zone does not rise quickly.

(3) Injecting inert gas to close the fire zone method

This method is one of the joint fire extinguishing methods, but also the safest and most effective fire extinguishing methods. In the closed fire zone at the same time, constantly inject inert gas (CO_2, N_2) into the fire zone. On the one hand, it can dilute the gas concentration in the fire zone and prevent gas explosion in the fire zone. On the other hand, the oxygen concentration in the fire zone drops rapidly, accelerating the asphyxiation of the fire source. However, this method requires a complete set of inert injection device, and there should be enough inert gas source into the fire zone.

(ii) **Determination of guidelines**

(1) The 3 times when the fire zone is closed

If the fire zone is not to be closed by inert gas injection, the method of closure should be selected based on the time for the flammable gas mixture in the fire zone to reach the lower explosive limit (T_1) and the time for the oxygen content to drop to the failure limit (T_3).

① T_1: the time when the combustible mixture in the fire zone reaches the lower explosive limit, i.e. the period during which the construction of the seal wall should be completed and the personnel evacuated to a safe area.
② T_2: the time when the combustible gas mixture in the fire zone reaches the upper explosion limit. Since explosions are not likely to occur even if the gas concentration exceeds the upper explosion limit, work in the closed fire zone is permitted after this time.
③ T_3: The time during which the oxygen concentration in the fire zone is reduced to the failure limit for combustion gases. This means that after this time, work can be carried out even if the gas concentration in the fire zone is within the explosion limit.

Between the periods T_1 and T_2, it is the most dangerous periods for gas explosions in the fire zone, so evacuate people to a safe location.

(2) Selecting the method of closing the fire zone

Two times should be selected based on T_1, T_3.

① If T_1 is much longer than T_3, i.e. the time for the oxygen concentration in the fire zone to drop to the failure limit is significantly shorter than the time for the concentration of the flammable gas mixture to rise to the lower explosive limit. There is no need to worry about a gas explosion during work in a closed fire zone. There is no requirement to inject inert gas or to maintain ventilation in the fire zone.
② T_1 and T_3 are similar, i.e. the time for the oxygen content of the fire zone to fall to the threshold of failure and the time for the gas to accumulate to the lower limit of the explosion is not much different, closed fire zone may occur gas explosion, which requires the construction of explosion-resistant seal wall, and to maintain the ventilation of the fire zone during work.
③ If T_1 is much shorter than T_3, there is a greater possibility of a gas explosion in the enclosed area, then an explosion resistant seal wall should be used and the fire zone ventilation should be maintained for closure. Finally, the ventilation holes should be sealed and the personnel evacuated immediately. Wait for a period of time until a gas explosion occurs and then send someone in to construct a permanent seal wall.

The three times T_1, T_2 and T_3 are variable for each fire zone. The choice of seal wall position allows the size of the enclosed space to be changed as well as the size of the gas flow in the fire zone to be adjusted.

4.2.5 Construction of the Seal Wall

(i) **Preparation work before the construction of the seal wall**

Time is the decisive factor for success, the faster the construction of the seal wall, the greater the chance of successful fire fighting. Therefore, all preparatory work should be carried out before the construction of the seal wall.

(1) Where significant surface leakage exists, advance knowledge of the location of these major fissures can assist in selecting the best location for the seal wall.
(2) When closing the fire zone, some of the original seal walls and dampers can be used. In addition seal wall materials should be prepared in advance and can be put in place quickly. These tasks should be carried out before a fire occurs.
(3) Measures to prevent explosions

Before the fire zone is closed, remove as much flammable material as possible, especially telephones and battery-operated sensors and electrical equipment in the fire zone, which can be easily ignited by flammable gases and can adversely affect the safety of future fire zone resumption. All circuits in the fire zone, including signal lines, overhead lines and metal pipes, should be disconnected and sections of track and conveyors should be removed to cut off conductive circuits. In addition, if there is a large amount of coal dust in the area to be closed, it should be covered with more rock dust for inerting.

(4) Sampling within the closed area

During the post-closure management and opening of the fire zone, the basis for understanding and making decisions on the combustion status of the fire sources in the closed area is mainly based on the sampling of the closed fire zone. Therefore, in the inlet and return air side, each seal wall should be equipped with a sampling tube. Sampling tubes should be extended as far as possible within the seal wall to minimize the effect of air leakage in the vicinity of the seal wall. Each air sample should be taken at least at three points—at the top, middle and bottom of the tunnel—to reflect the air composition within the confined area, particularly in inclined tunnels. In order to keep the sampling pipe as close to the fire zone as possible and to save time in constructing the seal wall, the existing air and water pipes can be used as sampling pipes by sawing them off near the fire zone. Isolation valves should be installed on the pipe to prevent the blast from being transmitted outside the seal wall through the pipe.

(ii) **Construction of fire protection seal walls**

After a mine fire, in order to quickly isolate the flow of wind inside and outside the fire zone, various research institutions and production departments at home and abroad have conducted a lot of theoretical discussions, experimental studies and field applications on the construction of efficient seal wall technology. The key issue in the construction of seal wall in closed fire zone is to reduce the leakage of air from seal wall and improve the quality of seal wall construction. Usually, the leakage of air from seal wall is mainly the perimeter of the contact between the seal wall body and the coal and rock of the side channel, which accounts for about 70–95% of the total leakage of air from the confinement.

At present, the domestic fire prevention seal wall mainly uses masonry or sandstone concrete seal wall, plus a variety of coating spraying. According to the summary of the previous experience of plugging, some plugging measures are proposed here to help reduce air leakage.

(1) In masonry concrete seal walls, glass fibre is mixed into the mortar to enhance the cementitious strength and adhesion of the mortar.
(2) The seal wall is most prone to leakage around the contact with the roadway. For this reason, some measures are taken at the bottom, the help and the top of the roadway respectively. ① Treatment of the bottom of the roadway: one method is to hollow out the slot and pour mortar mixed with glass fibre into the floor slot to form the base of the seal wall. Concrete blocks are then laid on top of the unconsolidated wall base. Alternatively, where rapid construction of a seal wall is required, the concrete block wall is built directly into the bottom of the artificially created fracture and water glass (sodium silicate) is poured into the hollow part of the concrete block, allowing the water glass to penetrate

the fracture floor. The water glass solidifies and forms a barrier, reducing the air leakage at the bottom of the seal wall. ② Corridor treatment: similar to the treatment of the bottom of the corridor, firstly hollow out the grooves in both corridors, then embed the concrete blocks and make them cemented to the corridor with mortar, plug all the gaps between the seal wall and the corridor with mortar, as well as then make a curved closed zone at the intersection of the wall and the corridor wall with mortar. ③ roadway top treatment: firstly, use a wooden wedge to punch 2.5 cm into the surface of the seal wall, stuff the mortar into the gaps, and also use mortar paste to form an arc-shaped closed belt within the intersection of the seal wall surface and the roadway top. If the seal wall has several layers of concrete block thickness, it can be handled in layers. ④ Seal wall peripheral treatment: in order to further reduce air leakage, spray polyurethane foam can be sprayed in the sealing dead corner where the seal wall is in contact with the roadway wall.

(3) In order to reduce air leakage and ensure the quality of the seal wall, the seal wall is built in the form of a sandwich seal wall, filled with cement and stone mortar or other colloidal sealing materials in the middle, as well as the upper void is filled and sealed with a grouting pump. In the upper, middle and lower parts of the external wall there are four pipe holes for the grouting main, the grouting pipe between walls, the observation pipe and the "U" drainage pipe. The grouting main should be left in advance and should be extended through the two walls as far as possible towards the working surface; the observation pipe in the middle of the wall should be extended through the two walls into the inner wall by not less than 0.3 m. When filling and grouting between the walls, it is necessary to wait for the walls to solidify before proceeding to prevent collapse accidents.

(4) Due to the high temperature in the closed fire zone and the influence of mine pressure, the seal wall will appear cracks, in order to further reduce air leakage, mud or other plugging materials should be applied to the wall surface regularly. In recent years, countries around the world have developed a variety of new fire prevention agents, plugging agents, such as: MEA plugging agent, rockxiu, nano-modified sealing plugging elastomeric materials.

The above measures will improve the strength and compactness of the seal wall.

4.2.6 Example of a Fire Zone Closure Incident

Figure 4.14 shows a mine with an upward flowing water–sand filled coal mining method, where a fire occurred in the mining area chute due to spontaneous combustion. There is a gas accumulation area of approximately 120 m^3 filled with gas near the source of the fire. The original plan was to use local containment measures, firstly in the three stone doors, but this was not possible due to the presence of the filling pipe in the tunnel. Finally, it was decided that the inlet air would be sealed in four

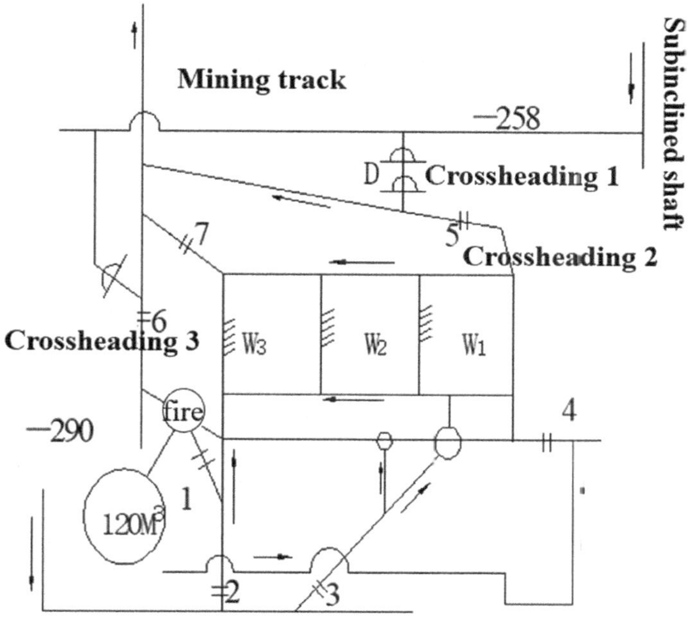

Fig. 4.14 Example of an incident where the air intake side was closed first I

places and the return air side would be sealed in three places, so that the fire zone could be closed to a larger extent for the purpose of extinguishing the fire.

The order of closure is 1, 2, 3, 4, 5, 6 and 7, i.e. the inlet side is closed first and the exhaust side is closed later. During the closure of No. 4, a gas explosion occurred when the water sand filling the confinement was about to be connected to the roof. The ambulance crew, who were preparing for work at No. 5, were also hit by the blast wave. There were nine consecutive explosions. When the 6th explosion, the second stone door in the air door D badly open, so that the wind flow short circuit, thus reducing the negative pressure of the three stone door, and the seventh explosion after the interval delayed for more than 7 h, the last three when the explosion power is also reduced.

The intervals between the explosions were in the order of 38, 11, 29, 95, 40, 3.30, 14.48 and 11.26 min.

The cause of the explosion is the first air intake side closed, so that the return air side of the negative pressure increases. Thus, the fire near the original 120 m^3 of accumulated gas outflow, through the fire zone and the explosion occurred. After the explosion due to the destruction of the shock wave, breaking the original gas accumulation state and the occurrence of a number of consecutive explosions. This is a great threat to rescue and relief work.

4.3 Determination of the Risk of Combustion and Explosion in Enclosed Fire Zone

4.3.1 Climate Change Patterns in the Closed Fire Zone

(i) **Breathing exercise in a closed fire zone**

As the fire develops and changes over time in the confined area, the gases inside and outside the confined area are constantly exchanged, with activity patterns similar to the human lungs performing respiratory movements. The current research on respiratory movements in mines is mainly focused on the mining area and the gas composition is mainly aimed at the gas in the mine.

The causes of respiratory movements are as follows.

(1) Influence of temperature inside and outside the fire zone

As the temperature of the fire zone is higher than the temperature outside the enclosed area, the temperature difference between inside and outside is formed, the temperature of the fire zone is high, the gas expands, the volume increases, the fire zone is in a closed state. In this limited closed space, the air expands and produces an outward expansion force, the air expands outward to form an air flow to the outside of the fire zone. This is called "exhaling". On the other hand, due to the high temperature of the fire zone, the gravity of the air in the fire zone decreases, the pressure decreases, and a pressure difference is formed inside and outside the fire zone, which is high and low, so that the external air flows inside the fire zone, which is called "suction". "breathing".

(2) Effect of changes in enclosed space

The coal mine underground closed natural fire zone, its space is constantly changing, causing the space to change for several reason.

Due to the fire zone to take slurry or inert fire extinguishing measures;

Water flow into and out of the fire zone;

The top of the roadway gang topping, the larger the closed space, the greater the change caused by its collapse, etc.

(3) Effect of changes in surface air temperature and pressure

When the surface air temperature decreases, the fire zone breathes in; when the air temperature increases, the fire zone breathes out. When the air pressure rises, the fire zone breathes in; when the air pressure falls, the fire zone breathes out.

(ii) **Changes in gas concentration**

During the fire zone closure, as the oxygen supply to the fire zone is reduced or cut off, the remaining oxygen in the fire zone is continuously consumed by combustion and mixed into the air together with the gushing CH_4, causing a continuous decrease in the relative oxygen concentration in the air under the premise of good closure performance. On the other hand, the CH_4 concentration in the coal rock seam keeps increasing due to the continuous influx of CH_4 after confinement. Therefore, the change of gas in the fire zone, especially the change of CH_4 and O_2, is the basis for determining the explosion risk of the gas in the fire zone.

(iii) **Changes in temperature and air pressure**

Variation of the relevant parameters during the confinement of the fire zone: (1) The temperature of the combustion zone quickly rises to 2–3 times the value of the temperature before the airflow is cut off; and the temperature of the flue gas on the section quickly tends to be homogeneous; due to the rapid rise in temperature, the volatile fraction and the amount of smoke produced increases; due to the lack of oxygen, the combustion weakens and the temperature rapidly decreases. (2) Cut off the wind flow when there is a large suction force on the dampers, and the pressure in front of the dampers is positive, after the dampers that the pressure on the side of the fire zone is negative; soon (about a few dozen seconds), the combustion weakens, the temperature drops, the expansion work disappears, and the pressure acting on the dampers is only a small fire wind pressure. (3) Smoke flow counter-flow layer quickly to the dampers transport. In a few dozen seconds, the temperature near the dampers from room temperature quickly rose to more than 200 °C.

4.3.2 Closed Fire Zone Gas Explosion Analysis and Determination

(i) **Fire zone gas explosion conditions analysis**

There are two main scenarios in which mine gas can cause accidents: high gas concentrations leading to a drop in oxygen concentration (e.g. when C gas > 57%, C oxygen < 9%) suffocating people; and gas concentrations within the explosive limit in conditions of high temperature heat sources.

According to the discussion of the mechanism of gas explosion in the previous fire zone, four basic conditions must be present at the same time for a gas explosion to occur, namely: (1) gas concentration within the explosion limits; (2) sufficient concentration of oxygen; (3) a certain temperature ignition source; (4) gas and fire contact time greater than the inductor.

(ii) **Analysis of Factors Influencing the Risk of Gas Explosion in the Fire Zone**

As can be seen from the fire zone gas explosion conditions, the only factors that determine the likelihood and danger of a gas mixture explosion are whether the concentration of combustible gases in the fire zone gas reaches the explosion limit, and the influence of the source of ignition. This can be analyzed from the following three aspects.

The first is the concentration of the gas mixture. When the gas and air mixture mixed with other combustible gases, if the concentration of combustible gases in the air is lower than the lower explosive limit, it will neither explode nor burn in the presence of open flame. If higher than the upper limit of the explosion, when encountering an open flame. Although it is not explosive, but it can occur combustion. Sometimes after a period of combustion and inhalation of air, so that the concentration of combustible gases in the air down to the explosion limit, it can occur; it may also occur after the combustion, combustible gases get a large supply, the concentration of combustible gases in the air did not fall, the formation of continuous combustion conditions, and thus it will not be transferred from combustion to explosion.

The second is the change in the explosion limit of the gas mixture. A variety of different combustible gases, due to their different physical and chemical properties, and therefore have different explosive limits. A combustible gas explosion limit is not fixed, they are affected by temperature, pressure, oxygen content, the diameter of the container and other factors.

(1) When the temperature rises, the lower explosive limit decreases (the range of explosive limits is positively correlated with temperature), making the explosion risk increase.
(2) When the pressure increases, the lower explosion limit also decreases, making the explosion risk increase.
(3) An increase in the oxygen content of the gas will also reduce the lower explosive limit, thereby increasing the risk of explosion.
(4) When inert gases are mixed in, the oxidation reaction activity is reduced, thus reducing the risk of explosion.

The oxygen content of the gas is reduced, the lower explosive limit will increase and the risk of explosion will be reduced. For flammable gases in general, combustion or explosion can be avoided if the volume fraction of oxygen in the mixture is reduced to 6–14%. The oxygen concentration in the gas is therefore an important indicator to monitor when monitoring the explosion hazard indicators in a gas mixture environment.

The third is the nature of the source of the fire on the explosion of the gas mixture also has a significant impact on the risk. If the intensity of the fire source is high, the hot surface area is large and the contact time with the mixture is long, the explosion boundary will be expanded and the risk of its explosion will increase.

(iii) **Analysis of the Impact of Relief Measures on the Risk of Gas Explosion in the Fire Zone**

Before the fire zone is closed, the flame spreads quickly. After the fire zone is closed, the fire zone space still has a certain amount of air, so that the fire zone can also maintain a period of oxygen-poor combustion. According to field tests, if no measures are taken, the temperature of the high temperature coal body is very slow to fall, as long as the coal temperature does not fall below the minimum ignition temperature to cause a gas explosion, the ignition source in the fire zone always exists. Fire zone closed, the gas concentration in the closed zone increases with time, but the rate of increase continues to slow down; due to the oxidation reaction between coal and oxygen, so that oxygen is consumed. The oxygen concentration in the confined

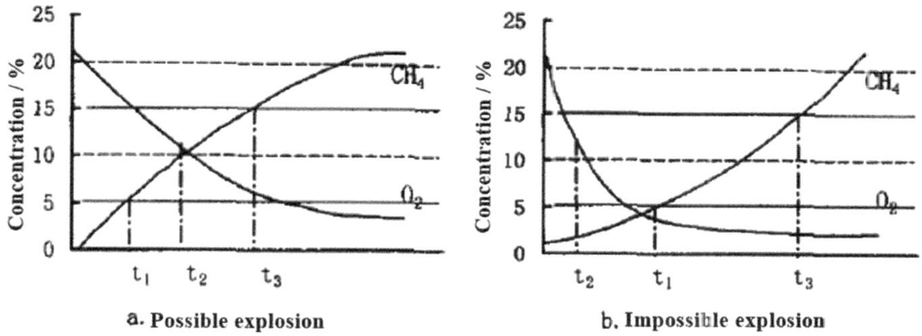

Fig. 4.15 Theoretical curves of gas and oxygen concentrations with fire zone closure time

zone gradually decreases with the extension of the confinement time, but the rate of decrease in concentration keeps decreasing.

As can be seen from the analysis of gas explosion conditions, the explosion risk of a closed fire zone depends primarily on whether the concentration of the gas mixture within the fire zone meets the explosion conditions, depending on the rate of change of the gas and oxygen concentrations in the specific case. As shown in Fig. 4.15a, when the concentration of the two gases change simultaneously to within their respective explosion limits, a gas explosion may occur. As shown in Fig. 4.15b, if the concentration of CH_4 rises to the explosion limit before the concentration of O_2 has fallen below the safe oxygen concentration, a gas explosion will not occur.

Assume that the upper and lower gas explosion limit concentrations correspond to a fire zone closure time of t_1 and t_3 respectively. Closure time is t, when $t_1 < t < t_3$, gas concentration in the explosive concentration range; fire zone gas failure oxygen concentration corresponding to the fire zone closure time t_2, when $t < t_2$, oxygen concentration to meet the explosion conditions; so possible gas explosion may occur at the time of $(t_1 < t < t_3) \cap (t < t_2)$, as in Fig. 4.16b. If $t_2 < t_3$, then the possible time of the explosion is $t_1 < t < t_2$, if $t_2 > t_3$, the possible time of the explosion is $t_1 < t < t_3$. If the intersection of the two is empty, then no a gas explosion occurs, as in Fig. 4.16 a and c. Otherwise, a gas explosion may occur.

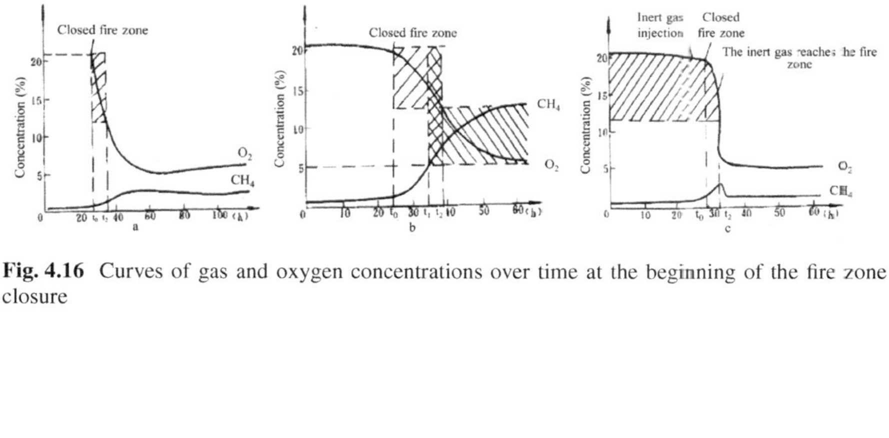

Fig. 4.16 Curves of gas and oxygen concentrations over time at the beginning of the fire zone closure

(iv) **Fire zone gas explosion hazard determination**

Whether it is a high gas mine or a low gas mine, the possibility of a gas explosion during fire fighting in the underground fire zone is the primary consideration when formulating fire fighting measures, so the determination of the risk of a gas explosion in the fire zone is of paramount importance. Many coal mining countries around the world have done more in-depth research on gas explosion determination methods. Common methods used to determine the explosivity of gas mixtures include gas mixture explosivity diagrams, the minimum combustion oxygen concentration method, and the comprehensive calculation method based on a single gas mixture in a gas mixture, etc. Each of these methods has its own advantages and disadvantages, but they can be used in combination to complement each other and improve the accuracy of the judgement.

When analyzing the explosive properties of gas mixtures in the fire zone, there are two broad types: explosive analysis of single combustible gases and explosive analysis of gas mixtures containing multiple combustible gases.

For the explosive analysis of a single combustible gas, experimental determinations and associated formulas and charts are currently used. Table 4.1 is the explosion limit for common combustible gases in coal mines based on experiments at a pressure of 101.3 kPa and a temperature of 20 °C.

Fire zone temperature is generally very high. Therefore, the actual calculation of the fire zone flammable gas mixture explosion limit, should be based on the measured fire zone temperature and the composition of the various combustible gases in the fire zone, the data listed in the table for temperature correction according to the following formula.

$$\begin{cases} N_{LT} = N_L \times \left[1 - \frac{721(t-25)}{10^6}\right] \\ N_{UT} = N_U \times \left[1 + \frac{721(t-25)}{10^6}\right] \end{cases} \tag{4.1}$$

Table 4.1 Upper and lower explosion limits for common combustible gases in coal mines (25°C)

Gas name	Chemical symbols	Blast boundary			
		In the air		In oxygen	
		Lower limit	Upper limit	Lower limit	Upper limit
Methane	CH_4	5.00	15.00	5.1	61.0
Ethane	C_2H_6	3.22	12.45		
Propane	C_3H_8	2.4	9.50	2.3	52.0
Hydrogen	H_2	4.00	74.20	4.0	94.0
Carbon monoxide	CO	12.5	75.00	15.5	94.0
Hydrogen sulphide	H_2S	4.32	45.50		
Ethylene	C_2H_4	2.75	28.60	3.0	80.0
Pentane	C_5H_{12}	1.40	7.80		

N_{LT}, N_{UT}—the lower and upper explosion limits for each combustible gas at t °C, respectively.

N_L, N_U—the lower and upper explosion limits for each flammable gas at 25 °C respectively.

By comparing the measured concentrations of flammable gases in the fire zone with the corrected upper and lower explosion limits, it is possible to determine whether the fire zone is explosive or not. However, this method is not simple and the simplest method is to use a mixed gas explosivity diagram. In most cases, combustible gases in coal mines are dominated by gas, as shown in Fig. 4.17, which can be used to determine the explosive nature of the gas mixture and the conditions under which it changes. When there is no rich inert gas (nitrogen as an example), the upper and lower limits of gas explosion are 5–15%. As the concentration of nitrogen in the atmosphere increases (and the concentration of oxygen decreases in relative terms), the upper explosion limit decreases and the range of explosion limits decreases, with the two sides of the upper and lower limits meeting at a point when the nitrogen reaches a certain concentration. Depending on the proportions of the components of the gas mixture, the gas mixture may be located in zones A, B, C or D. When the gas mixture is located in zone A, it means that the gas mixture is already explosive, but adding nitrogen can make the components of the gas mixture inert into other zones and lose their explosive properties. If located in zone B, the concentration of CH_4 is greater than the upper explosive limit and temporarily lost explosive. But after infiltration of the appropriate amount of air to reduce the concentration of CH_4 to the explosive limit, it will be explosive again. If located in Zone D, CH_4 concentration is low, less than the upper explosive limit and temporarily lost explosive, but infiltration of appropriate amounts of gas after the CH_4 concentration increases to the explosive limit will be explosive again. C zone is a mixture of gas failure zone. If the gas mixture component point in the upper part of the C zone, the gas mixture may be continuously added to the air first into the B zone and then into the A zone and explosive; if the component point of the gas mixture is to the lower left of zone C, the gas mixture may be explosive due to the continuous addition of gas first into zone D and then into zone A. The trajectory of the component points of the gas mixture follows the coordinates of the addition of air, inert gas and gas in a straight line towards the origin, 100% rich nitrogen and 100% CH_4 respectively.

Another commonly used application diagram to represent the explosiveness of a gas mixture is the Coward explosion triangle. As early as the 1950s and 1960s, H. F. Coward and G. W. Jenes and others carried out extensive tests on the explosive properties of flammable gases using an apparatus of the US Bureau of Mines. They determined the explosion limit values for methane at different oxygen concentrations (both with the addition of different inert gas concentrations) and plotted the data on a graph with the combustible gas concentration as the horizontal coordinate and the oxygen concentration as the vertical coordinate and found that the explosion area essentially formed a triangle called the explosion triangle, as shown in Fig. 4.18. Since then, the United Kingdom, Japan, Germany, Poland, the former Soviet Union, China and other countries have begun to determine the explosive nature of gas technology

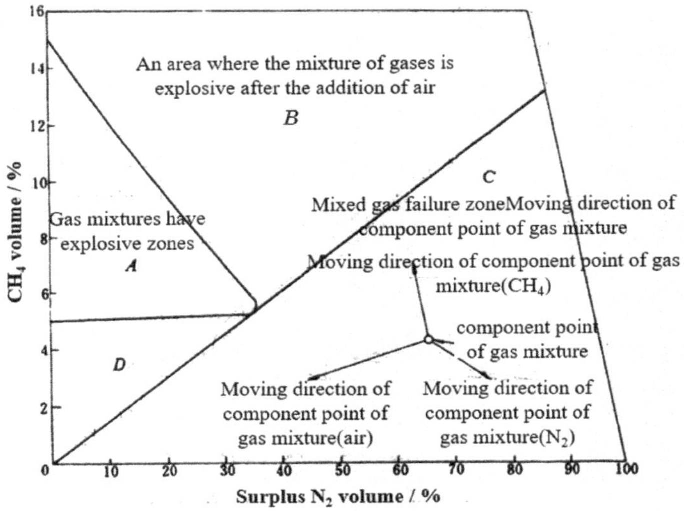

Fig. 4.17 Gas explosion concentration threshold map

research. At present, the world commonly used explosion triangle to determine the explosive properties of gas mixtures has no less than twenty methods.

Figure 4.18 in the marked methane an air mixture of gas in each combination of zones. Zone 1 is the methane-air gas mixture explosion hazard zone, Zone 2 is the non-combination zone, Zone 3 is the methane concentration of insufficient gas mixture failure zone, Zone 4 is the methane concentration of excessive gas mixture failure zone, Zone 5 is the oxygen-depleted asphyxiating gas mixture failure zone, DF—lower explosion limit limit, EF—upper explosion boundaries. That is, the use

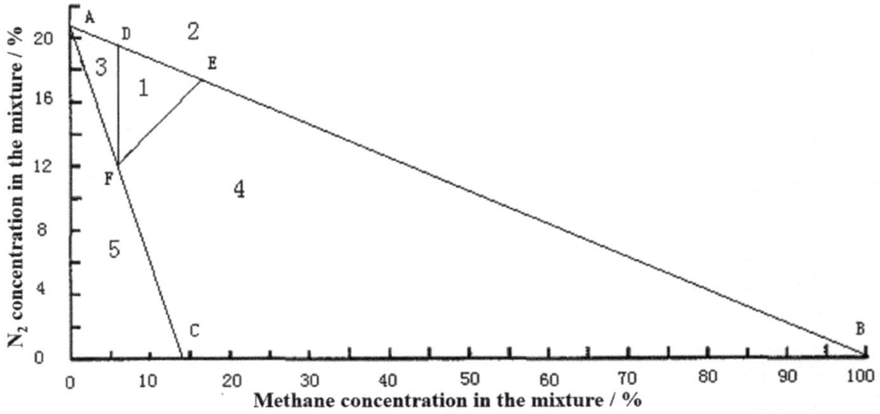

Fig. 4.18 Methane explosion triangle

of this chart to determine the explosiveness of the gas mixture, when the component points of the gas mixture is located in Zone 1, that is, △ DEF range, the gas mixture is explosive; when the component points are located in Zone 3, temporarily not explosive, but if the methane concentration increases, the component points will be transferred to Zone 1 again explosive; when the component points are located in Zone 4, temporarily not explosive. However, if the oxygen concentration increases, the component point will shift towards Zone 1 and become explosive again; if the component point is in Zone 5, it is not explosive. However, if the conditions are changed and air or gas is added continuously, the component point of the gas mixture may first enter Zone 4 or 3 and then Zone 1 as well as become explosive again.

For the explosive determination of gas mixtures containing multiple combustible gases, there are generally two methods. One is the dynamic triangle method, which is the direct use of LeChatelier's basic guidelines to calculate the explosion limit and critical point of the gas mixture, drawing the explosion triangle under the conditions of coexistence of multiple combustible gases, according to the location of the combustible gas component points in the explosion triangle to determine its explosivity. This method is convenient and flexible, but only considers CH_4, H_2 and CO, and ignores the influence of other hydrocarbons, so it is not convenient to predict the trend and state of gas changes. The other is the static triangle method. The explosion triangle for methane is first determined, then the corrections to the methane in the gas mixture by various other combustible gases are converted to the corresponding methane values and the component point coordinates are calculated. Finally, the explosivity is determined by comparison with the standard triangle for methane. This method has a clear physical meaning and describes the process of component change and combustible gas explosivity change clearly, but large errors can occur in the calculation process.

Although there are a variety of methods that can be used to determine the explosive properties of combustible gas mixtures, but the combustible gas explosion triangle is simpler to apply, easy to understand and judge, so the most widely used. At the same time, the use of inert gas fire, closed or open fire zone and fire zone rescue and disaster relief has important guidance. China since the eighties on the basis of foreign theory and technology, began the research of combustible gas explosion triangle and mine gas explosive monitoring technology, research direction is mainly based on the Polish fixed triangle correction value method to supplement and improve. The Fushun Branch of the General Coal Institute (formerly the Fushun Coal Research Institute), Xiangtan Mining Institute and Beijing Lingtian Century Automation Technology Co Ltd have all conducted in-depth research and developed equipment for this purpose. The Fushun Branch of the General Research Institute of Coal Science has developed a series of explosive gas measuring devices for coal mines, which are now widely used in mine disaster relief in China. Beijing Lingtian Century Automation Technology Co., Ltd. developed the BMK-II portable coal mine gas explosive tester, which is used in underground coal mines to use the mine explosion-proof and intrinsically safe portable instrument. It can be used in underground coal mines, carried by the mine rescue team into the disaster area to monitor the operating environment around the gas components whether there is a risk of explosion. In the process of dealing with a natural fire or fire zone, the instrument detects methane, oxygen and ambient temperature in the surrounding environment or in the air inside the confinement, displays the concentration of CH_4, O_2 and temperature values directly on the LCD screen of the instrument. It shows the graph of the "explosion triangle" and the coordinates of each gas component. When the alarm limit is approached or reached, the instrument simultaneously issues sound and light alarm.

Example: Let the composition and concentration of the flammable gas mixture in the closed fire zone of a mine are $CH_4 = 4.0\%$, $CO = 2.1\%$, $CC_2H_4 = 0.03\%$, $CH_2 = 0.04\%$, and the air temperature in the fire zone is 55°C. The explosive hazard parameters are analyzed as follows. The concentration of the flammable gas mixture and the percentage concentration of each component in the total concentration by volume are $CO_2 = 6.17\%$, $CH_4 = 64.83\%$, $CO = 34.03\%$, $C_2H_4 = 0.49\%$, $CH_2 = 0.65\%$. Explosion limits: the explosion limits of each flammable gas at 25 °C are found from Table 4.1 and corrected according to Eq. (4.1).

The upper and lower explosive limits of the flammable gas mixture in the fire zone are calculated to be 21.27 and 6.1% respectively in air, as well as 71.05 and 6.42% respectively in oxygen. This shows that the concentration of the flammable gas mixture in the fire zone is within the explosive limits in air and is explosive. In this case, if no correction is made for the explosive limit of the flammable gas, the lower explosive limit of the gas mixture is 6.24%, i.e. the concentration of the flammable gas mixture (6.17%) is less than the lower explosive limit concentration. As can be seen, it is easy to false alarm and the flammable gas mixture in this example fire zone is not explosive.

4.3.3 Determination of the Burning State of the Fire Source in the Closed Fire Zone

(i) **Reliability analysis of gas samples**

The main indicators of coal fires are CO, H_2 and hydrocarbons such as ethylene C_2H_4, propylene C_3H_6 and acetylene C_2H_2. Researchers have found that there is a clear correspondence between the changes in concentration of these gases and the coal temperature, among which the three gases O_2, CO_2 and CO have a strong regularity of change. The analysis technology is mature and can be realized in real time. The continuous monitoring and analysis of these three gases can be used to determine the development trend of the fire zone and the explosion risk, while the analysis of other gases can assist in determining the intensity of the fire [3].

The researchers also found that the indicator gases are produced and released in the sequence ethylene → propylene → acetylene, with a concomitant increase in temperature. When the temperature was abnormal, CO appeared first, followed by H_2 as the temperature increased; then C_2H_4, followed by C_3H_6, as well as finally C_2H_2 and other gases. The other gases are difficult to analyse in situ. Therefore, they have not been used to confirm the reliability of the monitoring data. Regardless of the temperature of the coal body and the oxygen concentration, the concentration of CO is much higher than the concentration of the other marker gases. A fire alarm should not be issued just because the presence of C_2H_2 or C_2H_4 is detected. Attention should also be paid to the relationship between the marker gases to avoid false alarms.

(1) Trend analysis of concentration changes

The most dangerous thing when fighting a mine fire is the inability to accurately analyse changes in the state of the fire source and abnormal changes, thus missing the best time to implement a rescue plan. At the same time, disaster areas pose a danger to rescuers. Gas trend analysis is an effective means of avoiding these problems, as it effectively eliminates human or equipment errors and removes any external environmental interference with the gas composition of the fire source, thereby accurately determining changes in the combustion state of the fire source.

Firstly, gas samples whose values do not correspond to the established concentration trend should be discarded. Generally speaking, as long as there are no drastic changes in the environment, such as explosions, serious collapses of the tunnel, destruction of the seal wall causing the inflow and outflow of water or air, or rapid changes in atmospheric pressure causing a large inflow of fresh air or CO_2 or CH_4. Then, the change in the gas composition of the fire zone is gentle and smooth. Since the gas concentrations and the ratios between them change exponentially, the detection data or ratios are best plotted on a logarithmic or semi-logarithmic scale. In cases where conditions do not vary much, the resulting curve allows a reasonable estimate of what has happened and a prediction of what is likely to occur. The analysis of fire conditions and trends would be much more difficult if the graphs were plotted in the usual linear coordinates.

(2) Trickett Ratio (Tr)

The Trickett Ratio Tr is a powerful tool for eliminating invalid gas samples and avoiding mis-judgements. It is based primarily on the fact that there is a certain interdependent ratio between the concentrations of the gases generated by the fire. When the ratio is abnormal, it means that the gas sample is somehow disturbed and invalid. When the Tr of a gas sample exceeds 1.6, that gas sample is not considered. If the main fuel of the fire is coal, a gas sample with a Tr greater than 1 is suspect. For example, in a mine fire example, a gas sample taken from the main ventilator exhaust contained 0.86% CO_2, 1.26% CO, 1.53% H_2, 0.94% CH_4, 19.8% O_2 and 75.6% N2 + Ar, which led to the decision to evacuate the direct fire extinguishing crew. The reliability of the gas sample is tested if the following defining equation for Tr is used.

$$Tr = \frac{CO_2\% + 0.75CO\% - 0.25H_2\%}{0.265(N_2\% + Ar\%) - O_2\%} \tag{4.2}$$

For this example: Tr = (0.86 + 0.75*1.26 − 0.25*1.53)/(0.265*75.6 − 19.81) = 6.4, as the Tr value was much greater than 1.6, the gas sample was unreliable and should not be used as a basis for decision making. It was later found that the reason for the unreliability of the gas sample was due to the fact that the gas sample bottle was not washed. By calibrating the reliability of the gas sample, the decision maker can be helped to reduce the possibility of incorrect judgement.

(ii) **Single indicator gas concentration change rate judgement indicator**

The various methods of determining the state of combustion are based on the changes in the various components of the gas mixture and their concentrations within the closed fire zone. The single indicator gas concentration change rate judgement indicator is mainly based on the trend graph of the various gas component concentrations in the closed fire zone over time rather than a single specific detection value, and calculates the rate of change of the gas sample concentration, especially after a continuous large change, to analyse the combustion status. It mainly includes changes in the concentration of gases such as O2, CO_2, CH_4 and CO, which is usually plotted with the gas concentration as the vertical axis and the time or date as the horizontal axis.

(1) Determination of the rate of change and preliminary cause analysis

Analysis of the interactions between the gases shows how they are produced, consumed or diluted. After making the concentration change curve as shown, the next step is to determine the trend and rate of change by calculating the rate of change in the percentage concentration of the gas from the following equation.

$$R = \frac{\log \gamma\prime - \log \gamma\prime\prime}{X\prime - X\prime\prime} \tag{4.3}$$

In the Equation,

X′, X″—the initial and final values of time during the analysis period.

γ', γ''—the percentage of the concentration of this gas corresponding to the time X′, X″.

R is negative when the change in gas concentration tends to decrease and positive when the change in gas concentration tends to increase. For the analysis of closed fire zone fire states, the rate of change of gas concentration within the time frame of the measured gas sample should be examined first.

A preliminary determination of the possible causes of the increase or decrease of gas in the fire zone is based on Table 4.2.

(2) Fire zone status analysis

The following are guidelines for analyzing the condition of the fire zone.

① Judgement criterion 1 (effect of rate of change of N_2 concentration): for most coal mines, if there are no significant roadway cross-collapses in the closed fire zone, the influx of CH_4 from the coal seam into the zone will dilute the other gases in the zone. When inert gas is injected, R_{CH4} must be replaced by $R_{(\text{Inert gas} + \text{CH4})}$, if the judgement criterion is applied.

Table 4.2 Possible causes of gas increase or decrease

CO_2 increase	Combustion (fire build-up), reaction of acidic water with acidic salts, slow oxidation (absorption reaction), natural emergence of coal rock	CO_2 reduction	Fire weakened, dissolved in acidic water (high wind speed, high alkalinity), adsorbed for carbon black, coke
CO increase	Combustion (increased fire), slow oxidation	CO reduction	Fire weakened, wood decomposed by fungi, absorbed by wet coal, adsorbed by carbon black, coke
H_2 increase	Combustion, water–gas reaction, decomposition of wet wood by fungal anaerobic-like bacteria	C_2H_4 reduction C_3H_6 reduction	Combustion, gas sample handling error
O_2 increase	Oxygen adsorption by the original coal seam (so that some coal seams contain more oxygen themselves)	O_2 increase	Combustion, slow oxidation, absorption or adsorption of O by coal or wood$_2$

② Judgement criterion 2 (comparison of the rate of change of O_2 and N_2 concentrations): where no combustion occurs and no or only a small amount of O_2 is adsorbed, the rate of change of O_2, R_{O2}, is approximately equal to the rate of change of N_2.

③ Judgement criterion 3 (comparison of the rate of change of gas concentrations associated with the fire zone): if the fire is extinguished, or if the main absorption reactions cease, then $R_{CO2} = R_{CO} = R_{N2}$ and the R-value of all gases will decrease while R_{CH4} will increase.

④ Judgement criterion 4: when the rate of decrease in O_2 concentration approximates the rate of increase in CO_2 and CO concentration, there are signs of fire development.

⑤ Judgement criterion 5: the fire is in a steady state when $\varphi(O_2)$, $\varphi(CO_2)$ and $\varphi(CO)$ concentrations are decreasing at a steady rate or their rate is approximately zero.

After the fire zone has been closed, a comprehensive analysis of the rate of change of various gases is carried out according to the above criteria to determine the status of the fire source and the development trend in the closed fire zone.

(iii) Carbon monoxide index judgment indicator

CO, the main natural coal fire indicator gas, is detectable at 60 °C from combustion generation and has a long trailing time, with its concentration increasing at a much faster rate than other fire gases. It differs from CO_2 in that CO only comes from the combustion or oxidation reaction of the fuel. Unlike CH4, it is also not consumed in a fire as its lower flammable limit is higher than 12%. So its concentration is more stable under fire conditions and less affected by other factors.

However, CO is also readily adsorbed by coke and carbon black, creating the illusion that the fire is extinguished. It can also be decomposed by some fungi or absorbed by wet coals. This decomposition and adsorption is only significant for CO generated by fires, which are more likely to occur in wet areas. For unknown reasons, CO generated by normal oxidation is not readily adsorbed or absorbed, so attention should be paid to the source of CO in the gas sample as well as the possibility of decomposition and absorption in the flow path, which is an important issue in determining whether the fire zone is open.

In addition, there are no clear regulations in China regarding the CO concentration value for coal spontaneous combustion warning, each mine needs to determine according to its own actual situation (Note: 24×10^{-6} is the health indicator, most mines use this indicator; Germany considers carbon monoxide concentrations in the return air stream above 50×10^{-6} as having a natural fire hazard; in the 1950s, the Soviet Union stipulated that the CO content in the gas samples collected every two days and nights was 0.001–0.002% or more, which was considered to be a sign of spontaneous combustion. (The disadvantage of this indicator when used alone is that the CO concentration varies considerably due to changes in air volume. So when using the CO concentration to determine the combustion status of a fire zone, the air volume at the sampling location must remain basically unchanged, otherwise the indicator will be inaccurate).

In order to overcome the shortcomings of CO concentration being influenced by wind volume, some coal companies in China use the absolute CO generation as an indicator to obtain a determination of combustion status. While for mines with CO gas in the normal airflow, a critical value for determination should be determined when using CO as an indicator to determine the combustion status of the fire zone. The following factors should be considered when determining the threshold: the background CO content in the normal airflow at each sampling location; the threshold should be determined to ensure a certain safety margin, i.e. to allow sufficient time for the implementation of fire fighting operations.

When using CO as an indicator for determining the combustion status of a fire zone, two points should be noted.

(1) The concentration or absolute value of CO should be greater than the critical value.
(2) There should be a steady increase in the concentration or absolute value of CO

The Carbon Monoxide Index (ICO), also known as the Graham Index, is the ratio of carbon monoxide to oxygen consumption ($CO/\Delta O_2$) produced by the oxidation of coal during natural ignition, which is proportional to the oxidation source temperature and oxidation time, reflecting the oxidation reaction of the fuel. The index can be used to predict the natural fire trend of the coal and also to analyse the change in fire conditions in the closed fire zone. An increase in the ICO value indicates that the spontaneous combustion of the coal continues to develop or that the air temperature in the closed fire zone increases. The value is calculated as follows.

$$ICO = \frac{\text{CO}}{0.265^{\star}(N_2 + \text{Ar}) - \text{O}_2} = \frac{\text{CO}}{\Delta\text{O}_2} \tag{4.4}$$

In the Equation: CO—concentration of CO generated after the wind flow passes through the ignition zone, %.

ΔO_2—concentration of O_2 consumed by the wind flow after it has passed through the fire zone, %.

O_2—O_2 concentration after wind flow through the fire zone, %.

A mine has determined from field and experimental data that the value of the carbon monoxide index (ICO) is in the range of 0.007–0.03, which is meaningful for evaluating the process of spontaneous combustion and re-ignition of coal with the following evaluation indicators.

ICO = 0.007, increased intensity of oxidation of the coal; ICO = 0.01, increased fire hazard; ICO = 0.02, significant heating of the coal; ICO = 0.03, open fire.

As the coal oxidizes and warms up, the ICO value increases more rapidly with increasing oxidation zone temperature. After the fuel has ignited and burned, the rate of increase in ICO slows down and in many cases will gradually stabilize. This is because during the open fire combustion period more CO_2 is produced compared to CO than during the negative combustion phase, so the rate of increase of CO slows down. When the fire zone is closed and the oxygen supply is reduced, the fire source is in a state of incomplete combustion and the proportion of CO generated relative to CO_2 increases, as its oxygen supply decreases, oxygen consumption also decreases. Although the overall trend of CO generation decreases, the rate of CO reduction is less than the rate of ΔO_2 reduction, resulting in the ICO value starting to rise again until the air temperature is similar to the ambient temperature. Thereafter, the ICO gradually decreases until it is close to zero.

When applying the CO index, the following points should be noted.

(1) When the intensity of combustion decreases, the oxygen concentration on the return air side increases and the CO concentration decreases, there may be a problem in using the ICO value to judge the fire situation. Due to incomplete combustion, the amount of CO_2 produced decreases and the amount of CO produced increases relative to CO_2, resulting in the rate of decrease in CO concentration may slow down and be less than the rate of decrease in ΔO_2, which eventually leads to an increase in ICO, whereby an incorrect judgement that the fire is increasing may be made.
(2) When the air component at the sampling point is similar to the ambient air component, other indicators should be used as a supplement to the ICO. In the closed fire zone, after the transition from the slightly higher temperature inactive combustion condition to the ambient oxidation stage, mineral desorption and ambient oxidation will also produce a very small amount of CO. The change in CO value at this time is very small and difficult to detect and should also be noted when applying the ICO value.
(3) The ICO value will be increased when there is infiltration of wind flow from other tunnels on the upwind side of the sampling point, when there is infiltration

of CO from non-fire sources, or when there is a certain concentration of CO in the incoming air flow on the upwind side of the fire source (e.g. reversal of wind flow); the infiltration of wind flow from other tunnels on the upwind side of the sampling point will cause the ΔO2 calculation to be incorrect and result in a misjudgement of the fire situation.

(4) Since the ICO value is closely related to ΔO_2, when the air entering the ignition zone contains excess or insufficient N_2, or more gases such as CO_2, CH_4 and water vapour, the application of the formula $\varphi(N_2)/\varphi(O_2)$ expressed in 0.265 will bring about a large error.

(5) At ΔO_2 (%) < 0.3, the ICO value will far exceed the theoretical possible value.

The correct application of the oxygen consumption indicator ΔO_2 gives the correct analysis. An increase in oxygen consumption means that the fire zone consumes more fuel. However, the oxygen consumption indicator also reflects an increase in asphyxiating gases such as N_2 and CO_2. When ΔO_2 reaches 22–26%, it means that the area is full of asphyxiating gases. This is due to the high consumption of O_2 to produce CO_2 during the oxidation process. Low values of ΔO_2 indicate that the gas is close to fresh air or that the oxidation reaction rate of combustion is low.

(iv) **Indicators for determining the ratio of carbon oxides**

The main combustion products of natural coal fires are CO, CO_2, alkane C_nH_{2n+2}, olefin C_nH_{2n} and hydrocarbon C_nH_{2n-2}, of which CO and CO_2 are the most abundant and can be detected at very low temperatures. As the increase in CO_2 concentration with temperature follows a similar trend to the increase in CO concentration with temperature in the absence of non-combustion generated carbon oxides, the ratio of carbon oxides (CO/CO_2) is the only indicator of all judgements that is not affected by wind flow, CH_4 influx and inert gas injection (N_2).

The conditions that cause misjudgements in this guideline are the generation of asphyxiating gases, the influences of other sources of CO_2 such as the influx of CO_2 from rock formations, the injection of inert gas (CO_2) and the reaction of acidic water with carbonates to produce CO_2. The infiltration of non-combustion generated CO_2 increases the amount of CO_2, decreases the CO/CO_2 ratio and tends to make the judgement more dangerous. In contrast, the reaction of alkaline water absorbing CO_2 decreases CO_2 and increases the CO/CO_2 ratio, increasing the margin of safety in the judgement. Adsorption has a similar effect on CO and CO_2. Therefore, it has a smaller effect on the CO/CO_2 ratio. As the intensity of the reaction changes, the relative amounts of CO and CO_2 also change, making the CO/CO_2 ratio a good indicator for monitoring natural fires.

The carbon oxide ratio (CO/CO_2) increases with temperature until the combustion flame appears. CO/CO_2 values vary with temperature for coals with different volatile contents, which are similar despite the fact that coal types vary greatly. After the appearance of a combustion flame, CO/CO_2 remains constant if $\varphi(O_2) \geq 5\%$, and decreases rapidly when $\varphi(O_2) < 5\%$; if the fire is high and the temperature is high, this indicates the risk of a fire transition to fuel-rich combustion.

When using the carbon oxide ratio (CO/CO_2) as an indicator of fire development in the fire zone, care should first be taken to distinguish between the sources of CO_2 and CO to avoid making incorrect judgements.

(v) **Temperature judgment indicators**

The air flowing through the fire heating is dissipated by heat transfer to the top and bottom slabs, both gangs, accumulated water and various materials during the continuous flow. Based on the fact that there is a certain regularity in the variation of each parameter with the environment, the heat transfer model within the closed fire zone of a mine fire is established by measuring the heat transfer coefficient of coal and rock, the temperature of the air flow and the volume of the closed space under reasonable assumptions. Through the study of the heat transfer law in the closed fire zone, the theoretical time for the opening and closing of the fire zone is deduced. In addition, the conditions for the opening of the fire zone are stipulated in the regulations. The air temperature in the fire zone must fall below 30 °C or be the same as the daily air temperature in the zone before the fire; the water discharge temperature of the fire zone must be below 25 °C or be the same as the daily water discharge temperature in the zone before the fire, and the temperature must remain stable for a certain period of time. The temperature indicator is therefore an important indicator to determine the burning status of a closed fire zone.

(vi) **Fire zone combustion state example analysis**

Example: A mine closed fire zone, after investigation, there is no large amount of wood in the fire zone, the water volume and PH value of the water body is not clear. The trend of various gases in the closed fire zone is shown in Figs. 4.19 and 4.20.

Judging indicators using the rate of change of a single indicator gas concentration:

(1) Calculate the rate of change of each gas

The rate of change for each gas was calculated over a period of 180 d (the rate of change for methane was the change in concentration between day 39 and 180 d).

$R_{N2} = -0.0004$; $R_{CO_2} = 0.0003$; $R_{O_2} = -0.008$; $R_{CO} = -0.001$; $R_{CH4} = 0.003$; $R_{H2} = -0.008$.

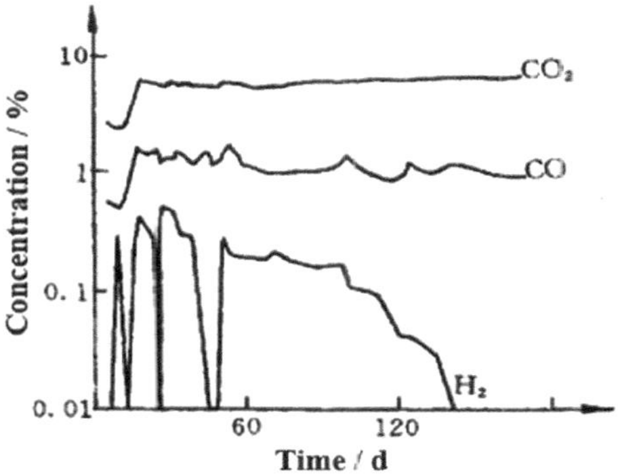

Fig. 4.19 Trend of CO_2, CO, H_2 in the closed fire zone

Fig. 4.20 Trend of N_2, CH_4, O_2 in the closed fire zone

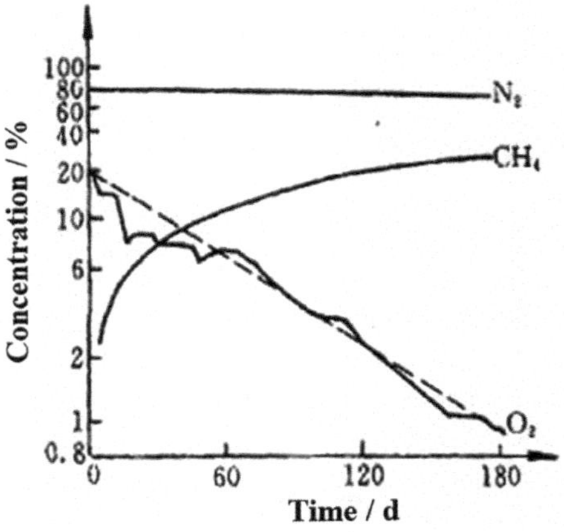

(2) Preliminary analysis of the reasons for the increase or decrease of gases such as CO_2, CO and H_2.

① Analysis of the causes of CO_2 changes, as shown in Table 4.3.
② Analysis of the reasons for the change in CO.

From Judgement Criterion 3, $R_{CO2} = R_{CO} = R_{O2} = R_{N2}$ holds for no combustion and slow oxidation. In this case: $R_{O2} = -0.008 > R_{CO} = -0.001$ (8 times) Slow oxidation has been excluded from the R_{CO2} analysis. Therefore, the change in CO concentration is due to combustion and the negative value may be due to CO being coke, which is absorbed by carbon black.

③ Analysis of the causes of the change in H_2

The fire zone has little wood and the H_2 production is due to combustion.

④ Other causes of gas change

See Judgement Guideline 1 and Judgement Guideline 2.

(3) The fire zone combustion status and development trend is derived from the judgment criteria

The fire zone combustion status and trends are shown in Table 4.4.

Table 4.3 Analysis of the causes of CO_2 changes

Indicator name	Indicator value	Phenomenon	Reason 1	Reason 2	Reason 3	Reason 4	Reason 5	Final conclusions
R_{co2}	0.0003	$R_{CO2} = 0.0003 > R_{CO}, R_{H2}$ (for negative values)	A. Reaction of acidic water with carbonate: the situation is unclear and cannot be judged	B. If the adsorption effect: CO_2 is more likely to be adsorbed than CO, R_{CO2} has a greater negative value than R_{CO}. However, in this case $R_{CO2} = 0.0003$ (positive) and $R_{CO} = -0.001$ (negative), indicating that adsorption is not the main effect	C. Slow oxidation: R_{CO2} should be around 0.001, in this case 0.0003. So: slow oxidation is not a major source of CO_2	D. Spontaneous combustion in the extraction zone: Initial $R_{CO2} = R_{CO}$, Development $R_{CO} > R_{CO2}$. In this case $R_{CO} < R_{CO2}$, so: Spontaneous combustion in the extraction zone is not the main source of CO_2	E. CO_2 enriched band surge, Rco_2 large, should have a brief, sharp increase. This example: Rco_2 is small, excluding this possibility	F. R_{CO2} increased, fire not extinguished due to combustion generation

Table 4.4 Fire zone combustion status and development trend according to judgement criteria

Fire zone fire source combustion status analysis			Fire zone fire source combustion state change process inference	
Judgemental guideline 1 (reasoning approach)	Judgemental guideline 2 (counterfactual approach)	Judgemental guideline 3 (counterfactual approach)	Judgement guideline 4	Judgement guideline 5
The CH_4 influx dilutes the gas and the N_2 concentration decreases at a certain rate. As shown in the graph, over a period of 180 d, methane concentrations increased continuously and nitrogen concentrations decreased continuously	If there is no combustion and no O_2 adsorption, the CH_4 dilution should make $R_{O2} = R_{N2}$. In this example: $R_{O2} = -0.008$ is 20 times more than $R_{N2} = -0.0004$, so there is combustion or O_2 absorption (adsorption). To further determine the cause, other gas change rates should also be analysed	If the fire is extinguished, or the slow oxidation reaction stops, the rate of change of the remaining gases will decrease due to the dilution of CH_4 so that $R_{CO2} = R_{CO} = R_{O2} = R_{N2}$ and R_{CH4} increases. This example: $R_{CH4} = 0.003$ but $R_{CO2} = 0.0003$, $R_{CO} = -0.001$ (decrease), $R_{O2} = -0.008$ (decrease) and $R_{N2} = -0.0004$ (decrease). So there is a combustion or slow oxidation reaction. The analysis of the cause of the change in CO_2 tells us that it was produced by combustion and the fire was not extinguished	When the negative value of R_{O2} is similar to the positive values of R_{CO2} and R_{CO}, the fire develops. For this example, month 1: $R_{O2} = -0.02$ and $R_{CO} = R_{CO2} = 0.03$. Therefore, the fire development deteriorated in month 1	R_{O2}, R_{CO} and R_{CO2} are steadily decreasing or their rate is approximately zero, indicating that the fire is contained and that this is the best time to establish a permanent firewall, plug air leaks and reinforce the firewall
Conclusions: A. After 180 days, the fire is not extinguished; B. $O_2 < 12\%$, not CH_4 burning: oxygen concentration is below 12% for most of this period, and since O_2 concentration has to exceed 12% for CH_4 to burn, it is not CH_4 burning; C. fire zone wood is too small to support 180 days of burning; D. coal is the main fuel; E. After 90 days, $O_2\%$ is reduced to 5%, at which point is a negative combustion, the fire is contained and will not spread; F. The fire is contained, the temperature of the fire zone decreases and the R_{H2} rate will be lower, in line with the actual diagram			Month 2 of this example: R_{O2}, R_{CO2} and R_{CO} fluctuate at position 0, while the concentration of CH_4 is steadily increasing, indicating a stable fire; however, there is a large air leak (due to R_{CH4}, the other gases R should drop and the air leak causes the concentration to fluctuate). The firewall should be reinforced and the air leak blocked	

4.4 Closed Fire Zone Management as Well as Opening and Closing Techniques

4.4.1 Fire Zone Management

After the fire zone is closed, the daily observation, testing and data analysis of the fire zone are collectively referred to as the management of the fire zone in conjunction with the fire fighting work. The specific contents are as follows.

(i) **Establishing a fire zone profile**

The ventilation department of the mine has a unified numbering system for fire zones and establishes a fire zone file to be kept. The contents of the fire zone file are as follows.

(1) Create a fire zone card, detailing the date of fire, cause of fire, location and extent of the fire zone.
(2) List of persons in the lead agency when dealing with a fire.
(3) The process of extinguishing the fire and the measures taken.
(4) The thickness of the coal seam, coal quality, lithology of the top and bottom slab, gas outflow, and the amount of coal confined in the fire zone at the site of the fire.
(5) Production information, such as the extent of the mining area, recovery rate, coal mining method and recovery time. Create a fire zone management card and map the location of the fire zone.
(6) Gas analysis and temperature changes before and after ignition.
(7) Ventilation before and after the fire (air volume, wind speed and direction).
(8) Draw a diagram of the fire zone of the mine. All previous fire zones and fire locations must be indicated on the diagram and numbered in chronological order. This is followed by the layout of the grouting boreholes and the direction of the wind flow around the fire zone, ventilation facilities, etc. Finally, the necessary sections are drawn.
(9) Location and number of permanent confinement, time of construction, material and thickness, etc.

The fire zone management card is to be completed by the mine ventilation department and bound into a book for permanent retention. The location of the fire source and the date of the fire should also be indicated on the mine ventilation system map for each fire. The date of the fire zone cancellation should be noted after the fire zone is canceled.

(ii) **Seal wall management**

(1) A fence shall be erected near each seal wall to prohibit the entry of persons and a sign shall be displayed indicating the date of construction of the seal wall,

material, thickness, gas composition inside and outside the seal wall, temperature, differential air pressure, date of measurement and the name of the person making the measurement.

(2) The air temperature outside the seal wall, the gas concentration, the differential pressure between the air inside and outside the seal wall and the seal wall itself must be checked once a day and the results of all checks must be recorded in the confinement log book. In the event of a sharp change being detected, at least one inspection shall be carried out per shift.

(3) The tightness of the seal wall determines to a large extent the effectiveness of the closed fire zone. So in addition to the above inspection, observation and alert system, the seal wall management should also strengthen the tightness inspection. Seal wall should be painted white with lime water to facilitate the detection of any air leakage. The sizzling sound from the seal wall can also be used as a sign of a leak in the seal wall and the seepage of fire gas. Any air leaks should be immediately sealed with clay, mortar etc.

(4) In addition, whether it is the seal wall on the air in take side or the return air side seal wall, good ventilation should be maintained on the outside and only personnel carrying good safety apparatus should be allowed to enter the area for observation and inspection.

4.4.2 Fire Zone Check

To keep track of changes in the fire zone, the gas composition and temperature in the fire zone should be checked regularly. Fire zone gas should be sampled at the return air side seal wall of the fire zone and gas samples collected through the observation tube on the seal wall. If the seal wall is far from the source of the fire, observation holes can be drilled close to the source of the fire. Sampling should be carried out regularly, once a day when the fire zone is not yet stable, and once a week or three days later. When sampling, the container should be cleaned of gas at the sampling site, and the location of each sample should be kept consistent, and the gas sample should be analyzed in time after it is taken out of the well to avoid human error. When using a mining temperature measuring instrument, the ground or underground observation holes can be used to measure temperature from a distance. Information on the composition and temperature of the gas in the fire zone should be collated in a timely manner, and gas composition and temperature change curves should be drawn to analyse the trend of the fire. If there is any deterioration, the causes should be identified and effective measures taken.

4.4.3 *Fire Zone Unsealing*

(i) **Fire zone opening conditions**

When a fire in a confined zone is gradually extinguished, the gas composition of the fire zone changes significantly, as do the temperature, pressure and natural wind pressure in the confined zone. These changes are used to determine whether the fire in the confined zone has been extinguished. The Coal Mine Safety Regulation states that a fire is considered to be extinguished when the following conditions are present in the fire zone

(1) The air temperature in the fire zone drops below 30 °C, or it is the same as the daily air temperature in the zone before the fire.
(2) The concentration of oxygen in the air in the fire zone is reduced to less than 5%.
(3) The air in the fire zone does not contain acetylene or ethylene, and the carbon monoxide concentration gradually decreases during the closure period and stabilizes at less than 0.001%.
(4) The discharge temperature of the fire zone is less than 25 °C or the same as the daily discharge temperature of the zone prior to the fire.
(5) The above four indicators have been stable for at least one month.

Improper handling of fire zones can lead to re-ignition and even gas explosions. A sealed fire zone can only be opened after long-term sampling and analysis to confirm that the fire has been extinguished. Before opening, an analysis of the risk of opening must be carried out, and safety measures and implementation plans must be drawn up and submitted to the supervisor for approval.

The main contents of the safety measures of the fire zone include the location and scope of the fire zone, the date and cause of the fire, the identification data of the fire zone, the fire extinguishing measures and the closure situation, the location map of the fire zone, the analysis and prediction of the fire consequences in the fire zone, and the fire zone unsealing scheme (including the unsealing method, the fire zone reconnaissance scheme, the order and process of opening and sealing, the ventilation method and the exhaust line, the ventilation equipment and its installation position, the ventilation method in the unsealing process, the action plan and route of the ambulance team, etc.). Analyze the hazards and possible hazards in the process of unsealing the fire zone, and the measures to prevent the hazards and hazards in the process of unsealing the fire zone (including the safety measures for the unsealing personnel, the safety measures for the personnel in the area affected by the fire zone gas emission, the safety measures for the ambulance team, the measures to prevent the gas and coal dust explosion, the treatment measures for the re-burning of the fire zone, and the safety measures after the fire zone is unsealed).

The main contents of the plan for the implementation of safety measures for the opening of the fire zone are: the time of the opening, the preparations for the opening (equipment, instruments, materials, fire-fighting equipment, water pipes, communication equipment, tools, etc.), the organization of the personnel for the opening (number and list of ambulance crew, number and list of gas detectors, number and list of safety inspectors, other personnel, division of labour and organization of personnel), and fire zone opening schedule.

(ii) **Fire zone status analysis**

The determination of the combustion status of the fire source in the closed fire zone can be analyzed in accordance with Sect. 4.4, Part 3 of this chapter. However, when a fire zone is closed, the seal wall cannot be very close to the fire source, which means that the gas sample obtained from the fire zone observation is not the original gas sample at the point of combustion. There is bound to be some error in determining the extinguishing status of the fire zone from this gas sample. For example, some fire zones have met the extinguishing conditions set out in the Coal Mine Safety Regulation, but it has re-ignited after being opened. As can be seen from the above, there are many factors that affect the status of a fire zone and it is not sufficient to judge the extinguishment of a fire zone based on the gas sample collected from the return air side of the fire zone, but a comprehensive analysis of all aspects of the fire zone is required to make a correct judgement.

(iii) **Preparation for the opening of the fire zone**

(1) Before opening the seal, be prepared to bring the return air from the fire zone directly into the return roadway; no personnel are allowed to work in the roadway through which the return air from the fire zone passes, and the power supply must be cut off.
(2) In mines where there is a risk of gas and coal dust explosions, rock dust should be spread in the tunnel connected to the fire zone or an explosion-proof water shed or rock dust shed should be installed.
(3) Prepare all materials, equipment and fire-fighting equipment necessary to open and re-seal the fire zone.

(iv) **Method of opening and closing the fire zone**

(1) Ventilation and sealing fire zone method

The ventilated open fire zone method is a method of opening the fire zone in one go. This method can be used if the fire zone is not very large and if it is confirmed that the fire zone is completely extinguished. Firstly, a member of the rescue team wearing a breathing apparatus enters the fire zone to reconnoitre and check for gas, and then opens the air intake side of the seal wall after the fire has indeed been extinguished. In order to gradually equalize the gas pressure in the fire zone, a small hole should be opened in the first seal wall and then gradually enlarged, it is strictly forbidden to open all the seal walls at once.

The air intake side of the seal wall is generally located in the lower part of the fire zone (fire zone in the upper mountain mining area), so special attention should be paid to the accumulation of CO_2 to prevent hazards caused by upwind flow when opening the seal.

After opening the inlet and outlet side seal wall, strong ventilation should be used. To prevent an explosion, all personnel must be evacuated to a safe place during this period and wait for 1–2 h before entering the fire zone for cleaning, spraying water to cool down and digging out the hot coal. The following precautions should be taken when using ventilation to open and close the fire zone.

① When opening the seal, it should be estimated that harmful gases such as fire zone gas and carbon dioxide are coming out.
② After opening the inlet and return air side confinement, strong ventilation should be taken within a short period of time to quickly reduce the gas concentration in the fire zone and thus prevent gas explosion, while personnel should be evacuated to a safe place and wait for at least 1 h before entering the fire zone for work.
③ Gas discharge from the fire zone shall be controlled to within the concentration permitted by the Coal Mine Safety Regulation.

(2) Air lock opening and closing fire zone method

The air lock method of opening and sealing the fire zone is a method of opening the fire zone in sections one at a time. In the case where the fire zone is large, it is difficult to confirm whether the fire source has been completely extinguished or a large amount of combustible gas may have accumulated in the fire zone, when using air lock to open and seal the fire zone, a temporary air wall 2 (air lock wall) with dampers is built 5–6 m outside the original seal wall 1 on the main air intake side as shown in Fig. 4.21. The air door is closed to create an enclosed space, which is entered by the ambulance crew. This is where the materials and tools needed to build a temporary seal wall are stored. Then open seal wall1, enter the fire zone for reconnaissance, confirm that there is no fire source within a section, a suitable location can be chosen (generally 150–200 m of the plateau seal wall) to build a temporary wind wall3 (air lock wall). After a quality check, remove air wall 2 and the original seal wall 1, use ventilation fan 5 for pressurized ventilation, discharge the gas accumulated in sections 1–3 and reinforce the support. The fire is gradually approached in sections until the seal wall on the exit side of the fire zone is removed and normal ventilation is restored to the entire zone. It is important to note that the dampers of the first wall are only allowed to be opened once the new seal wall has been constructed to ensure that the fire zone is closed and isolated.

Considerations for opening and closing the fire zone using the air lock method are as follows.

① Air lock work must be carried out in conditions free from the risk of explosion.

If abnormal conditions occur, such as a change in the direction of wind flow at the confinement or an increase in smoke, the work should be stopped immediately, personnel should be evacuated and observations made, and no danger should be posed before re-entering the fire zone.

Regardless of the method of sealing, the gas in the fire zone must be checked frequently during the sealing process. If CO or re-ignition is found, the sealing must be stopped immediately and the fire zone re-sealed. Within 3 days after the sealing of

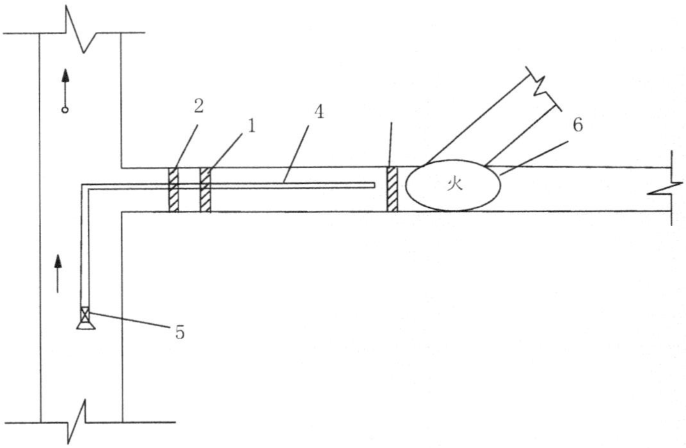

Fig. 4.21 Air lock opening and closing fire zone method. 1—original seal wall; 2. 3—temporary wind wall; 4—wind brief; 5—local ventilation fan; 6—fire source

the fire zone, the ventilation must be checked by the mine rescue team every shift and the water temperature, air temperature and air composition must be measured. Only after confirming that the fire zone is completely extinguished and the ventilation is in good condition can production operations resume.

4.4.4 Case Studies of Fire Zone Closure Incidents

(i) **Overview of the accident**

At 10:00 a.m. on 21 January 2005, a gas explosion was caused by spontaneous combustion of coal in the waste roadway of the auxiliary road in the west two east three sections of the Daming Mine (inclined shaft) −120 level, when 13 people were reinforcing the closed slurry spraying operation at the wind eye of the No. 2 tape conveyor road near the No. 4 closure of the No. 1 tape conveyor road in the west wing, resulting in nine deaths and four injuries.

(ii) **Accident history**

At 17:30 on 17 January 2005, a gas explosion occurred in the 15-storey closed abandoned tunnel on the east side of the west wing of the −120 m level of the Daming coal mine, destroying the closed confinement of three abandoned tunnels (three wood panel confinement and one material stone confinement) and the doors of the nearby 80 m tunnels (two groups of four doors). After a brief investigation by the relevant mine leaders, the decision was made to close the main roadway after the branch roadway had been closed in the absence of a clear source of the explosion, with the closure of the stone seal wall and the plank seal wall, see Fig. 4.22.

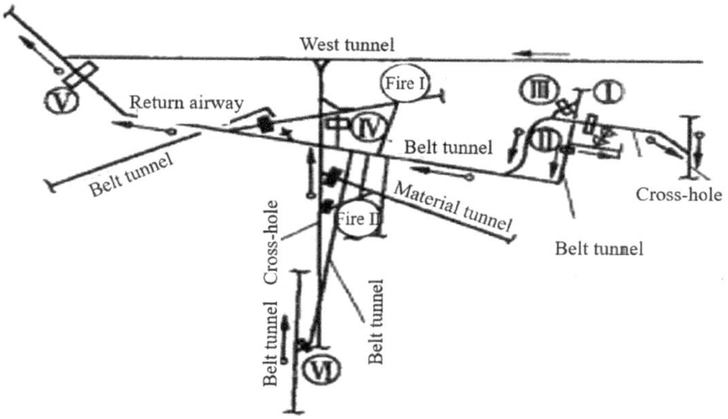

Fig. 4.22 Diagram of closure after the first explosion

On January 18, the second shift completed the construction of multiple branch roadway closure confinement. On January 19, the last moment of the closure confinement on both sides of the main roadway in the zero shift (Fig. 4.22 in II# and V# confinement) closed at the same time, a second explosion occurred and the closure work was temporarily stopped. Because of the small amount of gas involved in the second explosion, the strength of the explosion was weak, and only the upper 1–1.5 m high wall of the V# closure seal wall, which had just been completed, was knocked down. In the face of the two explosions, and taking into account the fact that part of the original plan has become a fact, the command did not change the original closure plan. But in order to reduce the leakage of air from the closed confinement, reduce the supply of oxygen to the danger zone, to ensure the effectiveness of the closure, increased the requirements of I#, II#, III# sealing surface spray concrete plugging.

At approximately 10:00 on 21 January, mine maintenance team personnel were working on the II# and III# containment closures to make up for the spraying operation as required when a third explosion occurred. As this time the amount of gas involved in the explosion was particularly large (it is presumed that the west wing 2# tape road, −120 m total return duct was filled with gas and participated in the explosion), the explosion was strong and powerful, destroying a number of closed confinements including II#, III# confinement, and personnel who were far away in the incoming air stream more than 700 m away to carry out replenishment spraying operations were killed and injured due to shock wave impact, seal wall material stone smashing. In view of the complex situation encountered in this rescue and relief, the command revised the closure program and was forced to expand the closure area again, and instead used the explosion-proof sand belt (outside stapled wood panel wall and sprayed concrete) to close the danger area, see Fig. 4.23.

When the implementation of explosion-proof sand belt to close the danger zone 88 h. That is, at 18:00 on 28 January, the closed area of the fourth gas explosion, this explosion only collapsed destroyed VIII# explosion-proof sand belt (the sand belt length of about 6 m, and the bottom plate pile of fallen rock up to 1 m thick,

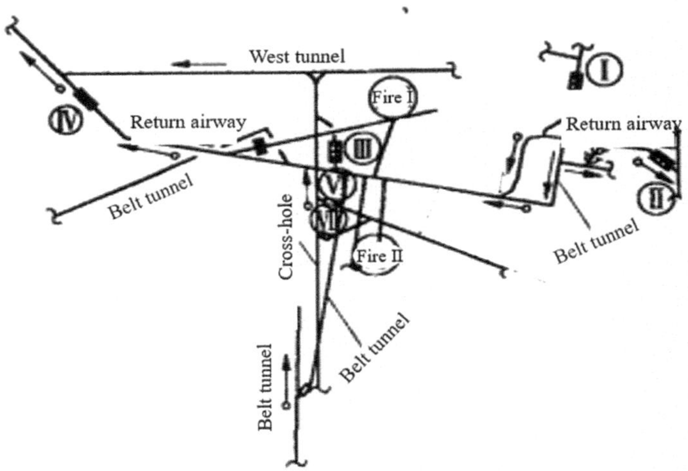

Fig. 4.23 Diagram of closure after the third explosion

poor blast strength), there are no serious consequences. After the fourth explosion, the command analyzed the dangerous area situation to confirm: because of the large enclosed area, the explosion fire point hidden, the enclosed area continued to gush gas, and there are multiple points of air leakage for oxygen, processing work should start from the elimination of gas explosion conditions; plugging air leakage, isolation of oxygen supply is the only way to eliminate gas explosion. To this end, crews injected nitrogen and three-phase foam into the blast area, along with sand filling, to submerge the explosive fire source and seal off VII#. In order to reduce the oxygen content in the blast area and eliminate re-explosion, the command made a decisive decision to retain the original nitrogen injection practice, to I#, II# anti-explosion sand belt before the use of wood panel wall plus wall spray concrete to establish a nitrogen chamber, through the pressure of nitrogen inside the chamber to eliminate air leakage into, see Fig. 4.24.

Through this creative nitrogen equalization technique, air leakage for oxygen supply was virtually eliminated, eventually reducing the oxygen in the blast area to less than 12%, completely eliminating the explosion and successfully ending the closure of the danger zone on 21 April 2005, turning the mine's safety around.

(iii) **Cause of the accident**

(1) Cause of the first explosion: the cause is unknown and the point of origin of the explosion is unclear.
(2) Cause of the second explosion: widespread closure of the fire zone.
(3) Cause of the third explosion: in a dangerous state to make up for the spraying of three (I#, II#, III#) surface spraying concrete plugging of leak-tight operations.

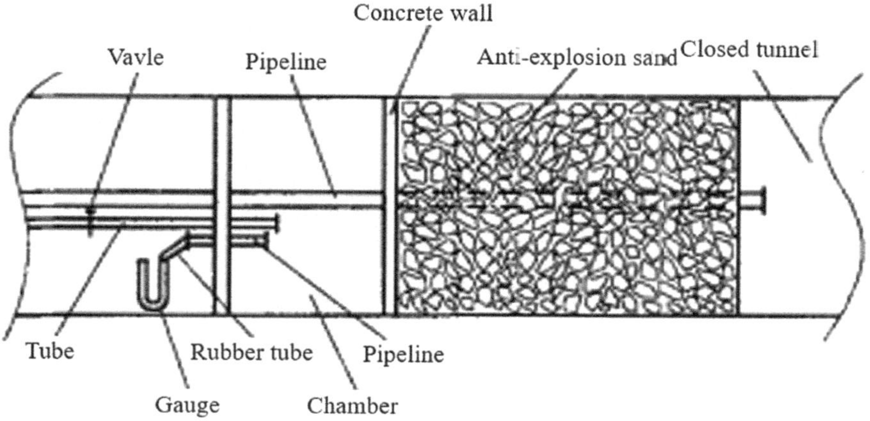

Fig. 4.24 Schematic diagram of the nitrogen equalization chamber

(4) Cause of the fourth explosion: the area was forced to be extended and the danger zone was closed off with a blast-resistant sand belt (with a wooden plank wall and concrete spraying on the outside)

(iv) **Precautionary measures**

(1) To strengthen the inspection and gas assay analysis of natural fires and enhance prediction and forecasting for each underground mining area confinement.
(2) Timely closure of abandoned roadways and improvement of the quality of ventilation facilities.
(3) When spontaneous combustion and natural fires are detected, well-thought-out measures will be formulated and dealt with in a timely and appropriate manner.
(4) When dealing with emergencies of "one pass, three defences", strengthen leadership, strict organization, comprehensive consideration and strict measures to prevent major accidents from occurring.

(v) **Lessons learned**

(1) The use of a wide range of closed gas explosion risk areas must be cautious. The obvious feature of this accident multiple gas explosions is: explosion a closed again and explosion again and again closed, not closed not explosion (except the first explosion). This fact seriously reminds us: this rescue and relief using a wide range of closed approach deserves deep thought. Large-scale closure of the gas explosion risk area, there is bound to be a large amount of air in the closed area, a short period of time the oxygen content of the explosion range can not be reduced to inhibit the explosion of the required safety concentration of 12% or less, even if the use of high-intensity nitrogen injection measures, the effect is also slow.

(2) Reasonable containment facilities must be used. The "1.21" accident initially used stone seal walls and plank seal walls to close the danger zone, which not only failed to achieve the purpose of closure, but also the stone became a deadly "killer"; when the rescue and relief efforts were replaced by explosion-proof sand belts, the purpose was gradually achieved, and the danger zone was finally controlled and closed. This not only failed to achieve closure, but also made the material and rock a deadly "killer".
(3) Nitrogen equalization to prevent air leakage is a good creative technique. The creative use of nitrogen pressure equalization to prevent air leakage played an important role in the ultimate success of the "1.21" accident.

References

1. Cao J, Shi Y. Argumentation of fire zone opening and closing parameters. Coal Technol. 2007;12(26):12–3.
2. Xie J. Research on locating the fire source in mine fire area and the safety analysis technology of closure opening and closing. Beijing: China University of Mining and Technology; 2009.
3. Shao S. Study on the variation of atmospheric conditions in confined fire areas and the optimal amount of inert injection. Beijing: China University of Mining and Technology; 2008.

Chapter 5
Fire Disposal Techniques for Different Locations Underground

Bo Tan, Haiyan Wang, and Liyang Gao

The handling of a mine fire is actually twofold. On the one hand, it is about extinguishing the fire, using various methods and tools, materials and equipment to extinguish the fire as quickly as possible. On the other hand, it is about saving lives, trying to minimize the casualties caused by the fire. It is therefore important that fires are dealt with quickly, safely and effectively, without missing the opportunity to fight them, and with reasonable tactics, otherwise even a small fire can develop into a large fire and cause significant damage [1].

The success of the mine fire process depends on the following factors.

(1) Location of the fire

This is the most important factor in deciding how to deal with a fire, and the method of extinguishing a fire varies from one location to another.

(2) Timing of fire fighting

Whether it is an internal or external fire, the initial stage of the fire is the best time to extinguish it, so the more timely it is, the easier it will be.

(3) Types of fire

B. Tan (✉) · L. Gao
School of Emergency Management and Safety Engineering, China University of Mining and Technology-Beijing, Beijing, China
e-mail: tanbo709@126.com

L. Gao
e-mail: gaoliyangcc@163.com

H. Wang
School of Civil and Resources Engineering, University of Science and Technology Beijing, Beijing, China
e-mail: b2146431@ustb.edu.cn

B. Tan et al. (eds.), *Mine Fire Prevention and Control*,
https://doi.org/10.1007/978-981-96-0466-1_5

Fire extinguishing methods vary according to the type of fire. For example, oil fires and electrical fires should not be extinguished directly with water.

(4) Fire-fighting equipment

Advanced and suitable fire-fighting equipment and gear facilitate fire-fighting.

(5) Correct command

Correct command is the most basic guarantee of mine fire handling.

(6) Quality of fire handlers

The personnel who carry out the firefighting have to be experienced in dealing with fires and have a good knowledge of mine fire. They must also have a brave and resourceful working style.

This chapter firstly introduces the techniques for determining the location of mine fires, the choice of fire-fighting methods, the techniques for detecting the affected area, the techniques for selecting a rescue base and the technical points for protecting underground personnel. It then focuses on the techniques for dealing with fires in different locations down-hole with examples.

5.1 Mine Fire Positioning Technology

The mine fire location technique refers primarily to mine external fire location techniques, which can also include external fire location techniques caused by internal fires.

The detection of internal fires is a technical difficulty and generally occurs in the following locations: (1) mined-out areas where there is a large amount of coal left behind that has not been closed in time or is poorly closed (near the mined-out area stop line, mined-out area collapsed coal left behind); (2) both sides of the roadway and the coal pillars left in the mined-out area that have been damaged by pressure; (3) floating coal piled up in the roadway or at the top of a coal tunnel where it has risen across the gang. (4) The place where it is connected to the old kiln, the place where the shallowly buried coal is connected to the surface, the outcrop coal, etc. In order to locate the source of the fire, a pre-buried temperature sensor is usually used in the mined-out area and a drill is used to measure the temperature to locate the source of the fire. At present, temperature sensors are seldom pre-buried in mines, which are mainly used to locate the source of fire after a fire has occurred.

After an external fire has occurred in the shaft, it can be difficult to determine the location of the fire source under the special conditions of smoke, reversal of wind flow and high temperature. Sometimes there can be errors in judgement and misleading treatment of fire accidents. Therefore, when a fire accident occurs, the basic situation of the fire area should be identified, and the location of the fire source, the nature, extent and direction of spread of the fire should be determined by the shortest route and the fastest method, so as to provide a basis for fire fighting work is very necessary.

The main current techniques for locating exogenous fires are empirical inference methods, qualitative analysis methods and quantitative inference methods. The empirical inference method is mainly based on empirical judgement through the abnormal phenomena such as smoke flow, water mist, odour or condensation droplets on the coal surface, temperature rise, etc. detected by the underground personnel from the wind stream, which is the current common inference method.

When the fire is small in the early stages, the source of the fire can be found against the direction of wind flow and according to the smell of fire gases or the direction of light smoke flow, provided that it does not prevent people from breathing.

When the fire has been going on for some time and the fire is large, the fire gases produced have reached a certain concentration which is harmful to people or the air temperature is high, the source of the fire should be sought by entering from the fresh air stream.

When the fire is very large, the fire wind pressure generated is high, wind flow reversal has occurred and fire smoke has filled the area, determine and find the location of the fire source from the direction of the reversed fire smoke flow. At this point, personnel must take extra care to find the right direction and route to enter and stay close to the point of origin of the fire to avoid being burned by the hot gases and avoid gas poisoning.

The following are examples of the method of determining the location of the fire source based on sensor alarms at the beginning of a fire and the method of analyzing the location of the fire source based on the direction of smoke flow movement during the life of the fire.

5.1.1 Steps to Locate the Location of Early Exogenous Fires

(1) Appropriate selection of the number, type and installation location of sensors, which is mainly a comprehensive consideration of the need for gas, wind speed, CO monitoring and timely alarming of disasters during daily production periods. The main purpose of the wind speed sensor is to determine if there is a significant change in wind speed in that roadway and possibly determine if the roadway wind flow has reversed. In addition to meeting the above requirements, the installation of the sensors should also take into account that in the unlikely event of a fire in a fire-prone area of the mine, the time difference between the alarm times of the sensors should be stretched in order to determine the location of the fire source.

(2) The smoke flow from the location of the alarmed sensor to the upwind side is reversed so that the sensor alarms through the roadway $A_{m} = (A_{m1}, A_{m2}, \ldots, A_{mn})$, where m is the number of alarm sensors and n is the possible fire branch (different m correspond to different n).

(3) Select the roadway corresponding to each alarm sensor as the common roadway $C = (C_1, C_2, \ldots, C_n) = \bigcup_{m=1}^{n} C_m$, where n is the number of common roadways.
(4) Work backwards from the upwind side of each non-alarm sensor location to the inlet shaft to obtain the relevant roadway $D_m = (D_{m1}, D_{m2}, \ldots, D_{mn})$ and find all non-alarmed inlet sides $D = (D_1, D_2, \ldots, D_n) = \bigcup_{m=1}^{n} D_m$.
(5) The common roadway corresponding to each alarm sensor is subtracted from the roadway containing the non-alarm sensor retrograde and the remaining is the suspected fire roadway obtained from the qualitative analysis $E = C \cap \overline{C \cap D} = (E_1, E_2, \ldots, E_n)$.
(6) Starting with each suspected fire roadway at E_m and assuming it is on fire, calculate the difference in alarm time for the smoke flow to reach each relevant sensor to form the combination of alarm time difference corresponding to a suspected fire roadway, $T_1 = (T_{11}, T_{12}, \cdots, T_{1n})$, $T_2 = (T_{21}, T_{22}, \cdots, T_{2n})$, ..., $T_m = (T_{m1}, T_{m2}, \cdots, T_{mn})$, where m is the suspected fire roadway and n is the roadway where the alarm annunciator is located.
(7) Based on the qualitative principle of fire source location, an alarm time matrix is established. The alarm time difference combination corresponding to a suspected fire roadway is further simulated by the dynamic fire simulation software, and compared with the measured alarm time difference combination one by one. The closest combination corresponding to the suspected fire roadway is selected as the presumed fire source location. In this way, a quantitative gray correlation analysis is used to further narrow down the suspected fire roadway.

5.1.2 Techniques for Determining the Location of the Fire Source in the Fire Area Based on the Direction of the Smoke Flow

(i) **Upward Ventilation Fire Source Location Analysis**

As shown in Fig. 5.1a, on a parallel branch where the wind flow has been reversed, the fire smoke is flowing in the direction of the main wind flow (no reversal) and merges with the fresh wind flow to return to the fire source, at which point the entrance to detect the fire source should enter from the direction of the fresh wind flow and follow the direction of the fire smoke inflow to the fire source.

As shown in Fig. 5.1b and c, after the reversal of a single wind flow in an angle-linked wind path, if the angle-linked wind flow AB is before the fire wind pressure is applied, it may be attacked by fire smoke even if the wind flow is reversed (B → A). In the case of a reversal of wind flow after the fire wind pressure due to the angled branch CD on the main flow branch, the fire will also flow from this branch to the normal branch (parallel point path) where no fire has broken out. In this case, the fire should be searched for in the direction of the original wind flow in the angled

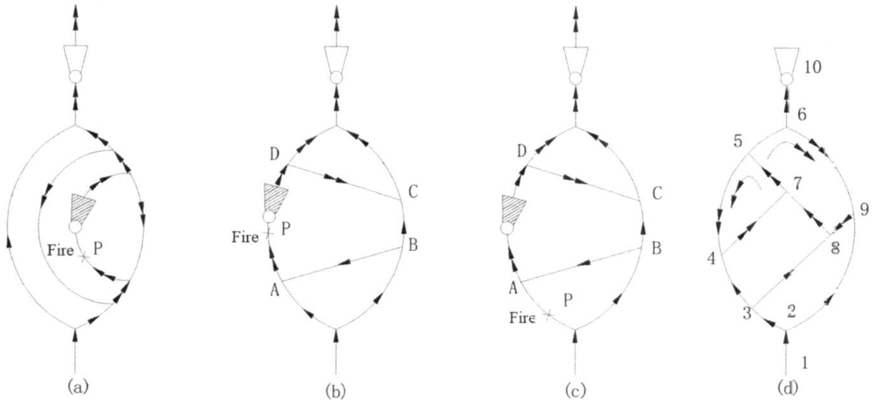

Fig. 5.1 Determining the location of the fire source in upward flow based on the direction of the smoke flow

wind path. When the fire wind pressure is high, the wind flow of the parallel branch BC may also be reversed, in which case the fire smoke will flow through the corner wind path AB to the place where the fire wind pressure is present.

Based on the pattern of fire smoke intrusion after wind flow reversal, it is possible to determine with some accuracy the location of the fire source or where the hot fire smoke is flowing through, i.e. where the fire wind pressure is generated.

As shown in Fig. 5.1d, when the direction of the fire wind pressure coincides with the direction of the main fan wind pressure, the main streams 1–2–3–4–7–5–6–10 and 1–2–3–8--7–5–6–10 that do not reverse. Analysis of the wind network structure and smoke flow of the diagram suggests that the fire either occurred in branch wind path 7–5 or within branch 5–6. The reversal of wind flows 6–9 and 9–8 would suggest that the fire could also have been in branches 8–7, 7–5 or 5–6. A reversal of wind flow 5–4 would indicate that a fire could have occurred in branches 4–7 and 7–5. Conversely, it would indicate that a fire could not have occurred in branches 5–6 and 8–7 and that the source of the fire must have been in branch 7–5.

It should be noted that the direction of the smoke flow is not the only basis for determining the location of a fire. Because fire smoke tends to remain stationary near the top of the roadway until the wind flow reverses, whereas in the lower half of the roadway the wind flow has already had a tendency to reverse.

(ii) **Downwards Ventilation Fire Source Location Analysis**

As shown in Fig. 5.2a, the main stream wind flow may be reversed and the fire smoke may flow from the main stream to the side stream. At this point, the fire smoke should be directed towards the roadway where it is flowing out to find the source of the fire.

As shown in Fig. 5.2b, if there is an angled wind path AB in front of the fire source, it is possible for the wind flow AB to reverse before the main flow on the BP section has reversed, indicating that there are signs that the main wind flow is about to reverse. When the angled wind path CD is after the fire source and has been

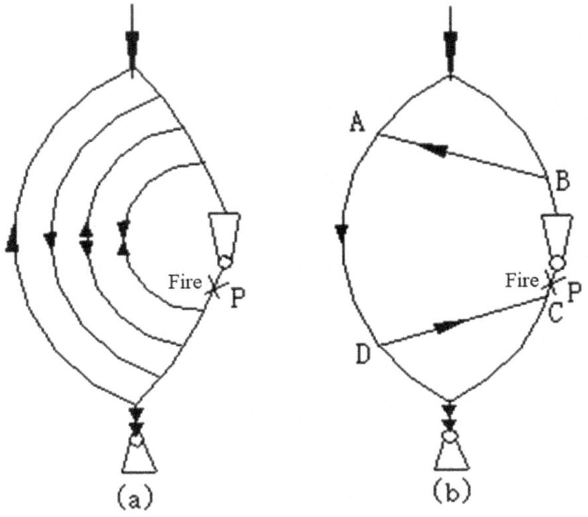

Fig. 5.2 Normal airflow diagram during a fire in the descending airflow stream

attacked by fire smoke at the beginning of the fire, the CD flow may reverse later when the fire wind pressure increases and may even flow over the fresh air flow.

In all cases, the location of the fire can generally be determined by the direction of flow of the fire smoke to find exactly where it occurred in the branch. In Fig. 5.3a, the fire is in branch 6–7; in Fig. 5.3b, the fire is in branch 4–6; and in Fig. 5.3c, the fire is in branch 4–3.

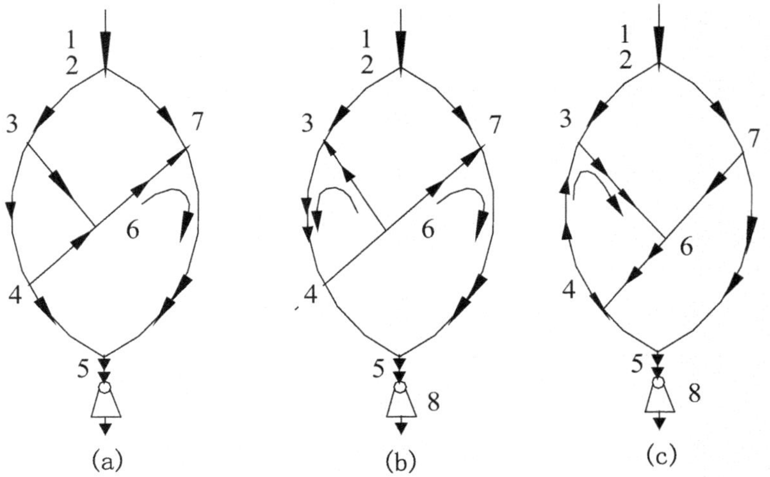

Fig. 5.3 Fire location in descending airflow based on the direction of smoke flow

5.1.3 Example of Determining the Location of the Fire Source from Sensor Alarms in the Early Stages of a Fire

Figure 5.4 shows a pressurized ventilation mine with gas and CO sensors set at four locations marked I, II, III and IV down-hole, where the CO sensors at locations I, II and III are alarmed at different times. The CO sensor at location IV is not alarmed, the process of applying the combined qualitative and quantitative analysis to deduce the location of the fire source is described.

(1) According to the three locations of the alarm sensor is located in the roadway, by qualitative extrapolation method to determine its upwind side of the smoke flow reverse retreat may flow through the roadway, corresponding to the three locations of the alarm sensor of the common roadway for the roadway 45, 2, 5, 6, 10.
(2) In the event of a fire in both roadways 9 and 15, which are located in the same area, there will be no smoke flow to the sensor at position I to cause an alarm when the initial air flow is not turbulent. So roadways 9 and 15 are non-public.
(3) Public alleys 2 and 45 in the non-alarm sensor corresponding to the IV position should be excluded. The final qualitative analysis obtained the suspected fire roadways as roadways 5, 6 and 10.
(4) It is then assumed that fires occur in roadways 5, 6 and 10, and care should be taken that roadways 6 and 10 are connected in series.

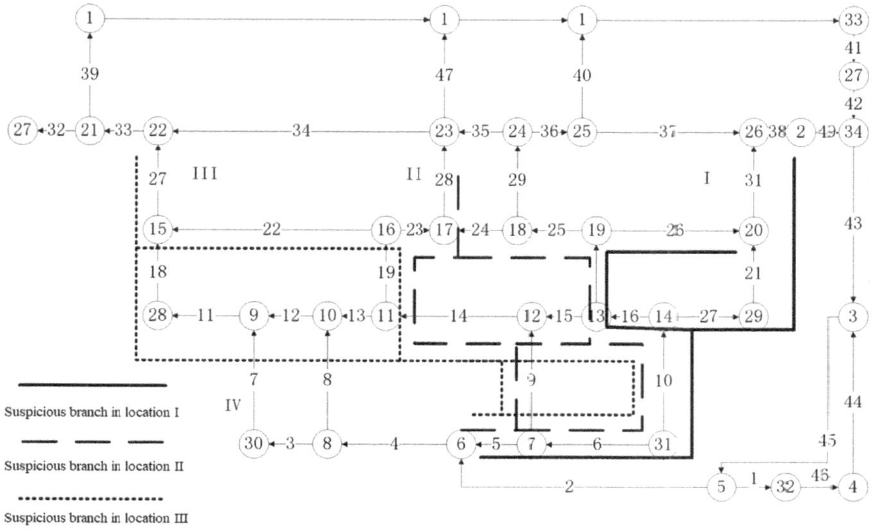

Fig. 5.4 Vardo mine ventilation network diagram

(5) As a result, the difference in alarm times between the sensors in positions I, II and III are the same for a fire in either aisle 6 or 10. It cannot be further inferred that a fire occurred in either aisle 6 or 10.
(6) However, suspicious fire roadway 5 is compared with roadway 6 or 10 and the corresponding sensor has a different combination of calculated alarm time differences due to the different smoke flow routes. The two combinations of calculated alarm time differences corresponding to roadway 5 or corresponding to roadways 6 and 10 are compared with the measured alarm time difference combination, respectively.
(7) If the calculated combination of roadway 5 comparisons is similar to the actual measurements, the exogenous fire occurred in roadway 5.

5.1.4 Example of Fire Source Location Based on Smoke Flow Analysis

(As shown in Fig. 5.5) A diagram of the ventilation system and ventilation network of a mine, the mine belongs to upward ventilation, 2–3 roadway is 1000 m away. When mine fire is found, all except 1–2 inlet roadway is filled with smoke, the following analysis of the fire source is located in which roadway.

As shown in the diagram, this is a fire that occurred in the upward flow, according to the 2–5–6–4 wind path analysis, it can be seen, before the fire the wind flow direction

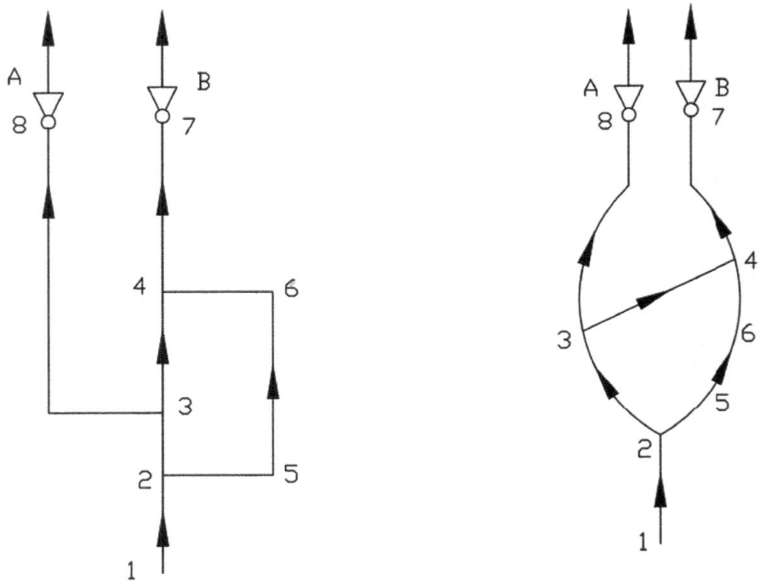

Fig. 5.5 Example of ascending airflow locating a fire source

is 2–4 upward fresh air flow, after the fire filled with smoke, but 1–2 no smoke, indicating that the 2–5–6–4 wind flow has been reversed into the 2–3 roadway. If the fire could only have occurred in the 3–4 corner joint wind-way, the fire wind pressure generated in the corner joint wind-way could easily have reversed the 4–6–5–2 wind flow, causing the smoke to fill all airways except the 1–2 inlet airway.

It is also possible that a fire broke out at the entrance to the 2–5–6–4 roadway and the smoke flow reversed back into the 2–3 roadway.

Incident management programme:

(1) The main fan cannot be stopped, neither A nor B can be stopped. In mines with multiple fans operating in combination, the operation of either main fan cannot be stopped at will without proper and precise measures.
(2) If the source of the fire is located in a 3–4 wind path, a cut-off containment on the upwind side of the fire is the most effective measure to control the fire and prevent reversal of the wind flow.

5.2 Selection Techniques of Fire Fighting Method

The correct choice of fire-fighting method generally requires consideration of the cause and ignition of the fire, the time of ignition and discovery, the location of the fire, the extent of the fire, the preparation time and the equipment and combat capabilities of the mine rescue team, as described below.

5.2.1 Causes of Fire and Ignition Sources

There are many different sources of high temperature fires. If they are found early, they are easy to identify. However, if they are found late and the scene is damaged by the fire, sometimes an investigation and analysis is required to determine them. Knowing the cause of the fire will facilitate the choice of fire-fighting methods.

Only a heat source with sufficient heat and temperature for a certain period of time can ignite combustible materials. In mines, spontaneous combustion of coal, gas, coal dust combustion and explosions, cannon work, mechanical friction, short-circuiting sparks, overheating from poorly functioning electrical equipment, smoking, welding and other open flames can all be sources of ignition heat. The main sources of ignition in mines are as follows.

(1) Open fires. Common open flames in metal mines underground include the flame of a calcium carbide lamp, a lit cigarette, an acetylene flame, etc. The flame of a quarry lamp used by miners for lighting is very hot and can easily ignite combustible materials such as broken wood and oiled cotton yarn. The heat from cigarette butts seems insignificant, but in fact examples of fires caused by throwing cigarette butts around are common. According to experimental

measurements, the centre of the cigarette burning temperature of about 650–750 °C, the surface temperature is also 350–450 °C, in dry, well-ventilated conditions, randomly thrown in combustible materials on the cigarette may cause a fire. Under the mine for cutting, welding metal acetylene flame, as well as the northern mine mouth heating stove, etc., may cause a mine fire.

(2) Arcing and sparking. Underground electrical lines, equipment short-circuiting, insulation breakdown, poor electrical switch arc, etc., will produce a strong arc or electric spark, the instantaneous temperature can reach 1500–2000 °C, enough to ignite combustible materials. Electrostatic discharge due to various reasons can also produce electric sparks, igniting combustible gases.

(3) Superheated objects. The hot surface of an overheated object is a common source of ignition for mine fires. The rotating parts of various underground machinery and equipment can become hot enough to ignite combustible materials due to frictional heating in poor lubrication, poor heat dissipation or other faulty conditions. With the increase in mechanization and automation in mines, there are more and more electrical equipment underground. If not properly used and maintained, electrical wiring and equipment can become overloaded and hot. In addition, electric heating equipment and incandescent lamps used underground are also sources of ignition that cannot be ignored. For example, when a 60–500 W incandescent lamp is lit, its surface temperature is about 80–110 °C, and the internal temperature of the hot tungsten filament can reach 2500 °C. In the case of poor heat dissipation and heat buildup, it can ignite nearby combustible materials. In addition, the high temperatures generated during blasting may ignite sulphide mineral dust, combustible gases or wood.

5.2.2 *Time of Fire and Discovery*

Fires that are short-lived, discovered early, and reported immediately and put out directly are easy to extinguish. Conversely, a fire that is found late or is not reported immediately but is walked away from, will expand the situation and make it very difficult to put out the fire directly.

The composition and temperature of the gases in the fire zone should be checked regularly in order to monitor the changes in the fire zone. The sampling location for the gas in the fire zone should be chosen at the firewall on the wind side of the fire zone, and gas samples should be taken through the observation tube on the firewall. If the firewall is far from the source of the fire, observation holes can be made close to the source of the fire, or ground boreholes can be used to observe the fire zone when it is not deep enough from the surface. Sampling should be carried out regularly, the fire area has not been stable stage, daily inspection sampling 1 time, later can be 3 days or 1 week inspection sampling 1 time. The fire area should be inspected and sampled by a full-time or ambulance crew. When sampling, the container should be cleaned of gas at the sampling site, the location of each sample should be consistent, and the gas sample should be analyzed in time after it is released from the well to avoid human error. The temperature of the fire area is usually determined by measuring the temperature of the gas in the fire area and the temperature of the water, which can be measured when the gas sample is taken. When using a mining temperature measuring instrument, the ground or underground observation holes can be used to measure the temperature from a distance. The data on gas composition and temperature in the fire area should be collated in time to draw gas composition and temperature change curves, analyse the trend of the fire. If there is deterioration, find the cause and take effective measures.

5.2.3 *Location of the Fire*

(1) Roadway use

Depending on the nature and use of the roadway, direct fire suppression should be sought for important roadway fires that serve the entire mine. Fires in the main shaft of the mining area, the flat shaft of the mining area (pan area), the stratified shaft and other shaft fires can be extinguished with a gradual focus on isolation.

(2) Roadway condition

Non-combustible material support of the roadway fire, easy to directly extinguish. Wooden supports or anchor spray off the roadway with exposed coal on fire, especially when the upper and lower gang has been mined and there is an old exit, direct fire extinguishing is difficult and should be switched to isolation fire extinguishing.

(3) Ventilation

If the fire site passes through a large amount of wind, it will ignite quickly and the fire will become large quickly. If the fire cannot be extinguished directly within a short period of time, it should be isolated and closed immediately; conversely, if less wind is passed, the fire will start slowly, which is conducive to direct fire extinguishing. Zafinor Coal Company is 650 m^3/min. After the wood stack fire only half an hour after the ignition of the belt, resulting in direct fire extinguishing ineffective, and finally had to be isolated and closed.

(4) Threat areas

The location of the fire is close to the mining area or the main mine exhaust duct, and a relative extension of the time for direct fire extinguishing is allowed after the post is set. If the return airflow from the affected area passes through places and alleys with a large number of staff, and direct fire extinguishing is judged to be ineffective or confirmed to be ineffective after implementation. The fire should be immediately isolated and closed. In addition, when the fire is located close to the pump room, substation and powder store or is about to threaten the main power supply line and main drainage line of the whole mine. The focus should be on using the isolation method.

5.2.4 Extent of Fire

The degree of ignition is divided into small, non-serious fires and serious fires, which can be determined on the basis of the following factors: distance of ignition, type of ignition involved, location of ignition, fire conditions, temperature of ignition, speed of ignition, etc.

(1) Distance to fire: the length of the roadway where the open fire started.
(2) Types of ignition materials involved in combustion: divided by one kind of combustion or many kinds of simultaneous combustion of cable electrical equipment, belts, coal dust, floating coal, supports, gang top coal, etc.
(3) Location of fire: classified by the location of the ignition material in the middle of the roadway, at the bottom, the two gangs or the roof.
(4) Fire conditions: divided into smouldering, flaming charcoal weak fires and large fires with embers.
(5) Fire temperature: determined by sensation located a few meters from the incoming air flow in the fire area.
(6) Rate of fire: estimated from wind speed, wind volume and fire size.

Once the extent of the fire has been determined, the direct fire extinguishing method can be used when the fire is small, while direct fire extinguishing techniques or isolation should be used when the fire is large.

5.2.5 Conditions for Selection of Fire Fighting Methods and Preparation Time for Implementation

Whichever method of fire fighting is chosen, careful consideration must be given to the availability of local conditions and the lead time required to apply the method. At the beginning of the fire, when the fire is not too serious, the fire extinguishing equipment and facilities available nearby can be used to directly extinguish the fire, the standby dry powder fire extinguisher can be used to extinguish the fire, and the safety helmet can be used to take water to extinguish the fire. The fire should be connected to the fire protection pipeline to extinguish the fire with a water gun, this method requires a special pipeline, sufficient water, a certain pressure and water gun, etc., do not have the above conditions, can not use this method. If you want to temporarily surprise the organization of human and material resources to create the above conditions, then you should estimate the preparation time and assess the development of the fire in the fire area, otherwise, you should give up this method of fire extinguishing. Although it is widely applicable, it is not suitable for non-large fires because of the long preparation time before implementation. When fires are serious and advanced equipment for direct fire suppression is not available, isolation and containment methods should be used. Emphasis should be placed on selecting a good location for the containment, arranging for the preparation and delivery of materials and the adjustment of filling lines. If the fire zone is connected to a large number of side wind flow paths, the number of containment should be increased and the required personnel and materials must be secured. Although some work can be split and worked in parallel, at least one hour is required at one location from the time the temporary confinement is made to the permanent confinement. The method of cutting fires by disconnecting belts and cables ahead of time is a means of reducing the length of belt fires and reducing the amount of direct firefighting work, but should only be used if there is a side channel in front of the fire area, the fire has not yet reached it and personnel can enter safely. The following methods can be chosen for different fire situations.

(1) Use water to extinguish fires.
(2) Extinguish the fire with sand or rock dust.
(3) Extinguish the fire with a fire extinguisher.
(4) Digging out the source of the fire.
(5) Grouting to extinguish fires.
(6) Inert gas fire extinguishing.
(7) Colloidal fire extinguishing.

5.2.6 Equipment and Combat Power of Mine Rescue Teams

China's mine rescue teams were first established during the "First Five-Year Plan" period, when the State formulated the "Trial Regulations for Rescue Teams". Subsequently, the State stipulated in the form of the "Mine Safety Regulations" that all coal mines must have a mine rescue team to serve them. A coal mine shall establish a mine rescue team or sign a rescue agreement with the nearest rescue team or jointly establish a part-time rescue team. Otherwise, it shall not be allowed to produce. Mine rescue teams must be certified by the State Mining Safety Supervision Bureau, which are managed by military and team rank systems. The overall quality and technical level of its members, the basic equipment of the mine rescue team, including individual protective equipment, underground rescue equipment, communication equipment, fire-fighting equipment, analysis and testing equipment and equipment tools, etc., are gradually configured in place and their modern technology level is gradually improved, basically adapting to the needs of rescue work in China's coal enterprises. To put out underground fires, the equipment and overall quality of the rescue team is very critical. A well-equipped, well-trained rescue team can quickly seize the opportunity to extinguish the fire. Otherwise, with poor equipment and low quality of commanders, emphasis should be placed on the isolation method.

Today's mine rescue teams are equipped in the following main areas.

(1) For personal protective equipment

Based on the introduction of American and German positive pressure oxygen respirator technology and the assembly of the BG4 positive pressure oxygen respirator, the HYZ-4 positive pressure oxygen respirator and the new PB240 positive pressure oxygen respirator were developed one after another at the end of the 1990s to meet the actual situation in China with more superior performance. It was used to equip front-line crews. China has also independently and successfully developed portable personnel locating devices, which allow ambulance teams to carry personnel locators that can search for miners in distress wearing miner location indication cards, adding an insurance policy for the safety of ambulance crews.

(2) In the area of underground rescue equipment

The ZDY4000S fully hydraulic disaster relief drilling rig developed by China to meet the target requirements of drilling rescue boreholes of 600 mm in diameter and 50 m in depth. The EBZ120QX type underground rapid rescue road heading machine, which is used to deal with the rock of the falling roof and to open up rescue channels for digging, became the main equipment for emergency rescue in underground mines. China introduced the world's most advanced search and rescue instrument, the DKL life detector, which was used for underground accident rescue life search and rescue.

(3) In the area of dedicated communications for rescue

Using Internet/Intranet/mesh network technology and GIS technology, it has researched and developed a decision support information system for rescue and disaster relief dispatching and command, as well as established a national coal mine safety information database, realizing automatic processing functions for accident response, disaster relief and emergency management and information sharing during accident rescue, improving the emergency handling capability of mine disasters and accidents.

(4) In the construction of rescue dispatching and command system

The technology and equipment used in the mine rescue and relief monitoring and command system was developed, which includes the ZLC series vehicle-mounted mine rescue command system, the mine rescue decision-making expert system, the ZS15 vehicle-mounted beam tube monitoring system, the low-illumination mining industrial television system, and the development of a national mapped website information management system. This achieves a short period of time to raise the level of individual technologies such as communication, information, decision-making, command and dispatch for rescue and disaster relief to a new level, forming a fast, accurate and efficient rescue command system.

Mine rescue personnel should be proficient in the standard requirements, operational requirements and operational skills for the operation of rescue techniques, including the following techniques.

(1) Hanging windbreak

This is a commonly used mining rescue technique, the technical principle is to avoid injury to mine rescue team personnel by establishing a temporary canvas barrier break in the pit to cut off the wind flow or high temperature into the smoke. The materials used are wooden squares, canvas and iron nails. According to the standard requirements, the hanging cloth, pressure strips and nails should be firm and solid. The entry should be quick and accurate and completed within 4 min.

(2) Construction of boarded up confined walls

This is the shaft rescue technology operation project, its technical principle is through the establishment of form-work closed wall, isolate the wind flow or high temperature smoke, avoid the injury to the mine rescue team commanders, use the scope and purpose and hang windbreak the same. Different is the wood board confinement wall confinement effect time longer, strong resistance to pressure. According to the standard requirements, the production of the use of wood and template, iron nails, the focus of the requirements is the board closed around the gap width shall not exceed into 5 mm, long into the degree of entry not into have to enter into over into into 200 mm. Column arrangement must be uniform, 10 min to complete.

(3) Construction of brick containment walls

The technical principle is equivalent to a plank confining wall, the materials used are red bricks and cement mortar, in a 4 m^2 non-combustible roadway, a sampling tube, a grouting tube and drainage holes are to be placed in the brick confining wall. The function is to close off the fire area, waste roadways etc. and is completed within 30 min.

(4) Installation of local fan and ducting

When there is gas to be discharged or directed ventilation to other locations, a local fan and a wind catcher barrel are required. The fan uses a 5 kW or 11 kW local fan and the switch must be explosion-proof. The connection barrel is a 400–600 mm diameter rubber duct with a double reverse side connection. As the fan is in a flammable environment, the installation requires that the connector for the power supply must be of the correct quality, installation and wiring. Short-circuit sparking phenomenon cannot occur and is completed within 8 min.

(5) Installation of high-capacity foam extinguishers

In the event of a large incoming fire in the mine shaft, the use of a high-capacity foam fire extinguisher is to be selected for a relatively flat roadway close to the source of the fire. Installation should be carried out in accordance with standard requirements, using explosion-proof magnetic starters, explosion-proof socket switches, wiring in accordance with explosion-proof requirements, foam fire extinguishers, chemicals, proportional mixers, firing nets, pumps, hoses, etc. should be placed in accordance with the prescribed distance locations. After installation and trial run, carry out fire-fighting operations. After the fire extinguishing is completed, the finishing work should be done.

5.2.7 Level of Safety and Security for Fire Fighters

Underground fires produce large amounts of toxic and hazardous gases, resulting in a large number of casualties. Improper adjustment of the wind flow during fire fighting can also cause reversal of the wind flow and cause explosive hazards resulting in casualties among fire fighting personnel. Injuries have also occurred in mine rescue teams. Therefore, there are problems with the safety and security of personnel in the fire-fighting and disaster relief process. The factors that cause casualties among fire-fighting personnel are as follows.

(1) Wind flow adjustment

Generally speaking, when a fire occurs in the general inlet airflow of a mine, a full mine counter draught should be carried out. In the main return flow, the original wind direction should be maintained. Sometimes, wind reduction measures can be taken, but not easily ceasing ventilation; and when it occurs in the mining area, the principle of disaster relief is to stabilize the wind flow. After the fire has started no further adjustment of the wind flow, if the wind flow must be adjusted, the firefighters should be evacuated to a safe place.

(2) Reversal of wind flow due to fire and wind pressure

Fire wind pressure can increase or decrease the amount of wind in a mine or local area, or even reverse the flow of wind, causing death or injury to firefighters in the upwind stream.

(3) Remaining fire in extinguished roadway

After sweeping through the focus of the fire with a water gun, it is difficult to press out the residual fire in the rear (upper) part of the stand and in the coal gap, producing more carbon monoxide to poison the personnel in front of the fire. In case of wind flow reversal, the explosive gas returned to this open fire, plus the appropriate oxygen, and the risk of explosion injury lurks.

(4) Over-fire roadway roofing injury and blockage

As a result of the fire, the strength of the roadway support (protection) after fire extinguishing has suffered damage. If it is found after the improper handling, it is easy to occur the roof injury or blocking accidents.

(5) Water and gas explosion

When using water to extinguish a fire, the jet should be gradually pushed to the centre by the edge of the fire source. If the sequence is not proper first shot to the centre, the resulting water and gas can not be quickly discharged due to the small wind flow, when the edge of the open fire there is a risk of explosion and injury.

(6) Gas explosions caused during isolation of enclosed fire areas

A person should be appointed to check the gas concentration and oxygen content at all times during the fire-fighting process and to watch for changes in the wind flow. In gas mines where the isolation method is used to extinguish fires, the constructed containment walls should be constructed on both the inlet and return sides at the same time, with the last agreed time being the simultaneous blocking of the vents reserved on both sides of the containment walls and the rapid evacuation of personnel. If reinforcement work is required, it should take 3 to 4 h before personnel can return. Otherwise, explosions caused by the accumulation of gas will result in casualties.

(7) Distance from the disaster area

Due to the different locations of the fire, the distance between the fire area and the external connected roadway with fresh air flow also varies. Short distances (within 100 m) make it easier to move around and provide a high degree of safety and security for fire-fighting personnel; conversely, long distances in the disaster area increase the danger to fire-fighting personnel.

Due to the rapid development of the fire, larger, the choice of a fire suppression method to open the confinement wall, may occur, such as gas and coal dust explosion, so that the measures to ensure the stability of wind flow may be disrupted. So before the fire is completely under control, to pay special attention to the safety of relief personnel, especially in the gas mine, even if the fire area has been confined, but still can not ignore the fire area may also occur in the danger of explosion. When building the main containment wall, the risk of explosion is most likely to occur in a gas mine, so it is best to make the main containment wall with a door. When the door is fitted and closed, leave as quickly as possible. When no explosion occurs after a certain period of time, go back for reinforcement and closure, but still finish as soon as possible and leave quickly, especially in a gas mine. All wind paths that lead to the danger zone should be set up with warning signs, and no other work is allowed within the warning signs except for work related to fire fighting. In order to keep abreast of sudden changes, a person must be responsible for checking the gas situation at the site. The inspector should be equipped with a portable detector for gas, oxygen and carbon monoxide as well as be ready to take samples and send them up the shaft for analysis.

In short, the factors that threaten the safety of firefighters come from many sources. Therefore, it is important to avoid prolonged fighting and delays when extinguishing fires, and to immediately isolate and close those that cannot be extinguished directly.

5.2.8 Communications in the Disaster Area and the Quality of the Commanders on Site

Timely and smooth communication in the disaster area and the high quality of the commanders on the scene are important guarantees for fire-fighting and disaster relief, which are also key in determining whether the chosen method of fire-fighting will be successful and verified as correct.

As can be seen from the above, the main factors in the choice of fire-fighting methods are basically dependent on the location of the fire, the extent of the fire and the degree of safety and security of the fire-fighting personnel, which are the three main aspects on which the choice of fire-fighting methods must be based.

The location of the fire is objectively inevitable after it has occurred, and it provides the environment and conditions for the choice of fire-fighting methods. The degree of fire predicts the ease of extinguishing the fire and directly determines the choice of extinguishing method; the degree of safety and security of the fire fighters is an important basis for achieving the purpose of fire fighting and disaster relief. Because in the process of disaster relief, it must not be allowed to lead to casualties in the event of another disaster. Therefore, it is a prerequisite for the choice of fire-fighting methods.

5.3 Reconnaissance Techniques of Mine Fire Disaster Area

After a fire has occurred in the well, such as a large fire, the rescue team must first enter the affected area for reconnaissance before dealing with the fire accident, to find out the specific location of the fire, the size of the fire, etc., before determining the specific treatment and fire-fighting program, do not enter the fire area blindly without understanding the specific circumstances of the fire. The following points must be made when reconnoitering a fire area.

(1) Before reconnaissance, prepare human and material resources and select people who are familiar with the situation and have experience in dealing with fires to undertake reconnaissance work, with a reconnaissance team of no less than six people.
(2) Essential equipment such as expedition ropes must be carried when entering the affected area for reconnaissance. Pay attention to concealed shafts, coal-slip holes, siltation, roadway support etc. when travelling. When vision is not clear, the expedition stick can be used to detect the advance and the team members should be linked with a contact rope. When scouting in fire areas, bad equipment and violations of rescue procedures can easily lead to serious accidents.

In June 1972, when a fire broke out in a mine, the squadron leader of the ambulance squadron led seven men into the disaster area to reconnoiter. When they returned to the base after reconnaissance, they found that one member was missing, so they went into the disaster area for a second time to look for him. Unfortunately, all six of them died because they ran out of oxygen. It was later found out that the missing team member had escaped to the ground during the reconnaissance. On the one hand, the reason for such an accident was the low quality of thinking of the escaped team member. On the other hand, the fact that the ambulance team had violated the rules during the reconnaissance by not carrying a contact rope and not checking carefully whether the amount of oxygen in the breathing apparatus was sufficient before entering the disaster area for the second time. Another example is that on 14 May 1992, when a mine in the Beijing Bureau was reconnoitering the disaster in the fire zone, the scene was so smoky that the visibility was zero, and a member of the ambulance team fell to the ground and died due to a malfunction of the breathing apparatus, while the ambulance team members in front of and behind him did not find it, which was also caused by the fact that the team members did not use a contact rope to link up with each other.

(3) A standby team should be set up at the underground rescue base to maintain constant contact with the reconnaissance team by telephone from the affected area.
(4) Before entering the disaster area, the reconnaissance team shall consider the measures to be taken if the retreat is blocked, specify the time to return and maintain contact with the well base by telephone from the disaster area. The reconnaissance squad shall return to the call by the original route at the specified time. Or if it cannot return by the original route, it shall be agreed by the commander who set the reconnaissance mission. If they do not return on time or if communication is broken, the standby squad should enter the rescue immediately.
(5) There should be a clear division of labour between the personnel of the reconnaissance team, who should check ventilation, gas content, temperature, roof and other conditions respectively, as well as keep records and mark the reconnaissance results on the drawings.
(6) When entering a fire area, the squad leader shall precede the line and the deputy squad leader shall follow the line, and the reverse when returning. When searching for persons in distress, the squad line should proceed diagonally with the centre line of the roadway.
(7) When reconnoitering over long distances and in complex roadways, several small teams may be organized to reconnoitre sub-sections. Persons found in distress should be actively rescued and escorted to the ventilation tunnel or the base of the shaft, and the place where they are found in distress should be checked for gas and well marked.
(8) Scouting should be careful and conscientious, so that all alleys must be visited, all the alleys taken should be marked and left with a name, and a diagram of the reconnaissance route should be drawn.
(9) Immediately after the reconnaissance, the squad leader should report the reconnaissance to the commander who set it up.

5.4 Selection of Rescue Base Techniques

When dealing with complex mine fires, ambulance bases must be set up for the timely supply of relief equipment and devices. Depending on the need, a surface rescue base and an underground rescue base can be set up.

The ground ambulance base shall have oxygen and other consumable materials maintained for at least three days and nights. The quantity of ambulance gear and equipment to be stored should be specified according to the nature of the incident, the extent of the impact and the number of participating ambulance teams. In order to ensure that the ground rescue base works properly and effectively, the mine rescue leader appoints the person in charge of the ground base.

The underground rescue base is the command post for front-line relief, the concentration of relief personnel and materials, the starting point for the rescue team to enter the disaster area and the temporary rescue station for people in distress, so the correct choice of the underground rescue base is related to the success or failure of the relief work.

The selection of the underground rescue base shall be determined by the general mine relief commander in accordance with the location, scope and type of disaster area, as well as ventilation and transport conditions, provided that the following requirements are met.

(1) The underground rescue base should be located in an area that is not threatened by the disaster area, or is not affected by further expansion of the disaster, but as close as possible to the disaster area to facilitate access for rescue crews to perform their duties.
(2) The underground rescue base should be located on the inlet side of the shaft where the wind flow is stable.

For example, on April 14, 1977, a mine in Liaoning in dealing with the 507 mining fire on the open fire, the underground rescue base in the return wind side adjacent to the disaster area, the result of the first gas explosion at 10:50, in the underground rescue base in the four mine level cadres were fumigated by carbon monoxide, lost the ability to command. Later, four gas explosions occurred in succession, killing 83 people and causing direct economic losses of 22.87 million yuan, which would have been greatly reduced if the underground rescue base had been set up correctly and commanded properly.

(3) A certain amount of space and area should be available to ensure relief activities and storage of relief equipment.
(4) The underground rescue base should be located in a location with easy access for transport and good ventilation and lighting.
(5) Do not choose an underground rescue base in a major transport big roadway with no connection to the affected area, in a corner jointed bypass or in a tunnel with excessive wind speed.
(6) Underground rescue bases are not required to be fixed in one place from start to finish, but can be pushed towards or away from the disaster area as the situation changes. Several alternative bases should be considered to facilitate selection.
(7) The underground rescue base should have a mine rescue team commander, a standby squad and an emergency doctor on duty, with a telephone to the surface rescue base and the disaster area, the necessary rescue gear and equipment, as well as clear lighting signs.
(8) The commander in charge of the underground rescue base should keep in constant contact with the relief command on the ground and the rescue teams working in the disaster area, pay attention to the ventilation and harmful gas situation in the base, as well as no personnel not related to the relief should enter the base.

(9) In the course of dealing with fire accidents, safety posts are set up as required in fresh air streams where the roadway where harmful gases accumulate crosses the fresh air stream, and the dispatch and withdrawal of the members of the station is decided by the ground command. A minimum of two ambulance crew members shall be assigned to the same post. In addition to a minimum of protective equipment, the team members should be equipped with various gas detection devices. Their main task is to prevent persons not wearing oxygen breathing apparatus or non-rescue personnel from entering the affected area; to introduce persons in distress into the fresh wind area and to provide medical treatment if necessary.

5.5 Safety Measures for the Protection of Underground Personnel

In the event of a fire in a mine, the first thing to think about in dealing with the disaster is to protect the safety of the people underground, to quickly evacuate people from the danger zone who are not involved in the rescue, and to take certain ventilation measures to prevent the wind flow from reversing and extending the disaster.

5.5.1 Safety Measures for Evacuating People from Danger Zones

The danger zone is the area directly threatened by the fire and the area adjacent to it, as well as the area through which the smoke flows to the wind shaft. Personnel working in these areas should therefore be evacuated first, except to participate in relief work, as well as also from areas where there is a risk of wind flow reversal as the fire progresses. It is at risk of being filled with smoke.

(1) As required by the Mine Disaster Prevention and Treatment Plan, consideration must be given to the shortest and safest route for evacuating persons in distress in the event of a fire anywhere in the shaft, the method of alerting them, and the measures to get them to safety.
(2) The routes for evacuating personnel must be well supported, properly lit and signposted to the exit refuge cavern, and the underground personnel must normally be made familiar with all the shafts along these routes, so that the names of the shafts are marked. They must also be informed at certain intervals of changes in the shafts along this route and evacuation methods must be devised accordingly to the changed situation.
(3) In the event of a fire in the mine, the appropriate personnel may be notified by telephone and lighting. Telephones must be installed in such a way that all personnel in the mine are aware of the location. Lighting signals are generally

used to notify the personnel concerned by cutting off the lighting power several times.

(4) Evacuate the danger zone in an orderly manner by choosing a safe escape route against the fresh wind flow and pay attention to the changes in the wind flow. When the evacuation route has been cut off by fire and smoke and there is a risk of poisoning, evacuees should immediately put on their self-rescue gear and pass through the nearby ventilation door into the fresh air flow as soon as possible. If the evacuees are sure that they cannot retreat, they should enter the nearby cavern or build a temporary shelter and wait for help. If there is a pressurized air line, evacuees should open the valve or try to cut the line and release the pressurized air to maintain breathing.

(5) If smoke is found to be discharged into the working face from the duct outlet, the duct outlet should be tied down immediately to trap the smoke and evacuate the personnel. When it is not possible to evacuate personnel, they should lie still in a smoke-free place in the tunnel and wait for rescue, as well as use clothing etc. as a temporary confinement where the fire and smoke may intrude.

(6) In areas of the shaft that are subject to smoke, if there are still people who have not been evacuated, or if it is not possible to know whether they have been evacuated, it should be considered that they may be in existing refuge caverns or temporary refuge caverns have been built, so that the pressurized air to these areas cannot be interrupted.

5.5.2 Safety Measures for the Protection of Disaster Relief Personnel

As fires generally develop quickly, opening up the confining walls before a particular method of fire extinguishing is chosen may result in gas and coal dust explosions etc., so that measures to ensure the stability of the wind flow may be disrupted. So the safety of the rescuers must be considered first before the fire is fully contained. Prevent rescuers from being scalded and burned by the hot gas, avoid fume poisoning accidents at the site and avoid gas and coal dust explosions at the fire site. In particular, consideration should be given to the possibility of an explosion occurring within the enclosed area even though the fire area has been closed.

When building the main containment wall, the risk of explosion is most likely to occur in a gas mine, so it is best to make the main containment wall with a door. When the door is fitted and closed, leave as quickly as possible. When no explosion occurs after a certain period of time, go back for reinforcement and closure, but still finish as soon as possible and leave quickly, especially in a gas mine.

In order to keep abreast of sudden changes, a person must be responsible for checking the gas situation at the site. The inspector should be equipped with a portable detector for gas, oxygen and carbon monoxide as well as be ready to take samples and send them up the well for analysis.

(1) All wind paths that lead to the danger zone are to be guarded and no work is allowed in the guarded area other than that related to fire fighting.
(2) A person must be responsible for checking the changes in gas composition in the main locations of the mine, especially in the case of a fire in a gas mine, it is even more important to check the atmosphere in the mine. Because changes in ventilation conditions in such mines may cause local gas overloads and are highly susceptible to gas explosions.
(3) When the gas concentration exceeds the limit, all relief personnel should be evacuated immediately and ventilation should be enhanced immediately to ensure the safety of relief personnel.

5.6 Fire Disposal Techniques for Different Locations Underground

5.6.1 Fire Treatment Techniques for Upward Flow in Inclined Tunnels

(i) **Treatment Techniques and Principles of Disaster Relief**

(1) Wind flow control

① Reduce the amount of wind to prevent the fire from spreading

Use the site conditions to actively put out the fire directly; to prevent the fire from expanding, hang windbreak on the upwind side of the fire source to reduce the wind flow; take measures to reduce the wind, put out the fire directly, and open the smoke exhaust channel to prevent the wind flow from reversing on the side branch; when taking measures to reduce the wind flow, prevent it from causing oxygen depletion and gas accumulation in the affected area.

② Keep the fan up and running

In the path of smoke evacuation, the main fan responsible for fire ventilation must not be depressurized or stopped in operation without threatening the central workplace or if personnel have been evacuated.

③ Local counter wind

Where possible, local counter drafts can be carried out to prevent the smoke flow from entering the mining area.

(2) Direct fire extinguishing

① Removal of combustible materials

When conditions allow, try to block the fire spread pathway. If the fire is small and the roof is intact, the wooden supports can be quickly removed to block the fire.

② Sprinkling water or aerial guns to cool down the fire

Open the water curtain on the return air side of the fire source point immediately, spray and sprinkle to cool down; the middle roadway and contact roadway can be used to directly extinguish the fire near the fire source. When the fire source cannot be approached, the mine car and skip can be used to lower the water sprayer into the roadway to extinguish the fire, or the high expansion foam can be fired for long-distance fire extinguishing.

③ Insulated fire extinguishing

When direct fire fighting is not successful, the fire area can be closed. The order of closing the fire area is generally close to the upwind side of the fire source first and then at a safe distance from the return airway. Try to keep the fire area as small as possible when closing it.

④ Other safety and security measures

During the handling of a fire, all junctions leading to the fire area should be fenced, marked with warning signs or guarded by personnel to prevent access by non-responders.

[Case 1] Example analysis of fire in inclined airflow roadway

[Course of the accident]

Figure 5.6 shows the ventilation system of a mine. A fire was caused by a cable fire at the lower opening F of the lower coal discharge eye in the uphill of the −150 level mining area. The workers in the W_1 working were quickly evacuated to the −150 inlet via the track uphill and D_1 ventilation door. W_2 and W_3 were standby working surface and were unoccupied at the time. 8 workers were trapped in the W_4 boring working surface and there were 60 people in the shaft at the time.

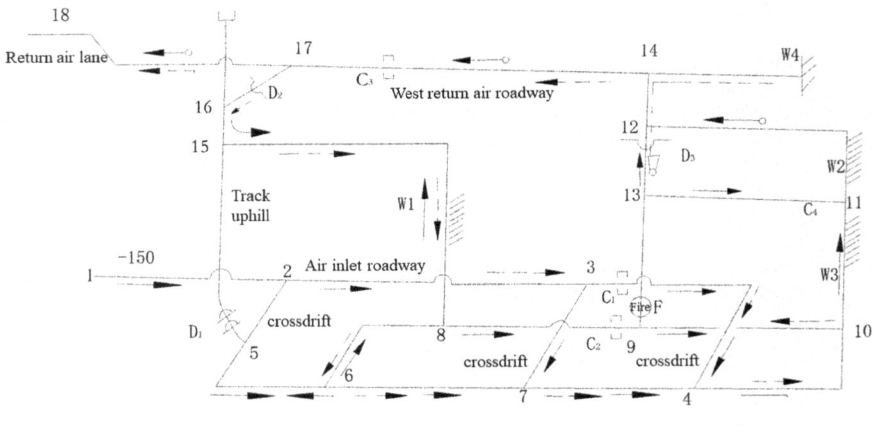

Fig. 5.6 Ventilation system diagram

After the accident, four mine leaders went to the upper yard of the track uphill D_2 regulating ventilation door to investigate the situation, while studying the rescue measures, as there were differences on the rescue plan, still discussing. At this time, someone informed to stop the main fan, about 15 min after the main fan was stopped. Under the action of fire and wind pressure, the smoke reversed to 14–17–16–15–5–2 (via the stone door and spread to −150 inlet big roadway 2–3, also spread to W_1 working face. At this time, the smoke diffused to all the underground roadway as shown in the diagram.)

Ventilation was restored 15 min later and C_1 confinement was built in 3–4 airways, C_2 in 8–9, C_3 in 11–13 and other confinements. The fire was controlled and over 60 people were saved from the shaft, but four mine leaders died near the D_2 ventilation door as they were in the middle of a reversing and reversing stream of toxic smoke for a long time.

[Cause of accident]

(1) After an accident, mine leaders should immediately be in place to organize rescue and relief work. They should not leave their posts or go down the shaft blindly.
(2) In the event of an accident, the mine manager shall be the main person in charge of the disaster relief work in accordance with the Mine Disaster Emergency Plan. One person shall be given the order, not more than one.
(3) It was wrong to stop the main fan, as a result of which a wind network structure dominated by fire wind pressure was created, thus contributing to wind flow disturbances in the whole mine.

[Technical measures for correct disposal]

(1) Maintain normal operation of the main fan to prevent the reversal of harmful smoke flow from infiltrating the roadway near the W_1 working surface.
(2) Firstly, temporary containment is built in the 8–9 and 4–10 wind paths to allow for increased air supply to the W_1 working face, reducing the air flow in the affected area and lowering the rate of harmful gas diffusion.
(3) Stop running the local fan of the W_4 excavation face and workers with self-rescuers are evacuated from 14–17-D_2-16–15–5–2. Depending on the situation, withdrawal from the 14–12–11–10–4–7–6–5–2 route is also possible [2].

5.6.2 *Fires in the Descending Airflow of Inclined Tunnels*

(i) **Treatment Techniques and Principles of Disaster Relief**

Descending airflow fires are more difficult to fight than upward flow fires because of the tendency to reverse the wind flow. The following measures can be taken.

(1) Increase airflow to prevent reversal of airflow

Measures must be taken to prevent the reversal of wind flow, to increase the amount of incoming and outgoing air and to reduce the return wind resistance. In the event of fire in the middle or lower part of the incoming slant hole, care must be taken that ambulance personnel are not allowed to flow into the well from the wellhead along the new air flow to prevent a sudden reversal of the reverse wind flow due to sthe fire wind pressure.

(2) Keep the fan up and running

In the path of smoke evacuation, the main fan responsible for fire ventilation must not be depressurized or stopped in operation without threatening the central workplace or if personnel have been evacuated.

(3) Wind flow short circuit

If the fire wind pressure is too high and reversal of the wind flow in the inclined roadway is unavoidable, it is advisable to find a short circuit path upwind of the fire source so that the reversed smoke flow is discharged to the return air system as soon as possible.

(4) Water sprinkling or aerial gun cooling

Open the water curtain on the return air side of the fire source point immediately, spray and sprinkle to cool down; the middle roadway and contact roadway can be used to directly extinguish the fire near the fire source. When the fire source cannot be approached, the mine car and skip can be used to lower the water sprayer into the roadway to extinguish the fire, or the high expansion foam can be fired for long-distance fire extinguishing.

(5) Safety and security measures

During the handling of a fire, all junctions leading to the fire area should be fenced, marked with warning signs or guarded by personnel to prevent access by non-responders.

[Case 1] Descending airflow multiple wind flow reversal example analysis

Figure 5.7 shows a fire that occurred in the descending airflow stream of an inclined tunnel, an example from abroad.

[Incident history]

(1) The source of the fire was at the lower F of the man uphill, as the fire area re-ignited causing violent and rapidly developing burning of the wooden supports. Ten minutes later, the wind flow in sections 5–6 of the fire roadway was reversed and reversed, with the smoke flow intruding into 5–18–13–14–6–5 to form a circuit loop, part of which was discharged by the east–west wind shaft. At this point, the rescue base was located at level 88.

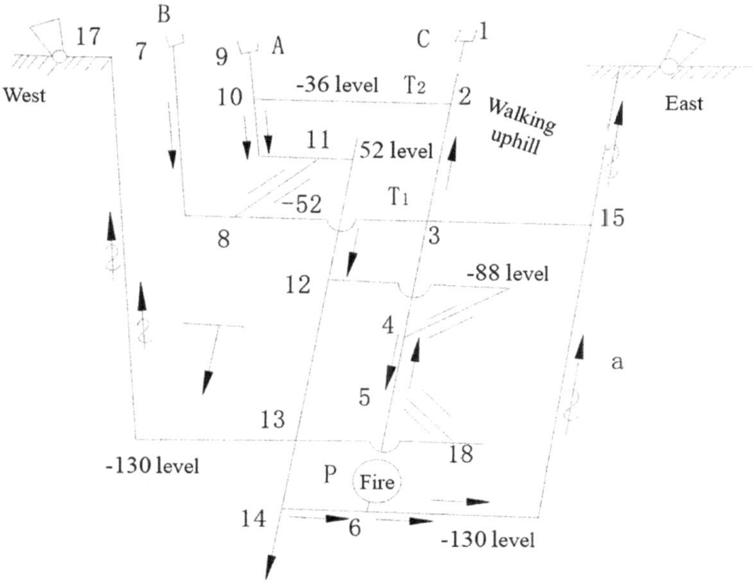

Fig. 5.7 Diagram of the ventilation system during a disaster

(2) Soon after a second wind flow reversal occurred in sections 4–5 and the smoke further intruded into the 4–12–13 roadway and flowed deeper via the inextricable rope uphill. The rescue base and personnel opened the lower confinement and withdrew to the −52 level.

(3) While the fire was being studied and analyzed, a third wind flow reversal occurred in sections 3–4 and the smoke flow was further expanded to invade the 3–8–11–12 roadway and flow to the deep level. At this time, only A and B shafts were free of smoke intrusion and personnel were concentrated at the bottom of A and B shafts to be rescued.

(4) When the fire further developed, due to the poorly sealed C_2, and the fourth wind flow reversal in sections 2–3, the smoke invaded the 2–10–11 roadway and attacked the evacuees at the bottom of shaft A. At this time, no one was spared from the smoke attack everywhere except the bottom of shaft B.

As a result, the evacuees at the bottom of shaft A were rescued by evacuating along the 11–8 stone door to the bottom of shaft B in a plume of smoke.

[Cause of accident]

Fires that occur in the descending airflow stream of inclined roadways often give the false impression that the wind flow in the upper part of the fire stream is fresh and safe from hazards, and fall for the deception. Depending on the magnitude of the fire air pressure, the airflow can be induced to reverse, with the following scenarios for reference.

(1) The fire occurs in a corner network circuit, as shown in this case. Wind flow reversal can occur section by section and is more likely to occur, with fire wind pressure at very little to reverse.
(2) In uphill and downhill where there is no parallel branching of the corner network circuit, there can be two cases.

① When the fire wind pressure is low and the ventilation pressure acting on the uphill and downhill is high (e.g. when the main fan pressure is high), wind flow reversal will not occur.
② When the fire wind pressure is high, a sudden reversal of wind flow will occur, and this sudden reversal of wind flow is extremely dangerous.

(3) When there is a parallel branch near the fire source, the fire wind pressure exceeds the ventilation resistance of the parallel branch, i.e. the wind flow can be reversed even if the fire wind pressure is small.

[Case 2] Descending airflow through the ventilation door to control the airflow to successfully save the disaster example analysis

[Course of the accident]

Figure 5.8 shows a fire in the lower part of the track downhill of a mine caused by the discharge of a cannon with a cable. The track downhill has an inclination of 18°, which is 270 m long. It is supported by wooden supports and the fire developed quickly. There are workers trapped at the working face, the gas detected at point A is 1.2% and the temperature is 42 °C. Due to the small cross section of the return uphill, high temperature and smoke, it is quite difficult and dangerous to rescue people from the return reconnaissance.

[Refer to correct disposal process]

(1) Someone check CH_4 and CO levels at point A.
(2) Send two people to control the two ventilation doors at D and strictly control the opening and closing of the ventilation doors as well as the degree of opening and closing in order to reasonably control the air flow.

① The effect and influence of fire and wind pressure must be carefully considered and the person opening the ventilation door must always act in accordance with the commander's orders. In one case, the fire suddenly increased in size when the ventilation door was closed by mistake because the command was not clearly understood.
② The wind flow reversal due to fire and wind pressure is always noted. On one occasion, when the ventilation door was opened to shorten the wind flow, the flames backed up for a short time by 8–10 m. As a result of the usual emergency drill preparation, the rescue team retreated backwards by lowering their positions, causing no danger to the personnel.

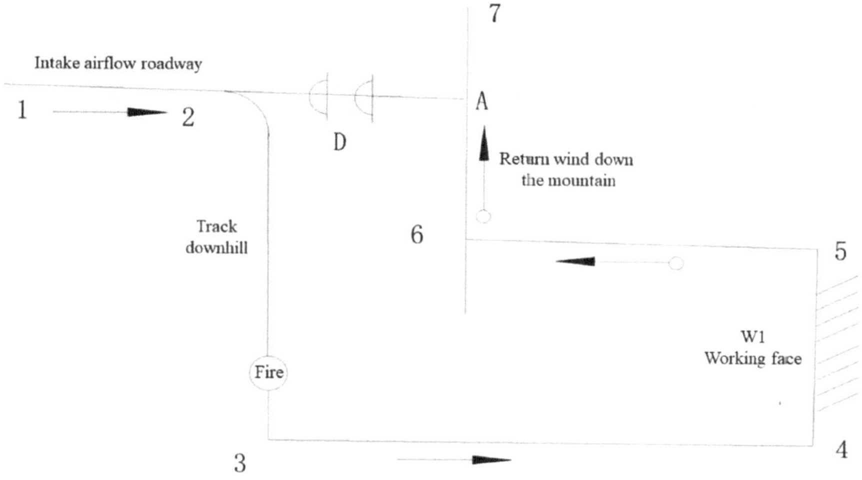

Fig. 5.8 Diagram of the ventilation system during a disaster

③ After the ventilation door was regulated, the downhill fire was significantly weakened and quickly extinguished with a dry powder extinguisher and fire pump, which basically extinguished the fire after 15 min.

5.6.3 Techniques for Dealing with Fires in the Vicinity of Underground Depots

(i) **Treatment Techniques and Principles of Disaster Relief**

In the event of a fire in the underground yard, the measures to be taken are as bellow.

(1) Main fan back draft or wind flow short circuit method

Take measures to rescue underground personnel by reversing the main fan or short-circuiting the wind flow so that the smoke from the fire is directly discharged into the main return airway, or stop the main fan to rescue underground personnel if the main fan stops and then reverses the wind flow in the mine. In mines with central parallel ventilation, the inlet and return air can be short-circuited to discharge the smoke flow directly to the surface; in the event of a fire at the bottom of the return shaft, the normal wind direction should be maintained, and the amount of air entering the fire area can be reduced on the premise that combustible gases will not gather to the explosion limit.

(2) Reduction of airflow

Reduce the amount of air flowing to the source of the fire in the underground yard by means of temporary containment and hanging air tents.

(3) Water extinguishing

Use all roads leading to the source of the fire and concentrate the maximum amount of human and material resources, especially applying the conditions of sufficient water supply in the yard at the bottom of the well to extinguish the fire directly and prevent its spread; to prevent the burning of the concrete supports and the wood chop above the masonry swan roadway, a water curtain may be set up by drilling or breaking the swan.

(4) Clean up combustible materials

After extinguishing the fire at the bottom of the well yard, strengthen the inspection of the top of the swan and the two gangs of the roadway (often with wooden stacks or left with floating coal, etc.). If abnormal temperatures are found, immediately take measures such as drilling or opening the concrete swan and digging the fire path to extinguish the hot or overcast fire in the top of the swan and the two gangs.

(5) Safety and security measures

During the handling of a fire, all junctions leading to the fire area should be fenced off, with warning signs hung or guarded by personnel, and underground personnel should be informed in a timely manner to ascend the shaft from a safe route to prevent non-rescue personnel from entering.

[Case 1] Belt inlet air slant hole fire accident example analysis

[Course of the accident]

As shown in Fig. 5.9, a mine with an annual production of 2 million tonnes is developed in a combined vertical shaft and slant hole, with seven inlet shafts and three outlet shafts, of which the east wind shaft is responsible for the air flow of the three inlet shafts. The belt slant hole and the vertical shaft main stone door form a parallel and complex angular network system at the bottom of the shaft yard.

During the construction of a lap joint between the upper part of the lower two belt sections (1334 m long) and the lower part of one belt section, gas welding to cut steel plates ignited the glue end and glue strips in the lap cavern. Due to poor fire-fighting measures, the timing was lost and the lower part of the first section of belt caught fire. The burning of the belt produced large CO, NO_2 and other nitrogen oxides and traces of cyanide, and the smoke spread down the shaft in the wind stream, posing a great hazard. There were 1,477 people in the shaft at the time.

As the fire got bigger and bigger, a strong fire wind pressure was formed and the incoming wind flow from the belt slant hole was suddenly reversed. At that time, two chief engineers and nine rescue team members did not understand or realize that the fire wind pressure would reverse the wind flow, and they entered the fire area from the belt inlet slant hole to reconnoitre and rescue, and all of them died.

[Cause of accident]

The slant hole was 357.2 m deep and the maximum temperature was estimated to be up to 1400 °C. The calculated fire wind pressure value was up to 3609.34 Pa

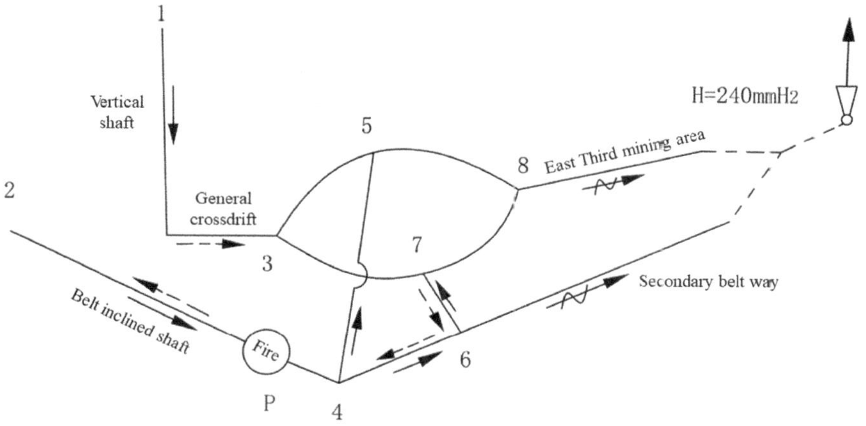

Fig. 5.9 Diagram of the ventilation system during the belt intake slant hole fire

(368.3 mm H_2O) and the fire wind pressure value had exceeded the total negative mine pressure of 2352 Pa (240 mm H_2O) for the main fan. But why did not the reversal of wind flow occur in the whole mine, but only the reversal of wind flow in the belt slant hole? The reason is that when the fire wind pressure exceeds the parallel ventilation resistance of the two inlet shafts (i.e. to the belt slant hole and the riser shaft a total stone door), the wind flow in the belt slant hole will suddenly reverse and reach the maximum fire wind pressure instantaneously. The reversed air flow will form a powerful high temperature smoke flow.

The ventilation resistance of the belt slant hole is approximately h = 630 Pa (64 mm H_2O). When the average temperature in the belt slant hole increases to 70 °C, the fire wind pressure (or natural wind pressure) can prompt the reversal of the air flow in the belt slant hole.

[Correct disposal process]

(1) Take the main fan back-draft so that the fire smoke is discharged directly into 2 places to exclude and rescue people down the shaft.
(2) In the event that a back-draft is not possible, rescue personnel will enter from the vertical shaft and set up a temporary containment at the 4–5 roadway and 6–7 roadway to prevent harmful gases from attacking the East 3 area. Be sure not to enter the fire area from 2 to 4.

[Case 2] Analysis of a fire accident in a belt conveyor roadway

[Course of the accident]

At 20:45 on 24 June 2009, a belt fire broke out in the main shaft transport belt roadway of the Huangcun Branch Mine in Yichuan County, Luoyang City, Henan Province. Twenty-eight people entered the mine during the shift, 25 were safely promoted and 6 were killed.

[Cause of accident]

(1) The accumulation of coal dust on the inside of the belt machine channel steel for a long time, the gradual gathering of energy and the oxidation of coal dust, causing spontaneous combustion of coal dust.
(2) The belt fire alarm system was not put into operation properly and the operator did not monitor the fire in time, resulting in the spread of the fire.
(3) Insufficient fire water pressure, inoperable rain shower system and fire hydrants, and failure of some fire extinguishers to spray properly, resulting in poor control of the fire at the scene and the expansion of the accident.

[Correct disposal process]

1. To strengthen the daily cleaning and inspection of the coal transmission system equipment and facilities, establish a perfect working system, implement personnel responsibilities, and prevent problems before they occur.
2. To deepen the safety hazard investigation and rectification work, comprehensively investigate the fire safety hazards of production equipment and facilities, formulate and implement rectification measures to prevent dust accumulation and spontaneous combustion of equipment and facilities, clarify management responsibilities, strengthen supervision and inspection of rectification implementation, achieve the five implementation of measures, responsibilities, funds, time limits and plans for rectification of hidden hazards, and form a closed-loop management of hidden hazards.
3. To rationalize the management process of the fire protection system and to implement the responsibility for the overhaul, maintenance and testing of the fire protection system.

5.6.4 Fire Handling Techniques in Cavern

(i) **Treatment Techniques and Principles of Disaster Relief**

The Coal Mine Safety Regulations set out a number of requirements for fire protection in underground caverns, as follows. The permanent central substation and other electromechanical equipment caverns in the underground yard should be built with swan; mining area substations should be protected with non-combustible materials; mining working surface distribution points should use special caverns and be protected with non-combustible materials; the central substation and pump room cavern in the underground yard must be equipped with outward opening fireproof iron doors. The following aspects should be noted when dealing with cavern fires.

(1) Main fan back-draft or short-circuiting of wind flow

When a fire occurs in a cavern located in a wing or in a cavern of the general air intake of the mining area, a back-draft or shortened airflow can be used to prevent the fire smoke from entering the underground working surface. If the cavern is located in a wing of the mine or at the connection of two roadways through which the general airflow from the mining area is fed, the cavern's special return airway regulating window or the ventilation door on the smoke pathway to the general return air should be opened where possible to short-circuit the smoke flow. When the conditions are right, a local back-draft can also be used; the cavern fire door should be closed in case of fire. When there is no fire door, the incoming air should be controlled by hanging wind tents or playing temporary panels.

(2) Removal of combustible materials

In the event of fire conditions permitting, the cavern should be cleared of combustible material as far as possible, especially when the powder magazine is on fire, the detonator will first be transported out. Then, other explosive materials will be transported out, such as the high temperature can not be transported, then close the fire door.

(3) Asphyxiation or hypothermia

Use the fire fighting equipment stored in the cavern, such as fire extinguishers and sand, for direct fire fighting.

(4) Other safety and security measures

In case of fire in the winch house, the mine car should be fixed under the source of fire to prevent burning the wire rope and causing injuries in the running field; in case of fire in the battery garage, to prevent hydrogen explosion, the charging should be stopped, ventilation should be strengthened and the battery should be transported out of the cavern in time. Cut off the power supply in the cavern.

[Case 1] Analysis of a major fire accident caused by burning wood shed of substation and main slant hole

[Accident history]

As shown in Fig. 5.10, a major fire accident was caused by the discharge of the transformer output cable at the bottom of a horizontal shaft yard in a mine, which ignited the transformer oil leakage, resulting in the burning of the substation wood shed and the main slant hole wood shed (11–10 into the wind tunnel). 25 people were killed on the return wind side (9–8 working area return wind side). During the rescue process, due to the reversal of wind flow, many more people were killed, making a total of 48 sacrifices. The accident was costly and the lesson learned was extremely profound. The mine was a low gas mine, with central parallel ventilation, and was a flat, dark slant hole with multi-level development and multi-stage winch hoisting.

[Accident management]

Segment 1: Fire incident at 08:30 on 9 August. The ambulance team entered from (2–5–6–7–8–9) to reconnoiter and rescue people and carry the victims. All 25 personnel were found killed in the second level return airway. As the fire in the main slant hole above the substation was serious and the roadway section was only 4 m^2, the route was more than 800 m long, it was difficult to carry the victims, so carrying was suspended.

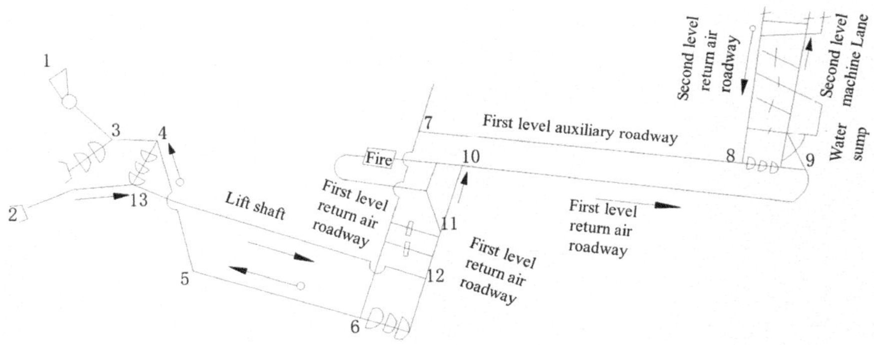

Fig. 5.10 Ventilation system diagram for a mine substation during a fire

The second section: at 2:50 on the 10 August, from the original program of transporting the victims first and then extinguishing the fire, to first extinguish the fire and then transport the victims; to reduce the fire stopped the main fan operation. First water to extinguish the fire near the substation, and remove the 10 m wooden shed, forming a fire separation zone, because of the fire wind pressure caused by wind flow reversal disaster area expanded to the incoming wind area and 23 people died of poisoning.

At 13:00, the back-draft started, as there was no water source in the main flat. That is, the fire engine was organized to transport water. At this time, the floating coal in both gangs of the shed column was also burning, the temperature rose to 70–80 °C, the rescue and fire fighting work was very difficult and dangerous. The fire was not clear, and the incoming and outgoing coal pillars were burning. At 15:10 on the 11 August, it was decided to withdraw all personnel from the shaft and to stop the operation of the fan.

The third section: at 19:30 on 11 August, an emergency meeting of leaders at all levels chaired by the governor was studied, and in order to prevent further expansion of the accident in the rescue and increase casualties, it was decided to temporarily close the wellhead and request that the fire be extinguished after one month before opening it to find the victims.

The fourth section: a confinement was built at the location of wind shaft 2 and the location of the main flat 1 respectively, then inert gas was injected into the fire area using a DQ-500 inert gas generating unit, injecting 450, 350 and 250 kg of inert gas into the fire area on the 14th, 22nd and 30th respectively.

The fifth section: On 15 September, the air locking method was used to enter the confinement for reconnaissance, with the temperature dropping to 21–22 °C and CO = 0. It was determined that the fire area was extinguished. Then, a sectional reconnaissance was used to restore ventilation to the first and second levels as well as the entire mine in sections.

[Analysis of the causes of accidents and correct handling measures]

(1) The substation is not used to form independent ventilation, if there is independent ventilation can be introduced into the return airway by short-circuiting the fire smoke through the wind flow, reducing casualties at the second level.
(2) Stopping the main fan operation causes a reversal of the air flow in the inlet side lifting slant hole.
(3) The mine counter wind measures were not timely. As can be seen from the system diagram, the source of the fire was close to the inlet shaft and the inlet route was a single route. So the accident could have been dealt with in time if the counter wind had been carried out at the initial stage.
(4) To ensure the safety of personnel in work areas 8–9, the 8–9 ventilation door can be opened to ensure the temporary safety of personnel in the work area.

[Case 2] Analysis of an accident in which an uphill belt was ignited due to an electrical short circuit in a substation

[Accident history]

Figure 5.11 shows a mine uphill panel substation that ignited the uphill belt due to an electrical short circuit, endangering the safety of personnel on the downwind side of the working face W_1, W_2 and the digging face W_3.

[Emergency response process]

At that time, measures were taken to short-circuit the air flow: the main ventilation door D1 in the air intake and return room and the ventilation door D2 between rooms 4–10 were opened, leaving the uphill pan area in a reduced air supply. As a result of the reduced air supply, the fire was weakened and the rescue team could approach the fire and enter the fire area to put out the fire, which achieved the desired objective.

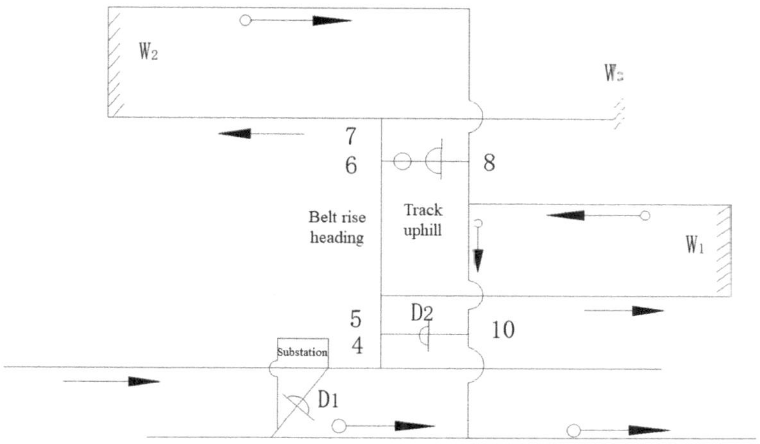

Fig. 5.11 Ventilation system diagram for a mine substation during a fire

[Accident experience]

This short-circuiting of the wind flow is a measure that must be monitored at all times for gas mines, analyzed and studied in relation to the timing of gas increases and fire suppression, so as not to cause gas explosions that could extend the scope of accidents and casualties.

[Case 3] Example of an oil and gas ignition accident at pump cavern due to irregular operation

[Course of the accident]

At around 08:00 on 4 December 2020, the retreating personnel of Shengjie Recycling Company arrived at the Hanging Water Cave Coal Mine shaft one after another to carry out retreating work. At around 16:40, when the retreating personnel were illegally using oxygen/LPG to cut the 2# and 3# pump suction pipes inside the −85 m pump cavern, the falling high temperature slag ignited the oil scale deposited in the suction well of the water bin. The burning of the oil scale and rock seepage oil produced a large amount of toxic and hazardous fumes, causing 24 people to be trapped down-hole (Fig. 5.12).

[Cause of accident]

During the evacuation operation of the underground retreat in the Hanging Water Cave coal mine, the retreating personnel illegally used oxygen/LPG to cut the pump suction pipe in the −85 m pump cavern, the falling high temperature slag ignited the oil scale deposited in the suction well of the water bin, the oil scale and rock seepage oil burned to produce a large amount of toxic and harmful fumes, which spread to the inlet roadway under fire and wind pressure, causing casualties.

[Correct disposal process]

(1) The air supply should be reduced by adjusting the nearby air windows.
(2) Pre-emptive control by nearby water sources when the fire is low.
(3) Extinguish the fire with equipment such as a foam sand box within a controlled area.

5.6.5 Fire Treatment Techniques for the Excavation Roadway

(i) **Treatment Techniques and Principles of Disaster Relief**

The treatment of fires in one-headed roadways is based on three main considerations: firstly, how to control the local fan; secondly, how to prevent gas explosions; thirdly, how to prevent the fire from expanding.

In recent years, there have been many cases where fires in the excavation roadway have been mishandled, resulting in extended accidents. Fires in the excavation roadway are difficult to deal with as there is only one route in and out of the roadway due to ventilation restrictions. This is particularly the case as the roadway stops

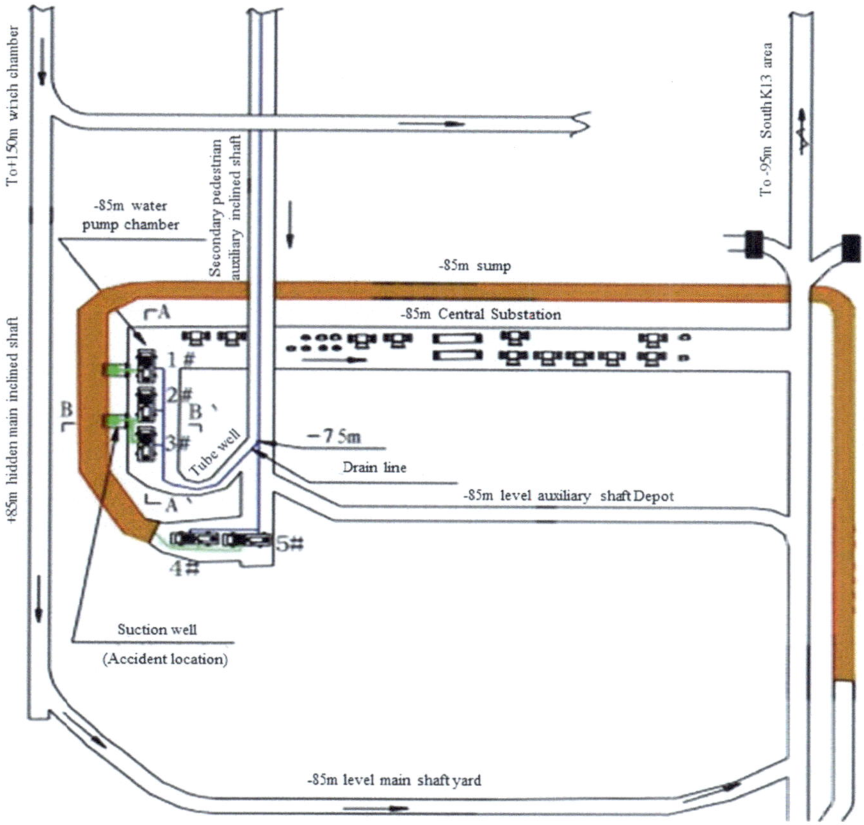

Fig. 5.12 Ventilation system diagram for a mine substation during a fire

running partially fan after a fire, the ducts are broken by fire, the ventilation in the roadway is destroyed by gas explosion and burning, the gas may reach the upper limit of explosion, the roadway is filled with thick smoke and fire, the temperature in the fire area increases, the wooden supports burn and lose their supporting capacity, and the hot roof plate falls. Whichever tactics and advanced equipment are used, they will bring varying degrees of danger and complexity to the fire-fighting effort. Therefore, exploring ways to extinguish fires at the head of the excavation and ensuring the safety of the firefighters themselves is an urgent problem to be solved.

(1) Keep the local fan running properly

In principle, direct fire fighting should be carried out while maintaining normal ventilation of the local fan. However, after arriving at the scene of the fire, attention must be paid to maintaining the original state of ventilation, i.e. the fan out of operation should not be opened casually, the fan in operation should not be stopped blindly, and disaster relief personnel should be assigned to guard the local fan, maintain normal operation and reconnoitre before determining the measures.

(2) Use dry powder extinguishers, water and other direct fire extinguishers

When the gas does not exceed 2%, the fire should be extinguished directly under ventilation. Dry powder fire extinguishers and water can be used to extinguish the fire directly under ventilation. After extinguishing the fire, the ignition point must be carefully inventoried to prevent re-ignition. If the gas concentration exceeds 2% and continues to rise, immediately evacuate the personnel to a safe place and close it from a distance. Take the direct fire extinguishing and sealing fire area, it should change the pressure air pipe and water pipe into strong pressure inert gas pipeline pressure injection CO_2 and N_2 and other inert gas; when using direct fire extinguishing, prevent the fire gas and water gas explosion; prevent a large amount of water steam can not be discharged and injury. Fire grenades etc. can be used first and then water can be used to remove the residual fire.

(3) Isolate fire extinguishing

① When the fire occurs in the middle section of the coal tunnel, the concentration of gas flowing to the fire must be detected during the extinguishing process to prevent the gas from passing through the point of origin of the fire and should be closed at a distance if the situation is unclear; if the fire occurs in the middle section of the uphill, the fire must not be extinguished directly and should be closed in a safe place.
② In the event of a fire in an uphill coal roadway, regardless of the location of the source of the fire, if the local fan has ceased to operate, it is strictly forbidden to enter to extinguish the fire or to detect it when no rescue is required, but to immediately evacuate the personnel in the vicinity and to close it from a distance.
③ When the source of the fire is at the head of the downhill coal roadway, if the condition of the fire is unclear, it should generally not be entered to directly extinguish the fire and should be closed.
④ When extinguishing an excavation roadway fire, the fire cannot be extinguished for a short period of time due to insufficient manpower and material resources, the gas in the fire area gradually rises, the gas in the fire point cannot be detected, or the coal dust flies due to the burning and falling of the coal roadway, etc. The direct fire extinguishing work should be stopped and measures to close the fire should be taken immediately.
⑤ In the event of spontaneous combustion fires in the central mined-out area of the excavation roadway and in abandoned roadways, the top of the pack should be filled with river sand, loam and fly ash to close the fire area. If this cannot be achieved, the roadway should be closed quickly and the fire should be extinguished by pressing inert gas, equalizing the pressure and filling.

(4) Other safety and security measures

If the local fan has stopped after the fire has started, do not turn it on during the fire treatment and assign someone to watch it. The decision to stop the local fan will be made after a detailed reconnaissance to determine the situation.

Check that there is no accumulation of gas on the inlet and return side of the fire escape. If there is, it should be closed first to avoid causing a gas explosion. In particular, if there are places where gas has accumulated on the return side of the firing roadway, there is a greater chance of an explosion and should be closed first. At the same time, pay attention to any faults, fracture zones, fissures, old roadways etc. around the firing roadway. If so, take extra care to prevent gas from coming out of these places and causing gas explosions or fires spreading around through these channels.

When a fire occurs in the middle of the excavation roadway and there are workers in distress at the working face, the mine rescue team, while fully engaged in extinguishing the fire, should open the compressed air pipe to increase the amount of compressed air or change the water pipe into a compressed air pipe to supply air to remove the gas and supply air in order to prolong their life saving time and prevent the accumulation of gas. However, care should also be taken to prevent gas discharge from causing an explosion. When carrying out rescue missions, the base of the shaft should be located in a safe area for the residual pressure of the blast wave. The minimum distance from the fire area is normally 500 m.

Where a fire occurs at the head of a horizontal, uphill or downhill working face to seize the initial moment, handling is relatively simple and the safety factor is relatively large. Generally maintain normal ventilation, gas gushing out with the flame and burning, not easy to constitute explosive conditions. The mine rescue team can easily extinguish such fires if they dare to break through the high temperature and smoke, using watering equipment and light fire-fighting equipment at the working face. In case of failure to extinguish, inert gas can be used to extinguish such fires when available. If there is no other good way to extinguish a fire, quickly close it and evacuate the fire area.

When a fire occurs in the excavation work, if the working face has been stopped under the conditions of ventilation, mine rescue work should first determine the explosive gases, according to the concentration of explosive gases to determine the course of action. If the concentration of explosive gases is low, the temperature is not high and the smoke is not large, entry reconnaissance can carry light fire-fighting equipment to extinguish the fire; when the temperature is high and the smoke is large but the concentration of explosive gases is not very large, ventilation can be restored to ventilate and cool down the fire, but measures to restore ventilation can only be implemented under conditions that confirm that no explosion can occur.

It is particularly important to emphasize that when dealing with fires in the uphill face, the residual fire source must be completely removed. Because the uphill head is most susceptible to gas build-up. Inadequate treatment of the ignition source can have unthinkable consequences. In the case of fires in downhill excavation workings, it is possible to remove the gas as it gushes out with the movement of the hot air stream.

[Case 1] Analysis of a fire caused by a cannon discharge at a digging face

A fire was caused in the excavation working surface of a mine due to a cannon discharge, as shown in Fig. 5.13. The mine is a high gas mine and workers reported

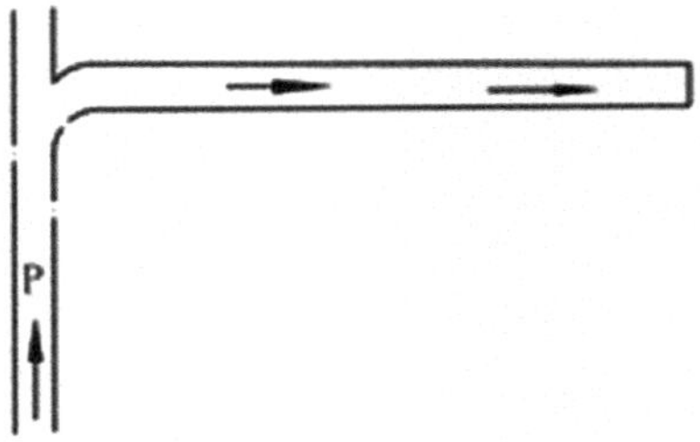

Fig. 5.13 Fire at a mine excavation face due to cannon discharge

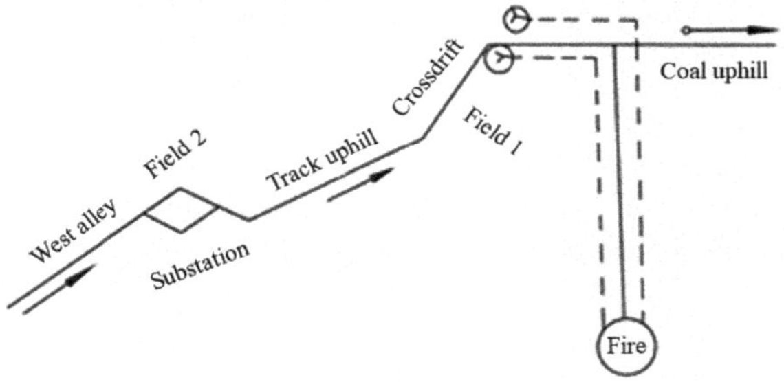

Fig. 5.14 Ventilation diagram during a fire

the flames at the working surface to the dispatch office immediately after the cannon discharge. As the flames spread against the wind stream, several sections of the air duct outlet burned and were damaged, causing a build-up of gas at the working face. When the ambulance team arrived at the accident site, an explosion occurred during fire-fighting killing a small party.

[Case 2] Analysis of successful start-up after partial fan closure of shutdown

As shown in Fig. 5.14, a case of zero-point ventilation (shutdown of local fan) was used to prompt a gas overrun to lose the fire.

[Course of the accident]

A fire was caused by a cannon discharge at the head of the downwind tunnel of a coal mining face in a wing of a high gas mine. 285 m of the tunnel had been constructed, with a section of 6 m^2, using two 28 kw local fans with 500 mm diameter fire retardant anti-static ducts pressed into the ventilation. The air volume at the head of the excavation was 200 m^3/min and the gas gushing volume was 3 m^3/min (obviously insufficient air volume), after the fire the workers all ran out without taking any fire fighting measures. However, an explosive box containing detonators and explosives for the shift was left 50 m from the head of the excavation. After the fire, the two fans did not stop and the smoke was so thick at the intersection of the excavation roadway and the return uphill that the scouts had to retreat after only 20 m of advance. It was not possible to enter the fire area and extinguish it directly.

[Accident management process]

It was decided to use a shutdown fan to stop the supply of air to the fire area, isolate the fire area so that the gas exceeds the upper explosive limit and the gas in the fire area reaches a loss of detonation before entering the fire area to extinguish the fire.

After 5 h reconnaissance check smoke disappeared, CH4 was 35% lost explosive. Temperature was 60–70 °C, digging headway has burned 35 m. After 7 days, the gas in the headway reached 100% and the temperature was 32 °C. Finally, measures were taken to drain the high concentration of gas and flush the working face and all the fire zone roadways with water. Finally, the confinement was removed and normal construction resumed. This was a case of stopping the air supply, causing oxygen depletion and gas loss in the fire area, and oxygen depletion and cooling to the point of extinguishing the fire, which provided experience in dealing with similar accidents.

[Case 3] Analysis of a local gas explosion caused by a fire at the digging head

[Course of the accident]

A mine excavation uphill caused a fire due to cannon discharge at the working face, as shown in Fig. 5.15; the mine is a high gas mine sharply inclined coal seam, the uphill eye and the return airway are 3–4 m apart and not dug through. When a fire was found at the working face after cannon discharge, personnel were quickly withdrawn, the local fan was put out of operation and reported to the duty officer in the dispatch room. The duty officer immediately informed the mine ambulance team to enter the well for reconnaissance, the head of the ambulance team together with the head of the supervision section led a small team to enter the well, upon arrival at the accident site, as the head of the supervision section had no breathing apparatus, he stayed at the fresh air flow of the local fan A point to stand by, the ambulance team hurriedly entered. When walking to the uphill entrance and was about to reconnoitre towards the uphill, a local gas explosion occurred.

When the blast shock-wave hit positively, all the personnel at point B were at the high step of the inner gang. As a result, the damage from the peak of the mechanical

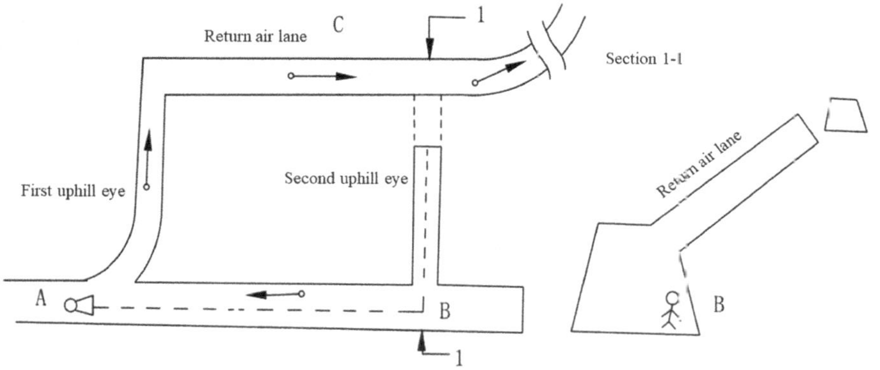

Fig. 5.15 Schematic diagram of the fire zone at the uphill working surface

shock-wave was avoided and the entire squad was injured. The men located at point A on the outside, however, were killed instantly on the spot. Several maintenance workers at point C were injured when the unexcavated portion of the excavation face and return airway was broken open by the explosion. The injured team members were protected from the poisonous gas as the explosion broke open the tunnel to form a complete ventilation system.

[Accident management process]

(1) Due to the destruction of the ventilation system caused by the explosion, allowing the excavation roadway to form an independent ventilation system with no more than 2% gas, direct fire fighting can be carried out on the upwind side of the fire source.
(2) Dry powder extinguishers and water can be used to extinguish the fire directly under ventilation. After extinguishing the fire, the ignition point must be carefully checked to prevent re-ignition.
(3) Monitor the roof, roadside gang, wind speed, methane and CO at all times during fire fighting and rescue.

[Lessons from the accident]

(1) In the event of a fire in a dug-in ventilation tunnel, direct fire extinguishing should, be carried out while maintaining normal local fan ventilation in principle. When it is not possible to extinguish the fire directly, the fire should be isolated and extinguished near the fire source.
(2) Solo-headed excavation roadway can avoid the harm of shock waves. As the concentration of harmful gases in the solo-headed roadway gradually increases, it is impossible to stay for a long time. After the first wave of impact, workers try to use self-rescuers as soon as possible to find a way to escape, so as to avoid casualties caused by poisonous gases and high temperature and the second shock wave.

[Case 4] Analysis of a fire at the downhill working face of a dugout

[Accident process]

A mine dug into the downhill roadway working surface caused a gas flaring accident due to a cannon discharge, as shown in Fig. 5.16, in which the gas ignited the wooden supports.

[Accident handling process]

In order to prevent the fire from expanding, the mine used boarded up walls to close off the entrance to the air shaft, thus causing the main and secondary slant holes to open up to high temperatures. In order to extinguish this fire, the local ambulance team entered from the secondary shaft and cut the pipeline at a small river to flood the fire from the surface water supply at point A. Due to the high temperature and large inclination, the ambulance crew could not physically support themselves while working and fainted when the small team retreated back, and the standby team entered

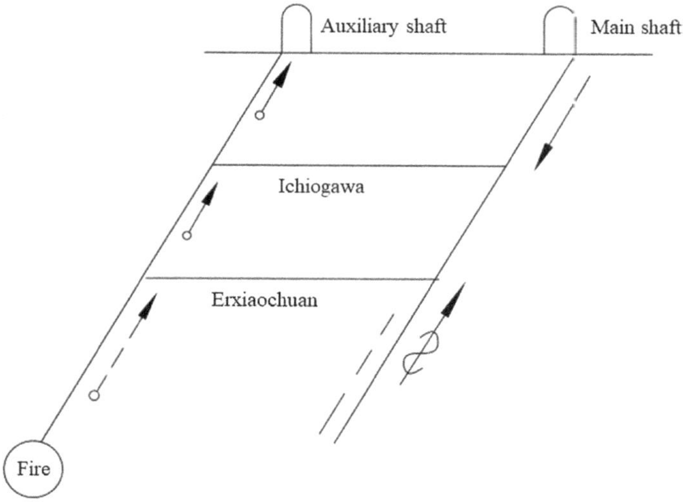

Fig. 5.16 Diagram of a fire at a sloping downhill excavation face

the shaft to rescue the same fainted. The nearby ambulance team arrived in time to take part in the rescue, and promptly opened the air shaft seal to restore ventilation and lower the temperature, which fortunately did not result in the death of the two squad members.

[Experience and lessons learned]

(1) When closing the fire area, keep the fire area as small as possible.
(2) Where available, inert gas can be delivered to extinguish such fires, and the delivery of inert gas should also prevent gas explosions from occurring.
(3) The temperature, gas concentration and roof conditions in the tunnel should be measured during the rescue. The rescuers should always pay attention to their physiological state to prevent fainting due to physical exhaustion.
(4) If the fire source is not very wide and located in the front of the working face, it can be considered to arrange the pipeline and directly use the water flooding method to eliminate the fire source.

[Case 5] Example of a fire in the middle of the excavation roadway

[Accident process]

In the process of the lower stratified excavation roadway, the regenerated roof was poorly glued from the outer to the inner part of the upper stratified layer at about 120 m due to poor closure of the return side of the upper and lower stratified layers, resulting in a collusion with the upper stratified mined-out area. In addition, the positive pressure ventilation in the excavation supplied oxygen to the mined-out

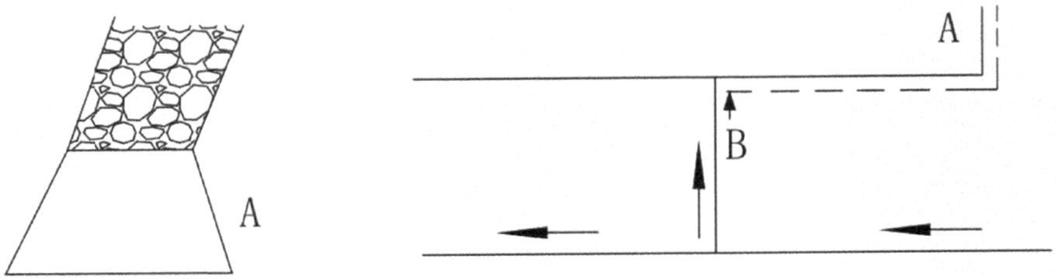

Fig. 5.17 Diagram of the fire in the middle of the excavation roadway

area, causing the floating coal in the roof of the mined-out area to spontaneously combust, as shown in Fig. 5.17.

[Disposal process]

The mine ambulance team used water jets to extinguish the fire ineffectively, personnel were withdrawn and the local fan operation was stopped. When the mine decided to recover the equipment without ventilation, the ambulance team arrived at the scene, it was very difficult to work with breathing apparatus and individual team members forced to start the fan. As a result, the gas gathered at the head of the excavation, in the process of removing the fire detonated.

[Experience and lessons learned]

(1) Positive pressure ventilation poses some problems for coal spontaneous combustion control.

① Fire spread has the characteristic of spreading in the reverse direction by absorbing oxygen along the upwind side of the more adequately supplied fire source, so that fires in the small mine mined-out area spread in the reverse direction to the adjacent large mine mined-out area, causing fires in the large mine mined-out area.
② In the case of spontaneous combustion in the mined-out area of the lower stratum, the positive pressure ventilation presses a large amount of CO generated into the upper stratum, and the CO concentration in the corner of the working face of the lower stratum or in the return airway is low, making it difficult to give timely warning and prone to sudden fires.

(2) Generally maintain normal ventilation, gas gushing out with the flame and burning, not easy to constitute explosive conditions.

When the local fan is stopped, gas tends to accumulate and with a high temperature fire source is prone to gas explosion. Therefore, direct fire extinguishing should be carried out while maintaining normal ventilation of the local fan in principle.

[Case 6] Analysis of spontaneous combustion in the middle of the downhill working face

[Course of the accident]

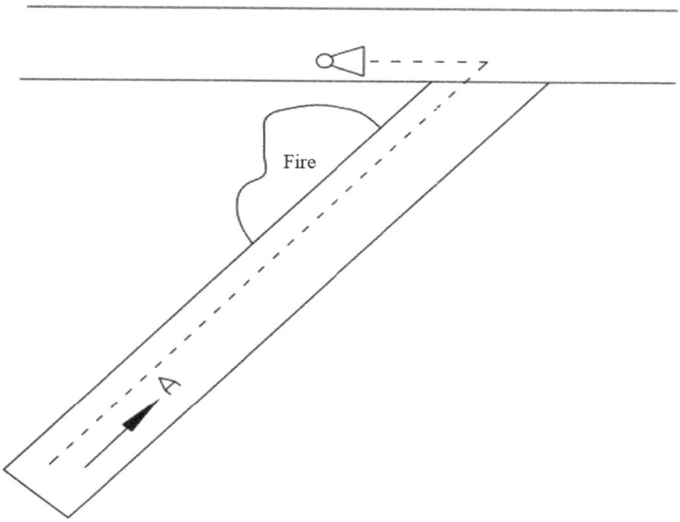

Fig. 5.18 Diagram of a high top fire in the middle roadway of the dug downhill

A mine was digging into the downhill face when a natural fire of floating coal occurred at the top of the middle roadway riser. This is shown in Fig. 5.18

[Disaster relief process]

While the rescue team was taking measures to extinguish the fire, the duct was disconnected in the middle of point A below the flames and was not detected by the fire-fighting crew. When another rescue team went down to reconnoitre, the duct was found to be disconnected and the duct was connected, at which point the gas stored in the working area was discharged with the wind stream and came into contact with the fire causing a local gas explosion.

[Lessons learned]

(1) In the event of spontaneous combustion fires in the middle of the excavation roadway and in the waste roadway, the fire area should be closed by filling with river sand, loess and fly ash. If this purpose cannot be achieved, the roadway should be closed quickly and methods such as pressing into inert gas, equalizing pressure and filling to extinguish the fire should be taken.
(2) When ventilation is resumed after the fan has ceased in the excavation roadway for a period of time, ensure that the source of the fire in the roadway has been extinguished.

[Case 7] Example analysis of insufficient fire extinguishing equipment

[Course of the accident]

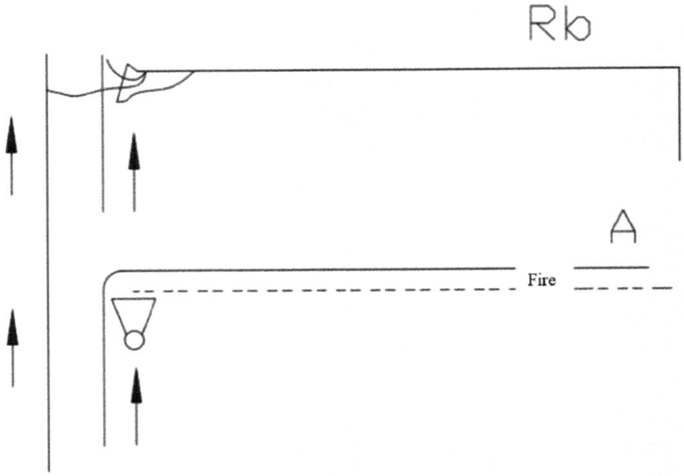

Fig. 5.19 Diagram of excavation roadway median fire

In a high gas mine, the excavation roadway was over 500 m long and the discharge caused gas flaring, which in turn ignited the duct 40 m from the head of the excavation, as shown in Fig. 5.19.

[Disposal process]

The underground workers immediately reported the accident to the mine's dispatching office. The comrade in charge called the ambulance team with light fire-fighting equipment to the underground to extinguish the fire and instructed that they should hold their positions underground and that the local fan should not stop running. The ambulance team arrived at the scene and immediately put out the fire. As the situation was unknown, the fire-fighting equipment was insufficient and a large amount of gas had accumulated at the burnt off duct, they also blindly carried out fire-fighting. An explosion occurred during the fire-fighting, resulting in many deaths and injuries to the ambulance team.

[Experience and lessons learned]

When fighting excavation roadway fires, due to the lack of human and material resources, the fire cannot be extinguished for a short period of time, the gas in the fire area gradually rises, the gas in the fire point cannot be detected, or the coal dust flying due to the burning of the coal roadway and falling should stop the direct fire fighting work and immediately take measures to close the fire.

[Case 8] Example analysis of a fire at the entrance of the excavation roadway

[Course of the accident]

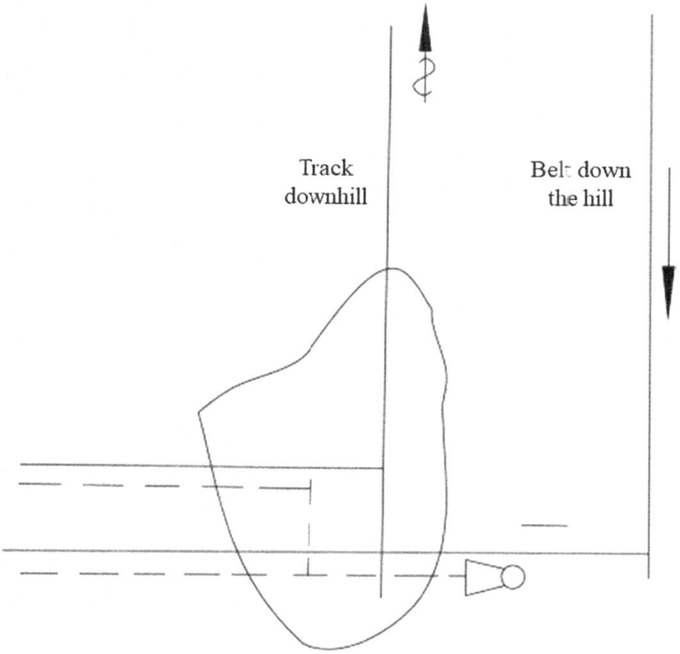

Fig. 5.20 Schematic diagram of the excavation roadway mouth fire

A mine excavation roadway entrance caused a fire due to a coiled stack of coal drill cables that were energized and heated up to produce electrical isolation, as shown in Fig. 5.20.

[Disposal process]

All workers on site were evacuated from the shaft after discovering the fire and did not report it to the relevant authorities, allowing the fire to spread into the track downhill fire.

As a result of the lack of active fire-fighting measures, gas accumulated inside the excavation roadway after the inlet side ducts burned off. When the gas build-up reached the explosion limit, a gas explosion and continuous explosion occurred, for which a long distance containment was forced and finally the flooding method was used to deal with the situation.

[Experience and lessons learned]

(1) In the early stages of a fire in a dug-out roadway, it should be normal in the local fan to carry out direct fire-fighting and try to put out the fire.
(2) If the initial fire extinguishing is a failure, isolation anti-extinguishing should be carried out and measures such as nitrogen injection anti-extinguishing should be carried out in confined areas.

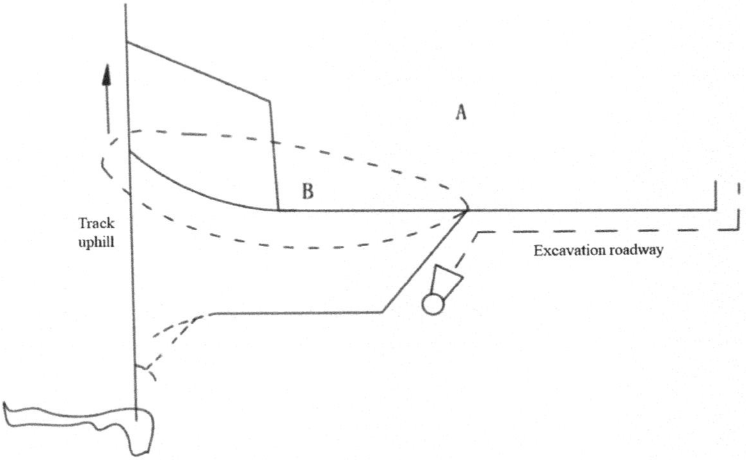

Fig. 5.21 Schematic diagram of excavation roadway crossing

[Case 9] Analysis of a fire caused by a short circuit in the cable at the local fan at the entrance of the tunnel

[Course of the accident]

A mine is a low gas mine due to a cable short-circuit fire at the local fan at the entrance of the roadway, igniting the roadway crossing, the fire burned down the wind stream and the ducts were burnt off, as shown in Fig. 5.21. Excavation roadway had 8 workers working, one of them found the fire and ran out himself in a panic, lost his way and died at point B. The other 7 people took shelter in the roadway pending rescue.

[Disposal process]

The mine rescue team actively extinguished the fire and after more than 10 h the fire was suppressed and the mine rescue team entered and gave the distressed person wearing a 2 h breathing apparatus a safe withdrawal. The post-event investigation flames burning roadway depth of only 4–5 m, temperature and smoke intrusion 40–50 m.

[Case 10] Analysis of a fire caused by gas protrusion

[Accident process]

A high gas mine, due to a gas protrusion in the excavation roadway, caused a fire from an external open fire and three explosions, as shown in Fig. 5.22.

[Disposal of disaster relief process]

After a few days of hard work by the rescue team to extinguish the fire, according to the site inspection, oxygen was supplied at the intersection of point A, all combustible

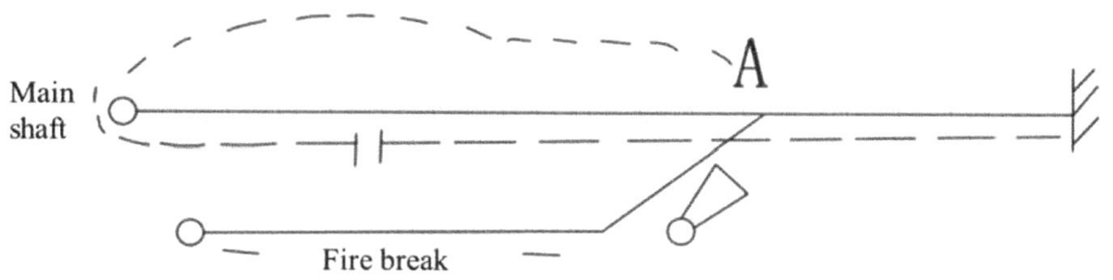

Fig. 5.22 Schematic diagram of excavation roadway crossing

material outside point A was burnt out, 30 m inside point A, and the temperature was lower than normal in the 45–50 m inside. In other words, the temperature in the high concentration gas filled tunnel was lower than the temperature in the rest of the tunnel, and the flame would not have burned farther in due to the formation of an oxygen poor zone in this tunnel.

[Accident experience and lessons learned]

This type of fire occurs at the entrance to a single headway, and the risk of causing a gas explosion is lower than that of a fire in an intermediate roadway and higher than that of a fire in a dug-out. Generally speaking, the gas gushing out from a fire in the roadway immediately participates in the combustion and does not accumulate to an explosive concentration. In contrast, a fire in the middle of the roadway is prone to gas build-up and detonation in its internal space. The fire at the mouth of the single-headed tunnel. Although the destruction of normal ventilation, but the accumulation of gas and detonation time is subject to the distance of the tunnel and the speed of gas gushing out two factors, characterized by the main ventilation tunnel close to the fire after receiving a thermal convection supply enough oxygen for combustion, so it spreads to the wind side of the combustion. When it reaches the ventilation tunnel, the fire takes advantage of the wind power and becomes a big fire. On the contrary, the airflow inside the fire point burns off the duct due to the fire at the entrance of the tunnel, breaking the normal ventilation of the digging work surface, for which the air forms a steady flow. The fire is close to the source of heat conduction and thermal radiation, a small amount of smoke and temperature to the inside of the slow conduction extension, this convection and conduction influence range of about 20–40 m. When the fire spreads inside, due to the oxygen supply is limited to its distance is only 20–30 m.

To illustrate this, the rescue of besieged persons in distress in the excavation roadway in low gas mines should be accelerated by fire-fighting, and the measures to organize entry for rescue and continued fire-fighting to prevent the expansion of the fire while ensuring that the entering personnel are not besieged by the flames are a double whammy.

When fighting such fires in high gas mines, the amount of gas gushing out and the total length, cross-sectional area and air volume of the tunnel should be calculated as well as the air supplied through a pressurized duct if necessary for the space in which people in distress will survive.

Excavation roadway occurs when gas protrusion causes gas combustion, when in the gas is greater than the air pressure in the roadway, gas ejected with the supply of air to burn; if the gas protrusion pressure is lower than the air pressure at the mouth of the roadway, the flame with the heat convection supply of air to the roadway spread. But this kind of fire generally does not cause long-distance burning, from practical experience, its maximum burning distance is about 20–30 m. Due to heat conduction and thermal radiation influence, distance is 20–30 m.

5.6.6 Handling of Fires in Coal Mining Working Surface

(i) **Treatment Techniques and Principles of Disaster Relief**

There is a large amount of combustible material in the back mining face, most of which is in the incoming air stream, and once a fire breaks out, the fire will rapidly develop in the direction of the working face. It often burns the roadway support, causing the roadway to collapse, so it is difficult to approach the fire from the outside of the fire source, for direct fire extinguishing. The "two lines" of the backing face are also prone to spontaneous coal combustion. Once spontaneous combustion occurs, it is bound to affect production and even cause casualties. When dealing with coal mining fires, it is very easy to cause gas and coal dust explosions. Therefore, the following methods should be followed to deal with working face fires.

(1) Wind flow control

① Fires should generally be extinguished under normal ventilation, and when there is a source of gas gushing upwind of the source of the fire to avoid the accumulation of gas causing a gas explosion, normal ventilation should be maintained as far as possible.
② Increase the air volume at the working face when gas flaring occurs, but note that as the air volume increases. The negative pressure decreases and care should be taken to prevent gas gushing out of the mined-out area.
③ Do not open and close the ventilation door on the return side at will during the handling of gas flaring to prevent the air from being impelled and transformed into an explosion.
④ When methods such as short-circuiting the airflow or closing the fire zone are necessary to control or reduce the development of the fire and to extinguish the fire close to its source, the gas should be diverted to the side airway or isolated from the fire passage as far as possible.

(2) Direct fire-fighting measures

① After evacuating personnel, first take direct fire-fighting methods from the inlet side: use fire extinguishers and water pipes for dust control, grouting and filling to extinguish the fire, or use high bubbles or inert gas to extinguish the fire if the fire source cannot be approached.
② When it is difficult to work from the inlet side, a partial back-draft may be used, with a water curtain on the inlet side followed by a back-draft.
③ In case of a fire at the working face of a sharply inclined coal seam, it is not allowed to extinguish the fire above the fire source to prevent water vapour from injuring people; nor is it allowed to extinguish the fire below the fire source to prevent the fire area from collapsing and injuring people; the fire source should be approached from the middle roadway or mined-out area direction when available.
④ Fires at the working face, especially the residual source of fire after gas combustion is mostly in the floating coal, which is very difficult to detect, mistakenly believing that the fire has been extinguished and easily triggering secondary combustion or gas explosion.
⑤ When the fire area is too large for direct extinguishing, the fire area can be closed off first and then treated by a combination of means when the fire has abated.

(3) Isolated fire extinguishing

When mine fire has developed to the point where it cannot be extinguished directly, it should be quickly isolated, i.e. the air supply to the fire area should be cut off by quickly constructing a confining wall in the tunnel leading to the fire area.

When sealing the fire area and extinguishing it with isolation and combined methods, the time difference between the reduction in air volume during the sealing process and the increase in gas volume should be analyzed to ensure safe working.

(4) Other safety and security measures

① When a fire breaks out in the coal mining face of a sharply inclined seam, the fire must not be extinguished above the source of the fire to prevent water vapour from injuring people; nor must it be extinguished below the source of the fire to prevent falling objects from injuring people in the extinguishing area, but from the side, i.e. the coal wall of the working face or in the direction of the mined-out area, using the protective table and protective cover to approach the source of the fire.
② When extinguishing gas burns in the upper corner, be careful not to force the flames into the mined-out area.
③ In the event of a fire at a coal mining face with downwards ventilation, effective measures should be taken, such as increasing the air flow at the face and keeping the return air system open, to avoid reversal of the air flow.

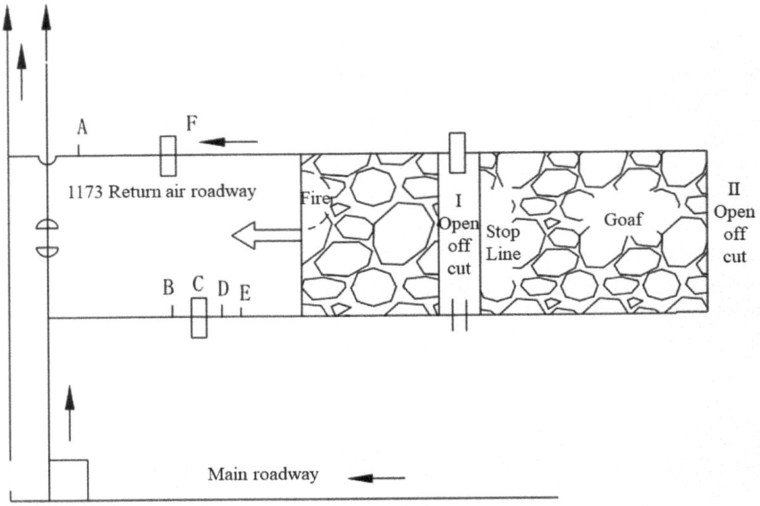

Fig. 5.23 Simplified diagram of the fire incident roadway at mining face 1173II

[Case 1] Example analysis of high bubble successfully extinguishing an externally caused fire

[Accident situation]

As shown in Fig. 5.23, a mine with coal and gas prominence has a relative gas outflow of 61.7 m^3/t. The working face is 240 m in strike and 60 m in inclination, with a 2.4 m mining height mono-hydraulic support, metal articulated roof beams, and cannon drop coal, with an airflow of 563 m^3/min. 1173II mining face has an external fire caused by gas combustion. CH_4 is 3% and CO is 1000 PPm at point A. The temperature is around 50 °C.

[Disposal measures]

(1) First seal 1173 into fan alley and build containment E in order to fire high bubbles to the source of the fire at the mining face.
(2) At the same time, set up an anti-aircraft fire extinguisher in position B to foam the fire area and then close the 1173 return roadway after the fire area has been stabilized.
(3) Ensure that the oxygen supply to the fire area is minimized after closure. 1173 into fan alley confinement with sack construction sandbag wall and a wooden section of yellow mud wall outside it to confine D. A total of 350 kg of anti-aircraft gun chemicals were fired for a total firing time of 4 h 26 min.
(4) Further reduce the oxygen supply to the fire area, it was decided to build an additional brick confinement C outside the D wood section confinement.

(5) Monitor on the return air side: increasing CO_2 concentrations, CH_4 concentrations have exceeded the upper explosion limit. CO concentrations and temperatures have decreased.
(6) Finally, where safe to do so, a wooden section of yellow clay containment F is built in the 1173 return roadway. The fire area is then closed off for treatment.

[Discussion]

This high foam fire extinguishing method has a simple process, a large safety factor, a short foaming time, a good fire extinguishing effect, a fast speed and an easy opening and closing to restore ventilation.

[Case 2] Analysis of Example 2

[Accident process]

As shown in Fig. 5.24, in a coal mining face of a mine shaft 1366, the iron wind pipe was buried by the collapsed gangue overlay. When the iron wind pipe was dragged with the pillar puller, sparks were generated by the friction between the pillar puller hook and the iron wind pipe, igniting the gas, which then caused the burning of the wood shed, bar piece and coal, resulting in a serious exogenous fire.

The coal seam is 2.75 m thick on average, with an inclination of 24–26°, and is prone to spontaneous combustion. The working face has a strike of 419 m and an average inclination of 96 m, with wind picks falling on the coal, metal friction pillars, articulated roof racks and barred sheet backing. At the time of the fire, the working face had advanced 88 m, 27 m from the small uphill, the air supply to the working

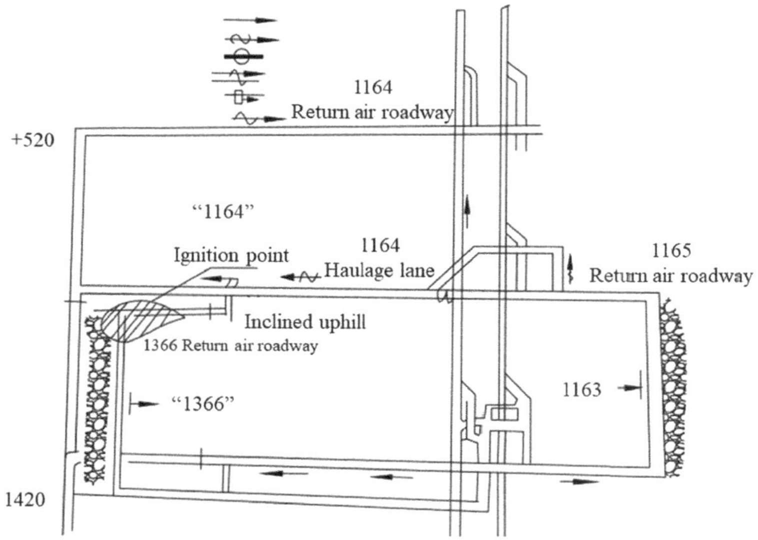

Fig. 5.24 Diagram of the 1366 fire incident in a shaft of a mine

face was 450 m^3/min, and the gas in the return airway often exceeded the limit. For this reason, a local fan was installed in the upper corner to draw out the gas.

[Disposal process]

The incident management programme was based on direct fire fighting and preparation of the fire area for closure. It was carried out in six stages over a total of six days, eight hours and thirty-one minutes.

Stage 1 (at 13:44–20:00 on 7 January)

(1) Removal and posting of personnel to stand guard at access points and normal operation of the fan.
(2) Carry a fire extinguisher into the well to extinguish the fire directly, as the action was slow and the fire had already spread, losing a favourable time for the initial extinguishing.
(3) Use hang windbreak to change the air path
(4) Power failure, cutting cables with electricity without the consent of the leading group, creating sparks and a near accident.

Stage 2 (at 20:00 on 7 January to 11:00 on 8 January)

The water was filled from the return side, as the water pipe was not put into position, the water flowed backwards for nearly 1:20 before it was put into its intended position until water flowed out of the lower down-chute. At this time, the upper part of the return wind chute was white with smoke and large water vapour, CH_4 was 10%; the lower part had no smoke and CH_4 was 0.14%.

Stage 3 (10:00 on 8 January to 10:00 on 9 January)

Due to the increasing gas, in order to reduce the gas to avoid accidents, take measures to increase the amount of wind, from 10:00 on 8 January to 00:53 in the morning to remove a total of CH_4 504.63 m^3, back to the wind tunnel smoke from white to yellow, so stop increasing wind.

Stage 4: (at 10:00 to 21:25 on 9 January)

In order to thoroughly clarify the fire situation and provide conditions for the rescue team to extinguish the fire, the working face was reversed and the ascending airflow was changed to descending airflow, while it was decided that the whole well would be evacuated and the rescue base was evacuated from +442 to +335 level. After reversing the airflow, the fire was extinguished by river sand in the 13 m tunnel. However, it was not possible to move forward to extinguish the fire because the wooden support was burnt off and the roof was falling.

Stage 5 (at 21:25 on 9 January to 09:20 on 10 January)

It is no longer possible to enter from the return wind to extinguish the fire, and the back-wind is long and may lead the fire to the whole working face. So at 21:25 on 9 January, it was decided to resume normal ventilation. Before resuming ventilation, the river sand was laid and a water curtain was set up. At 05:45 on 10 January, the wind was restored. At 09:20, the wind was restored to its original path.

Stage 6 (at 9:20 on 10 January to 1:35 pm on 14 January)

After ventilation was restored, it was originally intended to clear the roadway from the incoming direction and then flood the return roadway. However, due to several roof rises on the working face, the ventilation was blocked and the oxygen content was reduced. It was difficult to continue clearing the roadway, and if we delayed, the oxygen level might increase and lead to a gas explosion. Therefore, it was decided to take the final measure of "closing the working face". A 5 m water bag wall was built in each of the upper and lower chutes, and then a yellow mud wall of 2 m thickness at the bottom and 1.8 m thickness at the top. The sealing was carried out simultaneously. After the closure, in order to prevent accidents, another brick seal was built in the return wind chute on 16 January for permanent closure.

[Experience and lessons learned]

1. Delay in rescue and relief operations, missing the opportunity to fight the initial fire.
2. The explosion-proof switch was not completely closed and the cable was cut, which fortunately did not cause an electric shock or gas explosion.
3. The commanding officer did not go deep enough into the site, the decision was not decisive, the site was not well organized, there was not enough containment material, work was stopped for material and the plan was not fully implemented.

5.6.7 Wellhead Building Fires

(i) **Treatment Techniques and Principles of Disaster Relief**

A fire in the vicinity of the inlet shaft is bound to produce a large amount of smoke and harmful gases. These gases are likely to enter the shaft due to the fan power of the mine and pose a direct threat to mine safety and the lives of workers. In the event of a fire in a building at the entrance to the shaft, measures should be taken to prevent fire gases and smoke from entering the shaft: ① quickly extinguish the source of the fire; ② immediately reverse the wind flow or close the fire door at the shaft, and stop the main fan if necessary. For example, the Coal Mine Safety Regulations state that "wood yards, gangue hills and furnace ash yards shall not be less than 80 m from the inlet shaft." Then, its regulations state that "permanent derricks and shaft-head houses in new mines, with the joint building centred on the shaft-head, must be constructed with non-combustible materials."

(1) Wind flow control measures

Close the fireproof iron door of the inlet shaft to cover the shaft, install a temporary seal, reverse the wind flow of the main fan or short-circuit the wind flow, or stop the main fan from running, etc. to prevent burning fumes from invading the shaft;

in the event of a fire in the inlet shaft building, measures should be taken to prevent fire gases and flames from invading the shaft, i.e.: the wind flow should be reversed immediately or the fireproof door of the shaft should be closed, and the fan should be stopped if necessary.

(2) Direct fire extinguishing

After evacuating personnel, first use direct fire extinguishing methods from the inlet side. Use fire extinguishers and water pipes for dust control, grouting and filling to extinguish the fire, or use high bubbles or inert gas to extinguish the fire if the fire source cannot be approached. If it is difficult to work from the inlet side, a partial back-draft can be used, where a water curtain is first set up on the inlet side before the back-draft.

(3) Isolate fire extinguishing

The relatively large size of the building fire space makes it impossible to implement isolation fire-fighting technical measures normally.

(4) Other safety and security measures

① Guiding underground personnel out of the shaft.
② The need to wear an oxygen breathing apparatus when fighting a fire at the surface of a wellhead.
③ When fighting a fire in a shaft building, the fire brigade should be called in promptly, while the ambulance team should assist the fire brigade in extinguishing the fire.

[Case 1] Analysis of a fire accident caused by a short circuit in the cable of a simple coal processing plant on the ground

On 21 August 1958, a fire broke out in the Longfeng Mine of the Fushun Mining Bureau in a simple coal preparation plant on the ground (built with combustible timber) due to a short circuit in the cables. The fire was not extinguished in time and was not effectively controlled. The fire spread rapidly on the wooden coal processing trestle and on combustible materials such as coal dust and cables. The fire went straight to the main air intake shaft, which was the main hoisting shaft and air intake shaft with an air intake of 12,000–13000 m^3/min. A large amount of harmful and toxic gases from the fire could enter the shaft at any time, and the lives of thousands of workers underground were greatly threatened. The 64 m high hoist house and the washing equipment with a capacity of 2.4Mt/a were also at risk of being burnt by the fire. A blast was used to break the connection between the raw coal trestle and the inlet shaft of the simple coal processing plant, preventing the flames from continuing to burn and avoiding a mine fire.

5.6.8 Treatment of Fires in the Mined-Out Area

Fires in the mined-out area are generally endogenous, i.e. caused by the spontaneous combustion of coal. The risk of gas explosion is greater because of the tendency for gas to accumulate in the mined-out area, and the spontaneous combustion of the coal produces carbon monoxide that can endanger the safety of the operator.

(i) **Treatment Techniques and Principles of Disaster Relief**

Coal spontaneous combustion prevention starts from reducing the accumulation of broken coal, mainly by determining reasonable pioneering mining methods, reducing coal loss and improving the coal recovery rate. Starting from improving heat storage conditions, the main method is to carry out mine ventilation so that the heat generated during the self-heating process of coal is less likely to accumulate. Starting from the destruction of oxygen supply conditions is the most important method to prevent endogenous fires, mainly to eliminate air leakage or avoid coal-oxygen contact, the measures include isolation of mined-out area, equal pressure ventilation, yellow mud grouting, pressure injection of resistances, pressure injection of gel, injection of inert gas, etc.

(1) Wind flow control

Mined-out area wind flow control mainly refers to the control of air leakage, control to the mined-out area leakage methods is filling mined-out area, plugging method, equal pressure fire prevention methods. ① The best way to reduce wind leakage is to use all the filling method to fill the mined-out area; ② Plugging anti-extinguishing technology mainly refers to mined-out area sandbags in the upper and lower corner at certain intervals to seal or mined-out area plugging materials to seal the mined-out area wind leakage location cracks, etc.; ③ Even pressure anti-extinguishing technology refers to through fan or wind window, or adjust the ventilation system, so that the pressure between the source of air leakage sink to maintain a balance.

(2) Direct fire extinguishing

In the event of a spontaneous combustion fire in the mined-out area, liquid inert gas can be injected into the mined-out area, or a large quantity of slurry can be injected into the mined-out area.

(3) Isolate fire extinguishing

When the fire in the mined-out area cannot be controlled and starts to spread towards the ends of the working face, the mined-out area needs to be enclosed and isolated to extinguish the fire.

When mine fire has developed to the point where it cannot be extinguished directly, it should be quickly isolated, i.e. the air supply to the fire area should be cut off by quickly constructing a confining wall in the tunnel leading to the fire area.

When sealing the fire area and extinguishing it with isolation and combined methods, the time difference between the reduction in air volume during the sealing process and the increase in gas volume should be analyzed to ensure safe working.

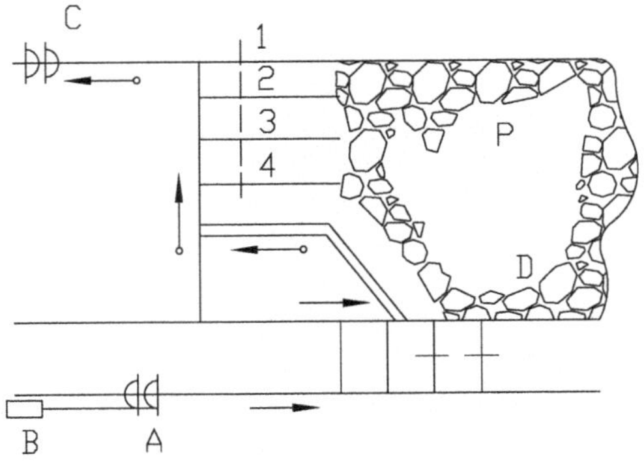

-regulation of ventilation door; P - fire source; D - mined-out area

Fig. 5.25 Uniform pressure to extinguish a mined-out area fire

(4) Other safety and security measures

Strengthen fire monitoring to prevent mined-out area fires from causing gas explosions.

[Case 1] Analysis of a fire caused by spontaneous combustion in a mined-out area

[Accident situation]

Figure 5.25 shows a schematic diagram of a fire caused by spontaneous combustion in the mined-out area of a mine. The carbon monoxide content of the airflow on the return side is 0.004%, the gas content is 0.45% and the temperature has reached 31 °C, proving that a fire has occurred in the mined-out area.

[Disposal method]

Firstly, the containment was put in place in roadways 1, 2, 3 and 4, but the carbon monoxide outside the containment wall still exceeded the permissible value and normal production could not take place. This proves that the fire in the mined-out area is still not extinguished and other methods of extinguishing the fire are required.

Considering the actual situation of the mined-out area, it was decided to use equal pressure fire extinguishing. Firstly, two ventilation doors were installed at A in the inlet roadway, a 28 kW local fan was installed at B and two regulating ventilation doors were installed at C in the return roadway to adjust the air pressure in the two roadways.

[Experience and lessons learned]

(1) Using equal pressure fire extinguishing, if the CO concentration is small does not indicate that the fire in the mined-out area has been extinguished, must accelerate the rate of advance, so that the source of the fire buried in the asphyxiation zone.

(2) In mines with complex air leakage or mines where the source of air leakage is difficult to identify, it is recommended that equal pressure fire prevention techniques are not used as far as possible.
(3) In the event of a spontaneous combustion fire in the mined-out area, liquid inert gas can be injected into the mined-out area or a large quantity of slurry can be injected into the mined-out area.

5.6.9 *Intake Shaft Fire Handling*

(i) **Treatment Techniques and Principles of Disaster Relief**

When a fire breaks out, the danger to the whole mine is relatively high. If the fire is large, the large amount of toxic and harmful gases produced will reach the various operating locations down-hole directly with the wind flow, causing poisoning to personnel, while the fire wind pressure generated will also directly affect the ventilation system of the whole mine.

The following methods can be used to deal with fires in the intake shaft in general.

(1) In the event of an intake shaft fire, the upwind side should be evacuated immediately to allow the main fan to back-wind.
(2) To close the fire-rated iron door and reduce the air supply if a back-draft cannot be achieved and it is not advisable to stop the main fan operation.
(3) Directing persons down the shaft along a safe route out of the shaft from another air intake or return shaft.
(4) When the slant hole is on fire, in the initial stage of combustion, the fire is not large, and it can enter the well-bore from top to bottom to explore and extinguish the fire. After the fire has increased, it is not possible to enter the shaft from top to bottom, but should be explored from bottom to top after a successful back-draft to extinguish the fire directly.
(5) In the event of a fire in a shaft, personnel must not be allowed to enter the shaft to extinguish the fire, and the shaft may be filled with high bubbles from a high bubble extinguisher at ground level.
(6) When ventilating a multi-wind shaft, the fan returning air to the shaft in the area of the fire shall not be shut down.

[Case 1] Case study of an accident during the restoration of mine ventilation

[Accident process]

(As shown in Fig. 5.26) It is an example of a foreign fire incident where the ventilation system was as shown. A year ago a fire broke out on the inlet side of working face I. The fire area was later closed with a containment and the main fan No. 1 was shut down; a year later, it was estimated that the fire had been extinguished and preparations were made to restore the fire area by opening the fire area containment. At the same time, fan No. 1 was restarted.

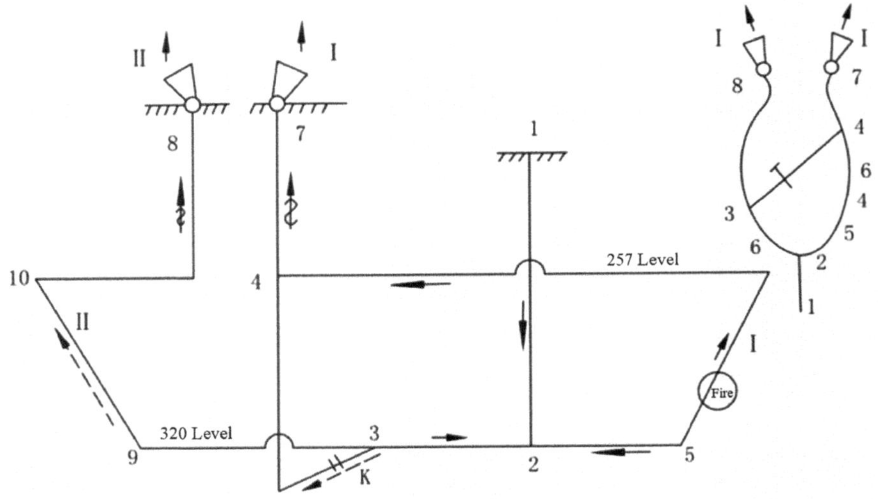

Fig. 5.26 Ventilation system diagram

[Incident handling]

However, the fire was not completely extinguished. When ventilation was restored, the fire area reignited shortly afterwards. The driver of No.1 fan at ground level found smoke flowing from the outlet air. Without receiving any report of the re-ignition of the fire area or any order to stop No.1 fan, the driver took the initiative to stop No.1 main fan when he saw smoke flowing from the outlet air. Soon afterwards, workers were poisoned at 320 level and at k ventilation door and in the inlet flat roadway of No.2 working face. The causes were analyzed as follows.

(1) With No. 1 main fan out of service and only No. 2 main fan operating alone, not only did air enter 1–2 at this time, but the return shaft of 4–7 also became incoming air. As a result, the fire gases in 5–6 had reversed into 5–2–3–9–10, poisoning the workers in that airflow.
(2) Because the fire occurred in the inclined roadway upward flow, when the fire wind pressure increases, from the diagram in the ventilation network schematic diagram, it can be seen that the wind flow in the corner joint air path 3–4 is easy to occur in the wind flow reverse by 4–3. At the same time, because k ventilation door is not strict, air leakage is larger, the same can make the toxic smoke flow from 4–3 successive leakage to 3–9–10 system.

[Experience and lessons learned]

How the main fan and wind flow is dispatched and controlled during the mine fire period is very important. If the wrong measures are taken, even for a short period of time, this can lead to a reversal of the wind flow, extending the scope of the disaster and causing unimaginably serious consequences. In addition, it is important to be unified in command and follow orders during the rescue and relief process to avoid mistakes.

5.6.10 Handling of Return Air Shaft and Big Roadway Fire

(i) **Return Air Shaft Fire Treatment Technology**

Although not likely to cause casualties, the return air shaft fire will affect the ventilation of the mine and the safety of the buildings and personnel at the entrance to the wind shaft.

(1) Return air shaft in case of fire, keeping the direction of wind flow unchanged.
(2) Control of the fire, the airflow may be reduced.
(3) Control of incoming fire doors, which may stop the main fan after the personnel have been evacuated.
(4) In the case of multi-fan ventilation in a multi-wind shaft, the main fan in the return shaft of the area where the fire occurred cannot ceasing ventilation.

(ii) **Treatment Technology of Return Airflow Big Roadway Fire**

In the event of a fire on the return airflow big roadway, the number of people at risk is small, so the fire should be dealt with by immediately evacuating the few people on the return airflow side and putting out the fire directly. The following points should be noted when extinguishing a fire.

(1) The direction of diversion cannot normally be changed, but the airflow may be reduced if necessary to control the fire.
(2) When there is a larger source of gas on the inlet side of the fire, it should be closed immediately to prevent an explosion from occurring when the gas concentration at the fire point increases.
(3) Immediately after the fire has been dealt with, the tunnel here should be cleared to maintain a proper mine ventilation system.

[Case 1] Case study of a fire accident caused by a fire in a big roadway cable

[Course of the accident]

At 20:30 on 15 March 2010, a cable fire occurred in the first bypass roadway of the west big roadway of the main shaft of Xinmi Dongxing Coal Industry Co., Ltd 31 people entered the well during the shift.

[Cause of accident]

(1) In early March 2010, during the technical reform of the Dongxing Coal Mine in Xinmi City, workers were privately arranged to enter the mine without a production safety permit or a coal production permit. The mine did not meet the safety production standards, and production was organized in violation of the law.

(2) A cable fire in the first contact roadway of the big roadway west of the mine rapidly expanded and ignited the roadway's wooden supports and coal seam, generating a large amount of carbon monoxide and other toxic and harmful gases, which entered the coal mining face along the incoming air stream, causing poisoning and asphyxiation.
(3) The management of Dongxing Coal Industry knew that the air compressors could not be used underground, but still used substandard air compressors for as long as three months. It was these long-term high temperature and high pressure operation of the air compressors that eventually led to a major underground fire accident.

[Precautions]

(1) The use of electrical equipment and materials etc. that do not meet the requirements of fire prevention and management is strictly prohibited underground.
(2) Regulations and rules to ensure safe and reliable operation of electrical equipment should be carefully implemented. Non-flame-retardant cables, belts, ducts and other materials should be prohibited from being used underground to prevent fire accidents.
(3) The reform of coal mining methods and support methods should be accelerated. The protection should be eliminated as soon as possible.
(4) Emergency drills such as anti-wind tests and disaster avoidance route drills should be organized as required to improve emergency response capabilities.

References

1. Lai KY. Correlation analysis and application of coal mine hazard warning parameters and influencing factors. Beijing: China University of Mining and Technology (Beijing); 2007.
2. Tan B. Research on dynamic simulation technology of mine fire based on physical field model. Beijing: China University of Mining and Technology (Beijing); 2007.

Chapter 6
Study of Gas Sampling Equipment and Methods in the Mine Fire Area

Bo Tan, Hongqing Zhu, and Liyang Gao

Manual sampling during the closure of the mine fire area is usually accompanied by differential pressure measurements inside and outside the confinement wall. The only way to help rescuers determine the state of development of the fire zone is to obtain gas samples that reflect the internal composition of the fire zone. Therefore, it is necessary to study and analyse the reasons for the fluctuation of the differential pressure inside and outside the confinement wall.

Gas transport in the mine fire area is influenced by both anthropogenic and natural factors. The human influenced factors mainly refer to the effect of inert injection on gas flow, while the natural factors include mechanical wind pressure, natural wind pressure and the thermal effect of the fire area.

During the closure of the mine fire area, as most of the mines have taken pressure equalization measures in the closed area to reduce air leakage from the fire zone, the pressure difference between the inlet tunnel of the closed fire zone and the outside of the confining wall of the return tunnel is small. It can be considered that the mechanical wind pressure is stable and its force does not cause fluctuations in the confining wall of the closed fire zone. The pressure difference between the inside and outside of the closed fire zone does not fluctuate. The irregular cyclical changes in atmospheric pressure at the surface tend to cause the gas in the closed fire zone to show an "expansion–contraction" breathing state, which makes the gas composition in a certain area inside the confinement wall susceptible to disturbance by external air leakage. Therefore, researchers generally consider atmospheric pressure changes

B. Tan (✉) · H. Zhu · L. Gao
School of Emergency Management and Safety Engineering, China University of Mining and Technology-Beijing, Beijing, China
e-mail: tanbo709@126.com

H. Zhu
e-mail: zhq@cumtb.edu.cn

L. Gao
e-mail: gaoliyangcc@163.com

B. Tan et al. (eds.), *Mine Fire Prevention and Control*,
https://doi.org/10.1007/978-981-96-0466-1_6

as the main reason for the fluctuation of pressure difference between the inside and outside of the closed fire zone [1].

6.1 Common Sampling and Testing Instruments for Gases in the Mine Fire Area

6.1.1 Portable Gas Detectors

Portable gas detectors are divided into pump and diffusion type according to the principle, and single gas detectors and composite gas detectors according to the type of gas detected. Pumped instruments have a built-in sampling pump, which allows the gas to be sucked directly to the sampling pump before detection, and are suitable for detecting gases in confined spaces or underground. Diffusion instruments are placed directly in the environment being tested to detect the gas and usually need to be carried by a person to detect it. Most are suitable for mine workers to carry, or for environments accessible to people. Coal mines are generally routinely equipped with portable gas detectors for portable CO, CH_4 detectors.

Portable mine gas detectors are often used in harsh environments and the instruments themselves are specially designed to be waterproof and dust-proof. The usual protection levels are IP65 to IP67 (this is the world's common industrial protection level, the last two digits indicate the level of dust and water resistance). IP65 means that the instrument is impervious to dust and protected against water injection; IP67 means that the instrument is impervious to dust and protected against water immersion. The protection level of a portable instrument depends on the sampling principle, with diffusion sampling being rated at IP67 and pumping products being at best IP65 as they are not completely waterproof due to the inhalation of the gas while sampling.

Portable gas detectors are generally used to detect combustible gases (methane, propane, ethane, butane and acetylene etc.), toxic gases (CO, hydrogen sulphide, ammonia etc.) and asphyxiating gases (carbon dioxide, nitrogen, hydrogen cyanide etc.).

Advantages of portable gas detectors: convenient, fast, flexible choice of measurement locations, possibility of recording data after measurement, therefore larger amounts can be measured; disadvantages: detection cannot be carried out continuously, hazardous environments are difficult to measure, the accuracy of measurement and the type of gas measured, the environment in which the instrument is used are limited.

6.1.2 Detection Tube

The detection tube (Fig. 6.1) is a tool for the rapid determination of airborne gas concentrations. Detection tube works by placing a fixed dose of detection agent inside a glass tube of a certain length and internal diameter, plugging it with an auxiliary material and finally sealing the two ports of the tube. The detector tube is then adsorbed onto the surface of a solid carrier particle. The choice of the detector and the chemical concentration of the detector allow the detection of different substances and their concentration levels.

Advantages of the detection tube: easy and fast, flexible choice of measurement location; disadvantages: continuous detection is not possible, hazardous environments are difficult to measure, the accuracy of the measurement and the type of gas to be measured as well as the environment in which it can be used are limited.

Depending on the actual needs, detection tubes have different measurement ranges and accuracies. The accuracy range of detection tubes produced by different manufacturers varies, and the commonly used gas detection ranges are shown in Table 6.1.

The detection tube should be used in conjunction with a manual gas sampler (Fig. 6.2), which should be flushed with air and tested for gas tightness before use. The gas sampler is used to take a certain amount of gas to be measured, the two ends of the detection tube are cut open and a short rubber tube is used to connect the end of the detection tube concentration scale "0" to the outlet of the sampler, the sampled gas passes through the detection tube at a constant speed according to the specified time. The gas reacts with the indicator gel and a coloured column appears.

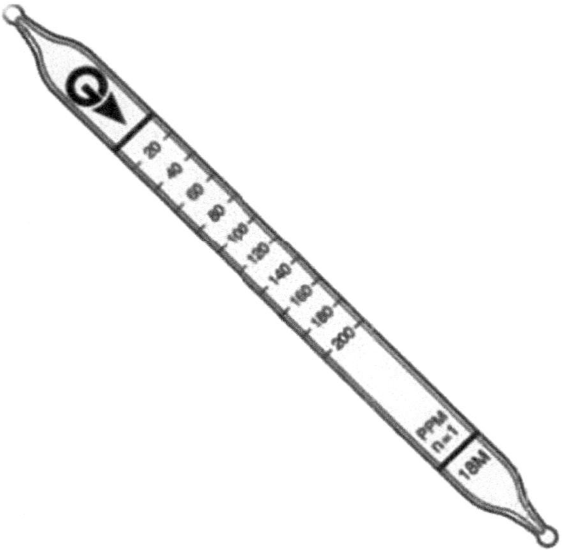

Fig. 6.1 Detection tube

Table 6.1 Common gas detection ranges

Test tube name	Unit	Measurement range
Carbon monoxide (CO)	%	0.00025–0.005
		0.001–0.05
		0.002–0.1
		0.1–5
		2–100
Carbon dioxide (CO_2)	%	0.03–0.5
		0.1–3
		0.5–20
		2–100
Hydrogen sulphide (H_2S)	%	0.0002–0.01
		0.002–0.1
		0.1–4
Hydrogen (H_2)	%	0.5–2
		0.5–4
Oxygen (O_2)	%	0.5–5
		1–20
		2–100

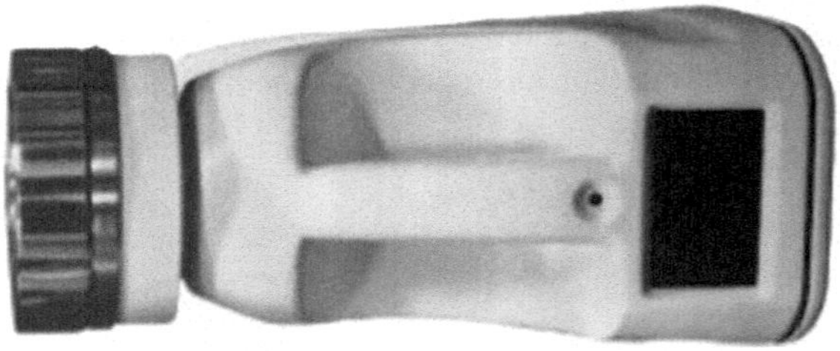

Fig. 6.2 Artificial gas sampler

The gas content can be read directly from the number indicated at the top of the column. It should also be noted that used detector tubes are corrosive and should not be discarded at will.

6.1.3 Optical Interference Gas Detector

The optical interference gas detector uses the principle that the difference in refractive index between the gas to be measured and the air to be measured creates an optical range difference and that the two beams of coherent light interfere in the space where they meet. To use the optical interference gas detector, it is first necessary to separate the underground gases. In the event of a fire, the composition of the

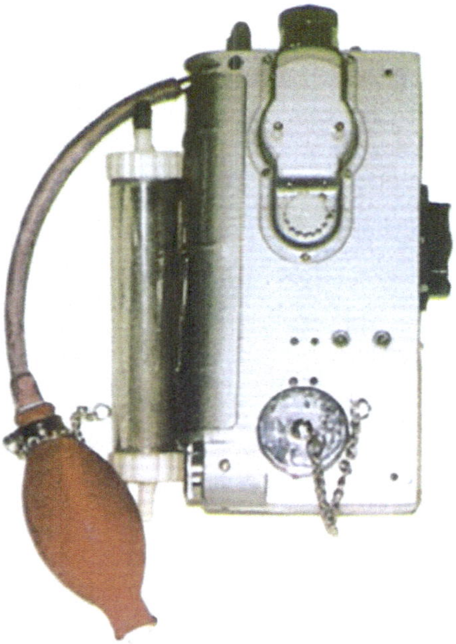

Fig. 6.3 Optical interference methanometer

gases in the underground mine will become even more complex, thus the use of the optical interference gas detector to detect the composition of the gases generated by a mine fire is extremely limited. Therefore, this instrument is basically not used. It is common in coal mines to use an optical interference methane detector (Fig. 6.3) to detect methane concentrations in mine air during normal production.

The use of the optical interference gas detector is introduced using methane as an example. The use of the instrument is divided into two main parts: preparation and carrying out the measurement.

(1) Preparation

① Loading: firstly, 2–3mm calcium chloride or silica gel, 4–5mm soda lime and other drugs are loaded into the absorber tube, where the calcium chloride or silica gel is assembled in the order of sponge pad → flower gasket → drugs (loaded to half way) → round gasket → drugs → flower gasket → sponge pad (the short absorption tube is filled with calcium chloride or silica gel, which is used to absorb moisture. The long absorption tube is filled with soda lime, which is used to absorb carbon dioxide. If the main measurement is methane and the moisture is large, it is preferable to have calcium chloride or silica gel in the long tube and soda lime in the short tube.).

② Check whether the balloon is leaking. Check method: the balloon on the rubber tube bending pinch the balloon after squashing the balloon, when loosened by the squashed balloon, the balloon expansion restore time of not less than 3min, that does not leak.

③ Check the smoothness of the methane gas system. Inspection method: use a non-leaking balloon to connect to the methane gas system, squash the balloon and release it immediately, the balloon will soon bulge up, indicating that the gas line is open.

④ Check the air leakage of the methane gas line. Check method: connect the gas line to a manometer of 700mm H_2O height for 1min. If the water column does not fall, it means the gas line is not leaking.

⑤ Clean the gas chamber with fresh air. The methane chamber must be cleaned before use in fresh air at a temperature close to that of the area of use (difference of no more than 5 °C). This will reduce streak shifts in the set zero position due to temperature changes.

⑥ Zero adjustment of the interference bar release.

(2) Conducting measurements

The measurement is carried out by extending the rubber tube connected to the gas sample population to the measurement site and then slowly holding the balloon under pressure five or six times so that the gas to be measured enters the gas sample chamber. Observe the interference fringes through the eyepiece and first read the number of fringes moving on the differentiation plate, e.g. read 3% if the fringes move to 3–4%. Then, turn the micrometer hand-wheel and back off the stripe selected for the zero position to the 3% mark, press the button on the micrometer illumination circuit and read the number on the dial. If the stripe is at 0.24–0.26%, it can be read as 0.25%. The result of this measurement is $3\% + 0.25\% = 3.25\%$. The dial should be returned to the zero position after the measurement.

There are a number of problems with the optical interference gas detector in its application. When using the optical interference gas detector to determine gas in a disaster area, special attention should be paid to the ambient oxygen concentration. If the oxygen concentration is too low, the instrument's readings will be inaccurate. Table 6.2 shows the gas concentration data from a mine fire using the optical gas detector readings and the gas concentration detected by sampling chromatography assays. From this set of data, it can be seen that there is a significant difference between the two types of detection data. In the standard condition, this is influenced by the oxygen concentration: for every 1% reduction in the oxygen concentration in the air, the optical gas identifier's gas dependence measurement is approximately 0.2% greater, which can be corrected for in the non-standard condition using the following equation.

$$C = 345.8T - V/P \tag{6.1}$$

In the Equation: C—true concentration.

V—concentration after correction with oxygen.

T—absolute temperature.

P—air pressure.

Table 6.2 Comparison of analytical checker results for fire gas chromatographs during the fire zone closure period

Sampling time	T/°C	O_2 concentration/%	CH_4 concentration/%		CO_2 concentration/%		Atmospheric pressure/Pa
			Chromatography	Checkers	Chromatography	Checkers	
22 June, 14:00	55	15.27	1.25	2.3	4.06	9	105,500
22nd June 22:00	50	11.10	2.12	4.0	6.16	8.4	106,500
23 June, 16:00	54	15.40	1.8	2.8	2.16	9	106,000
24 June, 16:00	62	10.40	1.23	3.2	6.85	7	102,500

6.1.4 Gas Chromatograph

Gas chromatography is widely used in various scientific research and gas analysis in many industries. It means that the stationary phase is a solid substance.

The chromatographic determination principle is based on the different partition coefficients between the mobile phase and stationary phase of the gas sample. After a certain length of time, the weaker adsorbed components are easily desorbed and are the first to leave the column and enter the detector, while the strongest adsorbed components are the least likely to be desorbed. It is therefore the last to leave the column. This allows the components to be separated from each other in the column and to leave the column in sequence into the detector (Fig. 6.4). The resulting ion flow signal is amplified and the peaks of the components are depicted on the recorder.

(1) System composition

The gas chromatograph has five main systems: ① the carrier gas system includes the gas source, gas purification, gas flow rate control and measurement; ② the injection system includes the injector, vapour chamber (capable of instantaneously vaporizing liquid samples into vapour); ③ the separation system mainly includes the column; ④ the temperature control system includes thermostatic control and low temperature device; ⑤ the detection and recording system includes the detector, amplifier, recorder, data processing unit, and workstation.

(2) Operational flow

In the meteorological chromatographic analyzer, the carrier gas is first released from the high-pressure cylinder, reduced to a certain pressure by means of a pressure reducing valve, and then purified and dried by means of a carrier gas purge and drying tube. Then, it flows through the pressure regulator valve and rotor flow meter, flows through the gasification chamber at a constant pressure and speed and mixes

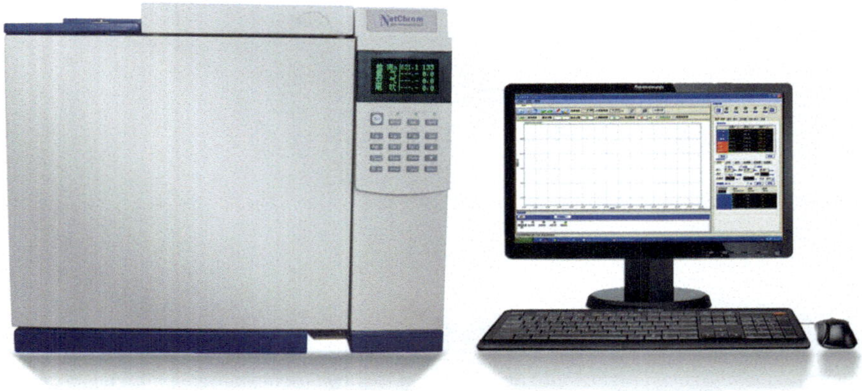

Fig. 6.4 Gas chromatograph

with the gasified sample, thus transferring the sample gas to the column for separation. The sample gases are transported to the column for separation. The separated gases are carried successively to the detector by the carrier gas, which is finally emptied. The detector converts the concentration or mass of the gas to be measured into an electrical signal, which is amplified and recorded using a recorder, thus obtaining the chromatographic outflow curve. Using the duration of the individual peaks on the chromatographic outflow curve, qualitative analysis can be done; using the peak area or the degree of peak height, quantitative analysis can be done. The gas chromatography process is shown in Fig. 6.5.

(3) Precautions for use

Care should be taken when using a gas chromatograph for gas sample analysis.

The gas chromatograph is calibrated with a standard component similar to the component of the gas sample to be analyzed.

① After the air compressor has been working continuously for 3 d, water should be released once, after the air compressor has finished working, with the pressure of water. At this time, there is a small amount of oil and water mixed liquid out is a normal phenomenon. When the water is released, connect the pipeline, put the water into the container, do not taint other facilities.
② When the liquid level in the hydrogen generator falls below half, distilled water should be added promptly.
③ The operator of the instrument must be well trained in the technique.
④ Special attention is required if CO is analyzed using a flame ionization detector or if concentrations of ethylene or methane of more than 10% are found in the analyzed gas sample.

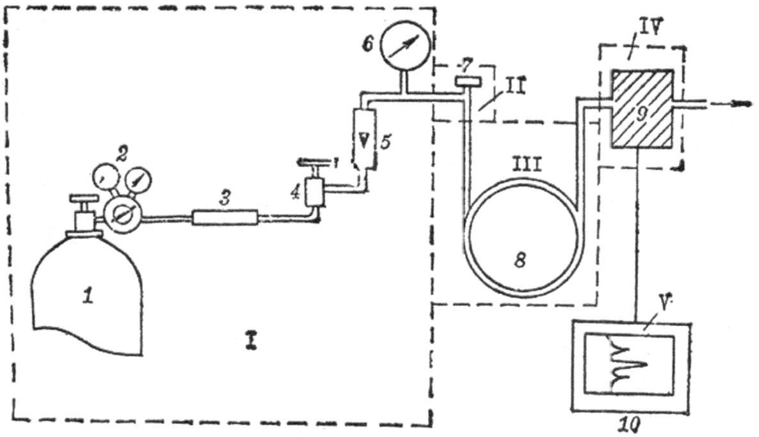

Fig. 6.5 Gas chromatography flow chart. 1—High pressure cylinder; 2—Pressure reducing valve; 3—Carrier gas purge and dry tube; 4—Needle valve; 5—Flow meter; 6—Pressure gauge; 7—Sampler; 8—Column; 9—Detector; 10—Logger

(4) Application of gas chromatograph in fire relief gas analysis

The gas chromatograph is one of the most accurate and reliable of all mine gas analysis instruments, especially when it comes to multi-component gas analysis. However, due to restrictions, the chromatograph is difficult to make into an explosion-proof or explosion-proof type for underground applications, which is therefore only used on the surface, limiting its use.

When using a gas chromatograph for the analysis of mine fire gases, the gas components O_2, N_2, CO, CO_2 and CH_4 can be measured within 3 min as well as the analysis of all gas components can be completed in about 15 min. The order of analysis of each gas component: ① O_2, N_2; ② CO, CO_2, CH_4, C_2H_6, C_2H_4, C_3H_8, C_3H_6 and C_2H_2.

When performing gas analysis, the possibility of error is higher if the composition of N_2 is determined by subtraction. In special cases, such as when analytical results are required as soon as possible, the concentration of Ar (argon) can be calculated by $\varphi_{Ar} = 12 \times 10^{-3}\ \varphi_{N2}$ in order to save time for the chromatographic analyzer to calculate the temperature, but it should also be noted the bellows. ① The concentrations of N_2 and O_2 in the gas sample to be measured; ② The concentration of 0_2 obtained from the analysis should be subtracted from the calculated Ar concentration, i.e. φ_{O2} (corrected value) = φ_{O2} (O_2 concentration obtained from the chromatographic analyzer)—φ_{Ar}. In the analysis of down-hole fire conditions, the N_2 concentration applied is actually the N_2 concentration + Ar concentration, so it saves time and avoids confusion to use the N_2 concentration + Ar concentration instead of the N_2 concentration in the analysis results.

6.1.5 *Infrared and Laser Gas Sensors*

When a substance is irradiated by infrared light, the molecules of the substance absorb some of the energy of the infrared light and convert it into the vibrational and rotational energy of their own molecules. In the absorption process, the vibration frequency of the molecules is closely related to their properties, and infrared radiation is only absorbed at wavelengths corresponding to the vibration frequency of the molecules.

For gases within the fire zone of a coal mine, if the absorption spectrum of a gas component is within the infrared human emission spectrum, then after the infrared light passes through the gas, an attenuation of the light intensity will occur at the corresponding harmonic line, and the absorption of infrared radiation by the gas follows the Lambert–Beer law, i.e.

$$I(\lambda)=I_0(\lambda)e^{-k(\lambda)LC} \tag{6.2}$$

In the Equation, I_0 (λ)—the intensity of the incident light.

I(λ)—the intensity of the transmitted light.

L—the thickness of the radiation passing through the gas layer.

C—the concentration of the gas being measured.

k(λ)—the absorption cross section, i.e. the absorption coefficient per gram of absorbing gas, as a function of wavelength λ.

As can be seen from the above equation, the concentration of a gas can be measured by the intensity of the incident and transmitted light. The selective absorption of infrared radiation by gas molecules is the theoretical and technical basis for infrared gas sensors.

For the non-split infrared analyzer, the optical system consists of three parts: the light source, the gas chamber and the detector; the electrical system consists of three parts: the preamplifier, the main amplifier, the temperature control and the power supply. The operating principle of the instrument is shown in Fig. 6.6.

The light source component modulates the continuous infrared radiation into a certain frequency of intermittent radiation, the radiation passes through the analysis side of the gas chamber and the reference side of the gas chamber. When there is no component to be measured in the analysis side of the gas chamber, the detector senses the reference side and the analysis side of the infrared radiation energy equal, the output signal is zero. The small current is transformed into a strong DC signal proportional to the concentration of the component to be measured by the preamplifier and main amplifier through impedance conversion, preamplification, main amplification, frequency selection, phase sensitive detector and filtering. The use of infrared and laser gas sensors for monitoring of fire zone gases in coal mines has the following advantages.

(1) Capable of measuring a wide range of gases. The vast majority of inorganic polyatomic molecules and organic molecular gases can be detected with optical gas sensors.
(2) Large measuring range. The upper limit of gas concentration can be as high as 100% and the lower limit as low as 1×10^{-6}. After processing, it is even possible to perform a 10–9 level of detection.
(3) Good sensitivity. It has a very good sensitivity and very small changes can be detected.
(4) High accuracy. Its accuracy is basically in the range of 3–5, and the best international instruments are available in the range of 2 and better than 1.
(5) Fast response. The time to react to 90% of the value is generally within 10 s, which is a significant advantage over other analytical instruments.
(6) Good selectivity. It has excellent selectivity, especially suitable for the analysis of a component to be measured in a multi-component gas mixture. When the concentration of one or several components in the gas mixture changes, it does not affect the measurement of the concentration of the component to be measured, which is a clear advantage compared with other analytical instruments.
(7) Capable of continuous analysis and automatic control. It is an automatic instrument with continuous feeding, continuous measurement and continuous display, and can continuously detect the gas concentration for a long time.

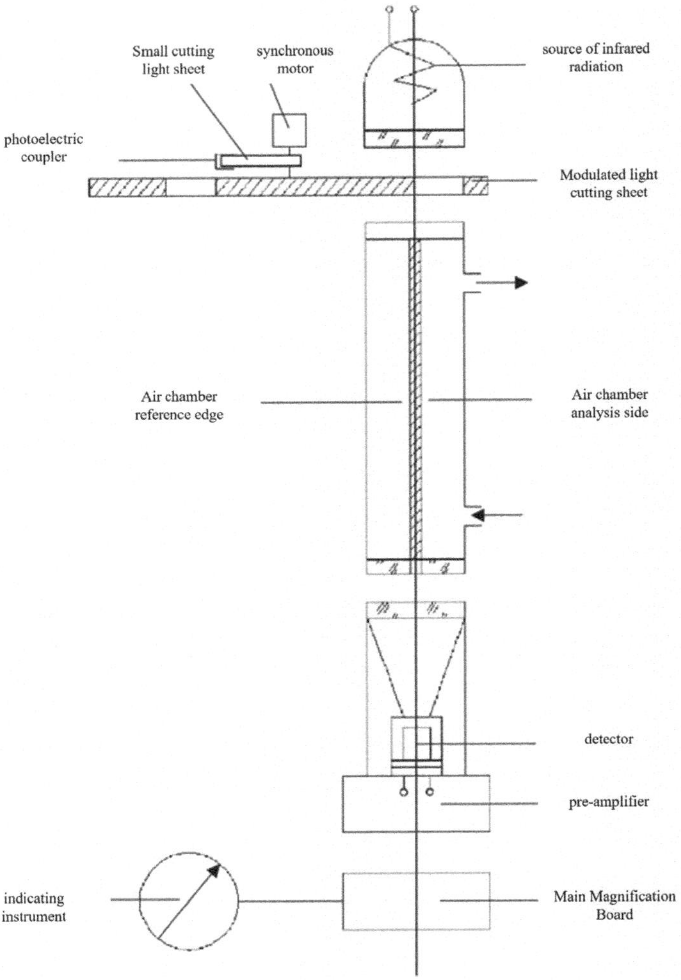

Fig. 6.6 Working principle of the infrared analyzer

(8) Non-poisoning. It has almost no "toxic" effect from harmful gaseous chemicals.
(9) Simple operation, easy maintenance and long life. The service life is basically more than 5 years.

6.1.6 Beam Tube Continuous Gas Analysis System

Continuous monitoring of mine fire area gas has the following advantages. It avoids sampling errors caused by human error and enables timely determination of changes in gas concentrations; it enables effective analysis of the influence of the external environment (changes in atmospheric pressure, etc.) on the concentration of fire zone gas; it enables the analysis of the correlation between the indicator gases of the fire zone, the analysis of the combustion state of the fire source and the analysis of the trend of the explosion hazard of the fire zone.

Earlier bundle tube continuous gas analysis systems were used to extract gas from a fixed location down-hole using a laid sampling tube path through a pumping unit set up on the well and then analyzed using a gas analysis instrument on the well to achieve the detection of gas from a fixed location down-hole. The disadvantage of this method is that although the system enables continuous sampling of down-hole gas, the gas analysis cycle is extended due to the long gas sampling tube path and the gas chromatograph takes 15 min to analyse each gas sample, which increases the measurement cycle when there are too many measurement points down-hole, making it difficult to guarantee the continuity of gas analysis. It is difficult to ensure the continuity of gas analysis.

Infrared and laser detection technology can measure a wide range of gases and has the advantages of a large measurement range, fast response time, no poisoning, high measurement accuracy, good stability and reliability, and can achieve online detection of gases. For gas analysis, it is only necessary to connect different infrared and laser gas analysis components in series to achieve rapid multi-component analysis with a single injection.

When the sampling location is far away, even if the surface type beam tube analysis system is equipped with a down-hole type positive pressure sampling pump, the sampling time is still too long and there are problems with pipeline maintenance. The down-hole type beam tube analysis system (JSG6N) is now a mature technology, which can avoid the monitoring and analysis errors caused by long pumping lines and improve the real-time monitoring and accuracy. The down-hole type beam tube monitoring sub-station can effectively analyse 6 gases CO, CO_2, CH_4, O_2, C_2H_4, C_2H_2 by using infrared and laser gas analysis technology.

The underground type bundle tube analysis system installs the mine gas monitoring substation underground and collects 24h real time gas from multiple sampling locations underground through the pump in the mine monitoring substation. The gas is filtered by dust and transported through the bundle tube to the substation, where it is filtered by water and then fed into the real-time monitoring substation for rapid analysis. The monitoring substation transmits the detection data to the ground control centre station through the underground communication line.

(1) System composition

A diagram of the composition of the JSG6N down-hole type fire bundle system is shown in Fig. 6.7.

① Down-hole gas sampling and analysis systems

It is mainly used for the collection of underground gases and the accurate measurement of gas sample components, which includes a mining explosion-proof and intrinsically safe fire tube monitoring substation (hereinafter referred to as the mining

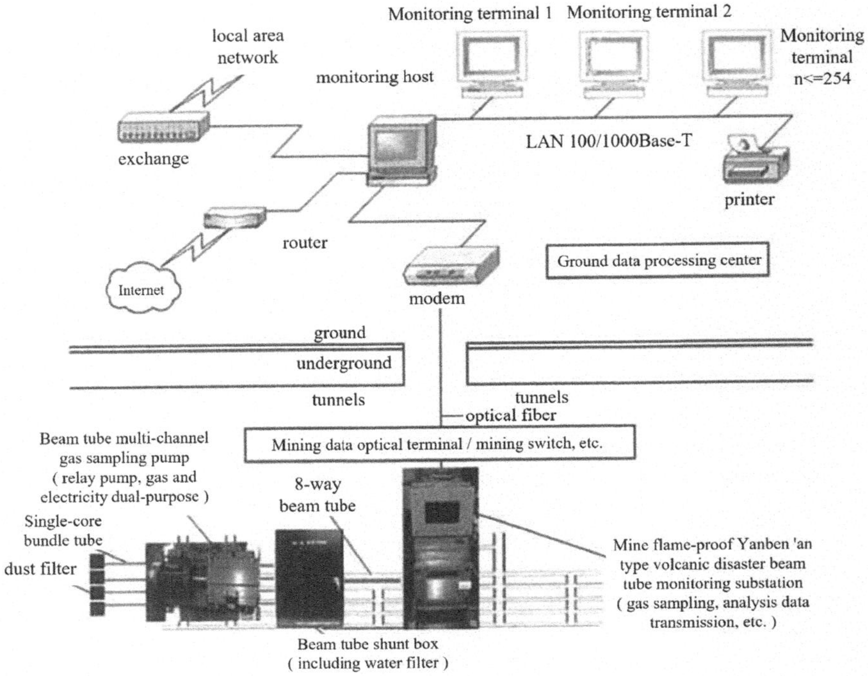

Fig. 6.7 Composition of the JSG6N down-hole type fire hose system

monitoring substation), a tube splitting box, a water filter, a filter cleaner, a beam tube, a fibre optic cable, an optical terminal and a supporting power supply.

② Ground data processing and sharing subsystem

It mainly completes the acquisition, storage, analysis and sharing of measurement data, including optical terminal, system industrial control machine, printer, workstation, system software and network software, etc.

(2) Mining monitoring sub-stations

The mine monitoring substation is mainly used for the determination of CO, CO_2, CH_4, O_2, C_2H_4, C_2H_2 6 gases, which is physically shown in Fig. 6.8. It is placed underground and consists of an explosion-proof case, analyzer assembly, sampling assembly, control assembly, automatic calibration and display assembly, data interface assembly, etc. The structure is shown in Fig. 6.9. A sub-station can monitor 8 monitoring points and the setting of the monitoring parameters (single circuit analysis, multiple circuit analysis, alarm limits etc.) for these 8 points can be operated from the mine monitoring sub-station or from a surface workstation.

The sampling gas is filtered by dust and transported through a bundle pipe to the sub-station, where it is filtered by water and fed into the mine monitoring sub-station for real-time analysis; after entering the sampling module, it is filtered by water in

Fig. 6.8 Mine monitoring substation

the second stage and automatically circulated (or at the operator's option) to collect "fresh" sample gas into the CO_2, CH_4, C_2H_4, C_2H_2, CO, O_2 analysis modules of the sub-station in turn. These components are connected in series through the gas circuit for multi-component analysis and the exhaust gas is discharged outside the station via the exhaust duct after the analysis is completed.

(3) Main technical specifications of the instrument

① Operating mode: 24h automatic analysis or manual setting.
② System capacity: 8 beam pipes per substation.
③ Measurement range: CO concentration: 0–1000 × 10^{-6} (resolution 1 × 10^{-6}).

C_2H_4 Concentration: 0–500 × 10^{-6} (resolution 1 × 10^{-6}).
C_2H_2 Concentration: 0–500 × 10^{-6} (resolution 1 × 10^{-6}).
CO_2 concentration: 0–30% (resolution 0.1%).
CH_4 Concentration: 0–10% (resolution 0.1%).
O_2 Concentration: 0–25% (resolution 0.1%).

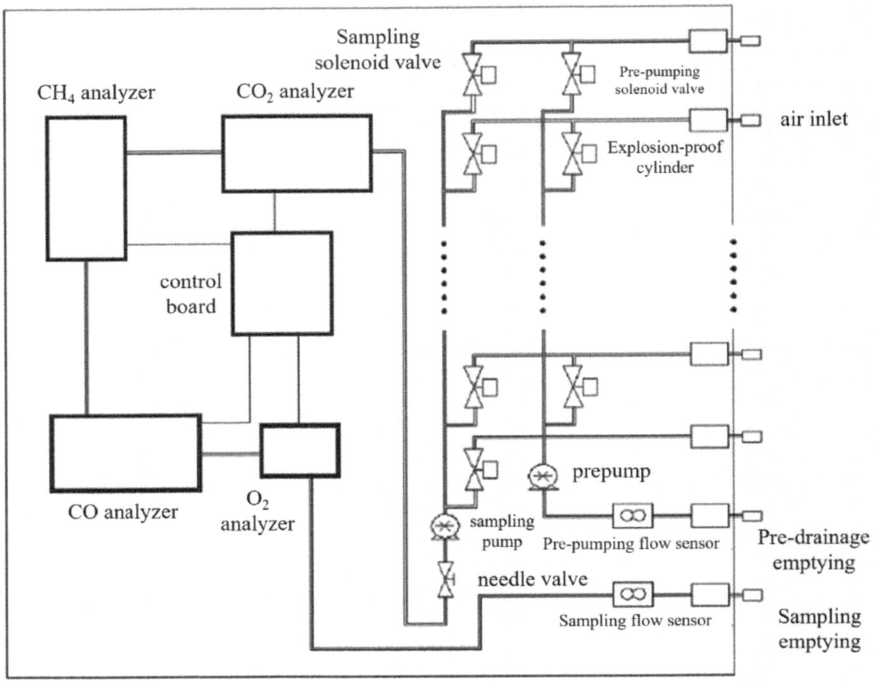

Fig. 6.9 Schematic diagram of the structure of a mining monitoring substation

6.2 Basic Principles of Gas Sampling and Analysis in the Mine Fire Area

Gas analysis is often used to analyse the combustion status of the fire zone during mine fires. But if problems occur during the gas sampling process, it can easily lead to misjudgement of the fire situation. Therefore, effective gas sampling of the mine fire area is fundamental to gas analysis, and the sampling process needs to ensure that each sample reflects the fire zone environment.

In order to make the collected gas sample is indeed a true and effective gas sample representing the fire zone fire development status, it is necessary to follow certain sampling principles, which are summarized as follows.

(1) Sampling bulbs or bottles may contain contaminants prior to sampling which may affect the analysis of the gas composition. In addition, sampling bulbs have a certain degree of adsorption. Therefore, it should be cleaned before use.
(2) Before taking a sample from the closed fire zone, the gas in the sampling tube must be pre-sampled for three times the time taken to complete a gas change in the sampling tube.

(3) When sampling, the "breathing" phenomenon of the gas in the closed fire zone caused by fluctuations in atmospheric pressure should be taken into account. Sampling should therefore be carried out during the time when the atmospheric pressure drops and the closed fire zone gas "exhales" outside the confinement. If conditions permit, it is advisable to measure the pressure difference between the inside and outside of the confinement wall at the time of sampling to ensure that a true and valid sample is taken to represent the fire conditions in the confined fire zone.

(4) In order to make an accurate determination of the location of the fire source and the extent of fire development, it is important to take as many samples as possible from the same sampling location or multiple sampling locations from each confined wall in the closed fire zone. Because the location and direction of spread of the fire is constantly changing until the fire is extinguished in the closed fire zone.

(5) Avoid taking samples too close to the point of origin of the fire. Because the air near the ignition point may flow slowly, which will cause the fire-generated gases to take longer to mix evenly with the air in the roadway. If the sample is taken too close to the ignition point, it will result in a high or low concentration of fire-generated gases in the sample taken. When sampling at the fire point downwind, the sampling location should be far away from the fire point, generally keeping a distance of more than 10 times the width of the roadway.

(6) Depending on the actual conditions of some mines themselves, it is again sometimes necessary to take samples as close to the point of origin of the fire as possible. For example, in some mines the coal seams themselves contain CO, CO_2, and CO, CO_2 may be generated by the slow oxidation of coal, which may cause the collected gas samples to contain more than the actual amount generated by the fire; some fires produce coke and carbon black which will absorb the CO, CO_2 generated by the fire, causing the collected gas samples to contain less than the actual amount generated by the fire. If there are factors such as these affecting the analysis of gas samples in the closed fire zone, the sampling location should be as close as possible to the point of origin of the fire. If there are difficulties in getting close to the point of origin of the fire, sampling can be carried out with the help of wind and water pipes in the fire zone.

(7) The sampling tube on a closed fire zone wall should be at least 5 m from the inside of the wall and the sampling tube should be installed in the centre of the wall; if the wall is to be opened at a later stage, the length of the sampling tube on the outside of the wall should be at least 20 m.

(8) If the wall is built on an inclined shaft, it is easy for fresh air to seep into the closed fire zone. To cope with this situation, it is advisable to install sampling tubes at different heights. In general, when constructing a confinement wall, it is necessary to install sampling tubes at three different heights on the wall, each at least 5 m from the inside of the wall.

(9) The preferred location for gas sample collection in the closed fire zone should be at the return air side of the confinement wall, taking care that the return air side of the confinement may be draughty due to fire wind pressure. Sampling on the return side should also ensure that the gas sample is taken only from the area to be measured. If there are air leaks that dilute the fire-generated gases, the CO/CO_2 indicator should be used for analysis.
(10) Samples should be taken at regular intervals from a fixed location in the closed fire zone, the number of samples taken and the frequency of sampling may be determined by the actual needs of the inspection. Repeat sampling is necessary for critical gas samples.
(11) When drilling for samples, if the air leakage from the ground is serious, the drilling should be abandoned first. The borehole should be resampled only after appropriate blocking measures have been taken to eliminate the factors affecting the air leakage.
(12) When sampling a borehole, attention should be paid to determining the pressure difference between inside and outside the borehole. Only when the pressure on the ground outside the borehole is less than or equal to the pressure inside the borehole can a true and valid gas sample be collected. The best time to take a sample is when the surface temperature is lower than the temperature inside the borehole, when the gas inside the borehole will flow towards the surface due to the natural wind pressure, thus increasing the effectiveness of the sample.

When looking at the records of gas sample analysis in all coal mines in China, there are often only the sampling location, sampling time and gas concentration But there is no record of the wind flow direction at the sampling point, the measurement method (portable gas detection or chromatographic analysis) and the pressure difference inside and outside the confining wall. The above content is very important and can effectively analyse the reliability of the measured data, such as the location of the gas sample taken to the gas sample can reflect the changes in the state of the disaster area has a critical impact. It must take the wind flow gas sample through the fire source in order to correctly reflect the state of the fire zone. Some people believe that the return air confinement sampling point can certainly take the wind flow gas sample through the fire source, as long as the emphasis on sampling in the return air area on the line. There is no need to write down the direction of wind flow at the sampling point. In fact, many coal mines fire zone often have a situation where the return air confinement is incoming and the incoming air confinement is outgoing. In some coal mines, the direction of the incoming and outgoing air in the fire zone changes several times a day. This is the result of the combined effect of the ground atmospheric pressure and the dynamic changes of the fire wind pressure in the fire zone. The air samples taken in the borehole should also be analyzed comprehensively to determine the actual direction of the wind flow in the whole area, including the inlet and return air.

The best way to analyse gas in a mine closed fire zone is to use an underground type beam tube analysis system to continuously sample gas from multiple locations in the

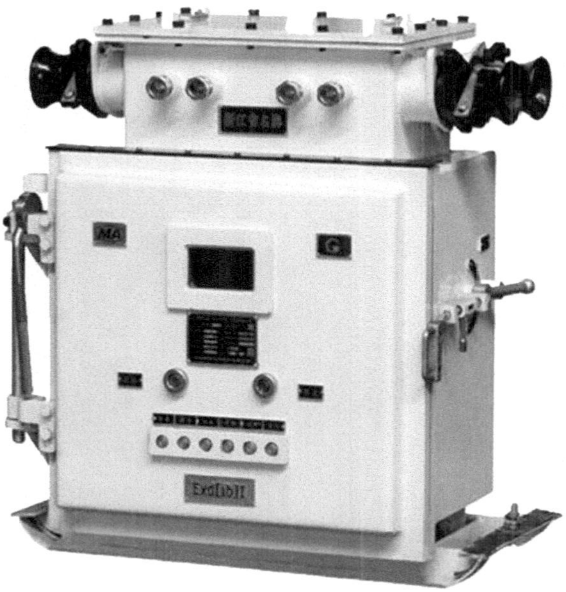

Fig. 6.10 Intrinsically safe differential pressure detector for mining

closed fire zone while continuously monitoring the differential pressure inside and outside the confining wall using a mine intrinsically safe differential pressure tester (Fig. 6.10). When the above requirements cannot be met at the fire management site, manual sampling is required, but to ensure reliable sampling, a number of conditions need to be met, which in principle can be divided into four main categories: personnel, sampling and monitoring equipment, location, frequency and time.

(1) Personnel

Sampling personnel need to be trained to understand the meaning and purpose of gas sampling and analysis, and to ensure that the sampling process is standardized. The sampling plan can be adjusted in time when the sampling environment changes during the down-hole gas sampling process.

When sampling a person wearing a breathing apparatus in the fire zone, to prevent interference with the gas sampled from the breathing apparatus. The sampling should be done by a single person, with other people waiting in the downwind.

(2) Sampling and monitoring equipment

If it is not possible to install an intrinsically safe differential pressure measuring device in the confinement wall, a U-shaped water column should be installed, and a continuous atmospheric pressure monitor should be installed on the ground or the weather bureau data should be consulted in time to record the hourly changes in atmospheric pressure on the ground.

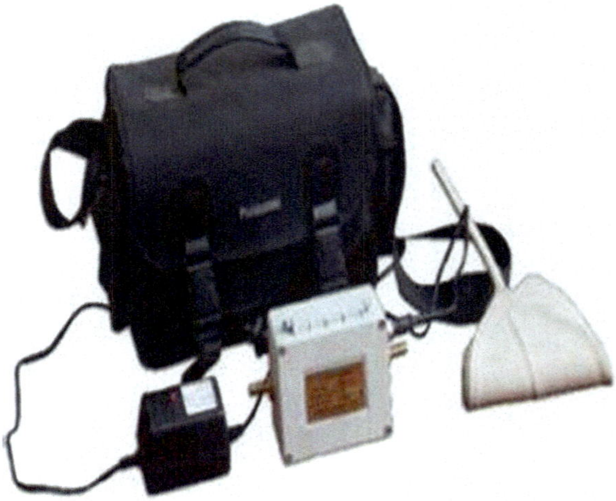

Fig. 6.11 CF222(A) automatic negative pressure gas picker

① The sampling bag must be easily closed to ensure timely closure after sampling to prevent leakage of the sampled gas. The sampling bag must be under vacuum before sampling.
② Check that the sampling bag is in good condition before sampling and rinse the bag at least 3 times with sampling gas.
③ When sampling with a certain length of sampling tube, flush the gas in the sampling tube at least 3 times with fire zone gas before sampling. When the length of the sampling tube is long, it is recommended to use a mine gas automatic negative pressure sampler (Fig. 6.11) for gas flushing and sampling to ensure reliable sampling.
④ When using a sampling bag for sampling, do not over-pressurize the sampling bag to prevent it from leaking due to excessive pressure, just fill the sampling bag by three-quarters of its volume when sampling.
⑤ The sample bag should be protected from bumps during transport to prevent leakage of the sample gas.
⑥ Sampling tubes should be made of steel, copper or suitable plastic tubes, not galvanized tubes, which will react with acidic water to produce H_2 and may cause interference with the gas components.
⑦ Each sampling bag should be labeled immediately after sampling. The label should include the location of the sample, the date and time of the sample, the number of samples taken at the same location at the time of sampling, the direction of gas flow at the time of sampling, the differential pressure inside and outside the confinement wall, other general descriptions and information about the sampling point.

(3) Location

The preferred sampling point for gas samples from a closed fire zone should be at the confined wall of the return air flow, and the direction of the air flow should be observed at the same time to ensure that the gas samples are taken from the area to be measured.

① During the fire fighting phase of a mine fire, sometimes the gas products generated at the point of origin of the fire are not well mixed with the other gases in the fire zone, so sampling points too close to the point of origin of the fire can result in too high or too low concentrations of gas products.
② According to the actual situation of the mine itself, sometimes it is also necessary to take samples as close as possible to the point of fire, such as some mine coal seams themselves contain CO, CO_2, the long distance gas sampling will lead to the concentration of combustion products gas exceeds the actual generation of fire zone; and some fires of relatively large fires produce coke, carbon black will absorb CO, CO_2, resulting in the content of these gases in the collected gas samples is less than the actual amount produced by the fire. If these factors affect the analysis of gas samples in the closed fire zone, the sampling location should be as close as possible to the point of origin of the fire. If there are difficulties in getting close to the point of origin of the fire, the sampling can be carried out with the help of wind and water pipes in the fire zone.
③ When sampling directly at the source of the fire downwind, the sampling location should be arranged at least 10 times the width of the tunnel from the source of the fire downwind, and sampling should be carried out at three heights in the tunnel: upper, middle and lower. Then, the average of the gas concentrations should be taken to analyse the burning state of the fire zone.
④ When carrying out a fire zone gas analysis, it is necessary to take samples of the wind flow on the upwind side of the fire point in order to compare and analyse the causes of gas changes.
⑤ To prevent distortion of the gas sampling results due to the "breathing phenomenon", a length of sampling tube should be pre-buried during the construction of the containment wall, the length of the sampling tube should not be less than 40 m.
⑥ When using surface boreholes for gas sampling, the surface air leakage should be determined first and gas sampling should only be carried out if the air leakage from the surface borehole to the shaft is eliminated.

(4) Frequency and timing

① The number of samples taken must take into account the urgency of the fire response and the rate of change of the gas sample. As these conditions are uncertain in the early stages of a teaching disaster, as many gas sampling analyses as possible should be carried out. When comparative analysis of results from different sampling points is required, the number of samples taken from these points should be the same; when used to analyse changes in the status of the fire source, the sampling frequency should be increased; when the results of the sampling analysis show that the fire zone is trending towards extinction. There is no other danger, and the sampling frequency can be shortened.

② Due to the limitations of the number of gas analyzers and the speed of analysis, sampling locations need to be determined prior to sampling. Rather, the change in gas trend at one valid sampling location is more useful for fire zone combustion state analysis than a large amount of gas data from multiple locations, and the sampling interval should be increased for valid sampling locations.
③ Fluctuations in atmospheric pressure can cause changes in the leakage of air from the outside environment into the closed fire zone, so in order to ensure that the sampled gases truly reflect the fire zone combustion state. Samples need to be taken when the atmospheric pressure at ground level is at its peak and trough.
④ When sampling a borehole, attention should be paid to determining the pressure difference between the inside and outside of the borehole. Generally, when the surface temperature is lower than the temperature inside the borehole, the gas in the borehole will flow towards the surface due to the natural wind pressure, so this time period is the best time for gas sampling.
⑤ To prevent degradation of the sampled gas and accidental leakage of the sampling bag, the sampler should deliver the sampled gas to the gas analyst as soon as possible.

6.3 Common Indicators for Gas Sampling in the Mine Fire Area

6.3.1 Indicator Gas Selection Principles of Mine Fire Area

The selection of an appropriate indicator gas is a key factor in the effectiveness of the gas analysis method for determining the combustion status of coal. It is generally accepted that the gases involved should have the following basic characteristics for the determination of indicators.

(1) Sensitivity: once the coal is burning or spontaneously combusting in the underground coal mine, or the coal temperature exceeds a certain value, the gas involved in the determination indicator must appear, and the trend of the determination indicator has monotonicity as the coal temperature rises.
(2) Uniqueness: the gas in question is only present as a result of combustion and is not produced in other cases.
(3) Regularity: there is a good correspondence between the change in concentration of the gas involved in this determination index and the coal temperature, as well as the repeatability is good. The initial temperature at which the gas is generated during pyrolysis of each coal sample from the same seam or mining area is basically the same or does not differ much, and has obvious regularity.
(4) Measurability: the gas involved in the determination is produced in sufficient quantities to enable existing instrumentation to detect the production and change of the gas in a timely as well as easily and accurately.

As there are no indicators that fully meet the characteristics of the above criteria in practice, it is necessary to choose the most suitable indicators or to use a combination of indicators for the determination, and to use some methods to solve the problems that may arise in the determination of the combustion status of a proposed indicator in the fire zone.

6.3.2 *Common Indicator Gases for Mine Fire Area*

(1) Carbon monoxide (CO)

At present, CO is commonly used as the main indicator for determining the combustion status of coal seam fires at home and abroad. Its advantages are as bellow. ① It meets the sensitivity characteristics, coal oxidation warming gas appears in the order of CO-H_2-C_2H_4-C_3H_6-C_2H_2 regulation, this is the most significant advantage of CO as closed fire zone combustion state determination index, CO as the determination index is more sensitive; ② It meets the regularity characteristics, coal in the low temperature oxidation process. The amount of CO generation and coal temperature is closely related with the coal temperature. As the coal temperature rises, the amount of change in CO concentration is most obvious; ③It meets the characteristics of measurability. CO detection is relatively easy and convenient. CO is released from low temperature, through slow oxidation, accelerated oxidation until the intense oxidation stage, and a strong regularity appears with temperature change.

However, there is a serious disadvantage of CO as an indicator of fire combustion status that can be easily overlooked, i.e. it does not have the unique characteristic that not only the coal heats itself and burns to produce CO. It is not only the self-heating and combustion of the coal that produces CO, but also the environmental factors such as the small amount of residual CO in the coal seam itself and the slow oxidation of the coal from underground blasting that can produce CO. These environmental factors will cause the CO concentration to increase. Carbon black and coke etc. in the fire zone will absorb it. These environmental factors will cause the concentration of CO generated by combustion to decrease. Therefore, the application of CO as an indicator for determining the combustion status of a closed fire zone must pay attention to the environmental factors and the effect of airflow CO on the concentration.

(2) Absolute production of CO

In order to solve the deficiency that CO concentration is greatly influenced by the air volume. Some coal enterprises in China use the absolute CO generation as the parameter for the determination of the combustion status of the fire zone, and name it as the natural firing coefficient, expressed by H. When the concentration in the inlet air stream CO is 0, H is calculated using the following Equation.

$$H = H_{CO}Q/100 \tag{6.3}$$

In the Equation, H_{CO}—the carbon monoxide content of the wind stream at the sampling location, %.

Q — air volume at the sampling location when taking samples, m^3/min.

If the carbon monoxide concentration in the inlet air flow stream is not zero, H is replaced by the difference between the absolute generation of CO in the return air and the inlet air flow.

For mines with carbon monoxide gas in the normal airflow, a threshold value should be determined when carbon monoxide is used as an indicator for determining the combustion status of the fire zone. The following factors are generally taken into account when determining the threshold value.

① The background content of CO in the normal wind flow at each sampling location.
② The threshold should be determined in such a way as to ensure a certain safety margin, i.e. to allow sufficient time for the implementation of fire-fighting operations.

It should be noted that when applying CO as an indicator of the combustion status of a fire zone, two points should be noted.

① The concentration or absolute value of CO is to be greater than the critical value.
② There should be a steady increase in the concentration or absolute value of CO

(3) Grahame coefficient (I_{CO})

In 1914, the British scholar Graham introduced the parameter of oxygen consumption (ΔO_2) as an indicator for determining the combustion status of a closed fire zone, which is easily influenced by the fresh air flow, and proposed a corresponding composite indicator, the Graham factor (I_{CO}).

The Grahame factor is the ratio of the amount of carbon monoxide produced by the combustion reaction in the closed fire zone to the amount of oxygen consumed, which is mathematically expressed as follows.

$$I_{CO} = \frac{100CO}{O_2} \tag{6.4}$$

In the Equation, $\Delta O_2 = 0.265N_2 - O_2$, if the ratio of O_2 to N_2 in the inlet side of the fire zone is not 0.265 due to factors such as inerting, the actual value should be used instead of 0.265.

CO, O_2, N_2—volumetric percentage concentration of gas in the return air side of the fire zone after the airflow has passed through the fire zone.

I_{CO} is very useful in predicting the trend of spontaneous coal ignition and determining the state of combustion in the closed fire zone, but the coefficient has some drawbacks. For example: I_{CO} can produce considerable errors in very small cases; I_{CO} can also produce inaccurate results due to non-coal spontaneous combustion or CO, CO_2, which can lead to an inaccurate determination of the state of combustion of the closed fire zone.

(4) Carbon dioxide ratio (CO/CO_2)

The carbon oxide ratio is the ratio between the amount of carbon monoxide and carbon dioxide produced by the combustion reaction in the closed fire zone. In the early stages of a fire, the value of this indicator increases with the development of the fire; in the full combustion stage and when the oxygen concentration is above 5%, the value stabilizes as a constant.

(5) Other indicator gases

Carbon dioxide index (ICO_2), oxygen consumption (ΔO_2), chain ratio, ethylene (C_2H_2), acetylene (C_2H_2), multi-gas combination, hydrocarbon index, Bestrong coefficient, gas component ratio, etc.

6.4 Example Analysis: The Effect of Atmospheric Pressure and Temperature on Gas Sampling

6.4.1 Laws of Atmospheric Pressure and Temperature Change

The Datong coalfield in Shanxi Province is an important coal energy base in China, which is also an area with more serious coalfield fires in China. It is located in the north of Shanxi Province, with longitude 112°34′–114°33′E and latitude 39°03′–40°44′N. It belongs to temperate continental monsoon climate with four distinct seasons. The natural fires in Datong coalfield are more serious and there are more closed fire zones. Based on the two years' atmospheric pressure time-by-time data from 2017–2018 in Datong, Shanxi, the atmospheric pressure fluctuation law is analyzed to provide reference for the development of gas sampling measures and gas analysis methods for the fire zone.

Atmospheric pressure is the pressure of the atmosphere, and the atmospheric pressure at a given point is equal to the total mass of the column of leaded air mine gas borne per unit area at that location. The distribution and variation of atmospheric pressure is related to weather changes and air movement. As the total mass of the atmosphere is roughly constant and the atmosphere is a continuous fluid, the mass of the air column in some places decreases the mass of the air column in some other areas must increase. The variation in air column mass from place to place is the trigger for the redistribution of air mass across regions caused by air movement.

When the temperature of the air mass in a local area falls, the air contracts and the air in the upper layers of the surrounding area converges from all around to the upper layers of the area, resulting in an increase in the mass of the air column, a rise in ground pressure and a dispersion of the air in the lower layers in all directions, forming a cold high pressure. When the temperature of the local air mass rises, the air column expands upwards and the air pressure in the upper layers of the region is higher than the surrounding air pressure, resulting in an outward dispersion of air and a reduction in the mass of the air column, forming a thermal low pressure. In the real atmosphere, the change in atmospheric pressure at ground level is influenced by temperature changes, but also by the atmospheric circulation, humidity and many other factors.

According to the long-term observation and collation of atmospheric pressure data by researchers in the field of meteorology, it has been found that the variation pattern of atmospheric pressure over time is divided into periodic and non-periodic variation, and the overall variation pattern is generated by the superposition of the periodic and non-periodic variation patterns. Statistical studies show that non-periodic changes in atmospheric pressure are directly related to the movement and evolution of atmospheric pressure systems, and that non-periodic changes in atmospheric pressure are much greater than periodic changes, such as cold wave outbreaks and thunderstorm processes, which can disrupt the periodic pattern of atmospheric pressure.

The non-periodic variation is usually more pronounced at middle and high latitudes than at low latitudes, as can be seen by using the one-day atmospheric pressure variation values as a comparison, which are usually up to 1000 Pa at high latitudes, about 400 Pa at middle latitudes and about 100 Pa at low latitudes (except for severe weather). Therefore, at middle and high latitudes, the cyclical pattern of atmospheric pressure is often masked by non-cyclical variations, and the overall characteristics are usually a cyclical. At lower latitudes, the non-periodic pattern of atmospheric pressure is often weaker than the periodic pattern, thus the periodicity of pressure changes is more pronounced. In addition, the annual fluctuation pattern of atmospheric pressure varies from region to region, which can be basically divided into continental, oceanic and alpine types according to the region.

(1) Annual variation of atmospheric pressure and temperature

Data for the atmospheric pressure study were obtained from the National Meteorological Information Centre, see Table 6.3 for information on the geographical location of the Datong meteorological station measurement points, with data collected in January 2,017,131 at intervals of 6–10 h, with an atmospheric pressure resolution of 10 Pa and a temperature resolution of 0.1 °C.

Through the collation and analysis of Datong's atmospheric pressure and temperature time-by-time data information, it was found that the collected data completeness

Table 6.3 Information on the geographical location of the Datong weather station measurement points

Location	Monitoring site area station number	Latitude/(°)	Longitude/(°)	Altitude of the observation site/m	Sensor altitude/m
Datong, Shanxi	53,487	113.25	40.05	1067.2	1068.4

rate was above 99.4%, with 104 missing data for atmospheric pressure and temperature for two years respectively. Since most of the missing data are single point data, the linear interpolation method can be used to fill in the missing data and then plot the time-by-time change curves of atmospheric pressure and temperature in Datong from 2017–2018, respectively as shown in Figs. 6.12 and 6.13.

It can be observed that the amplitude of fluctuations in atmospheric pressure is large and the daily periodicity is difficult to judge directly, while the amplitude of fluctuations in temperature is small and the daily periodicity is more obvious. There are

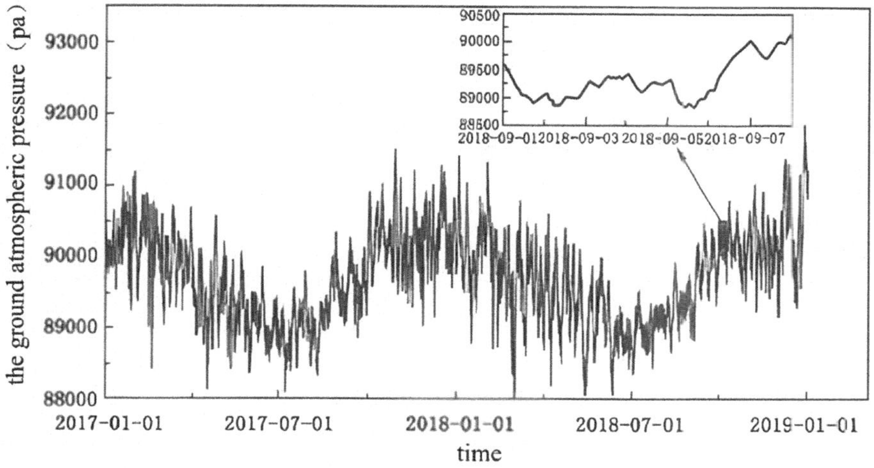

Fig. 6.12 Hour-by-hour variation curve of atmospheric pressure in the Datong area, 2017–2018

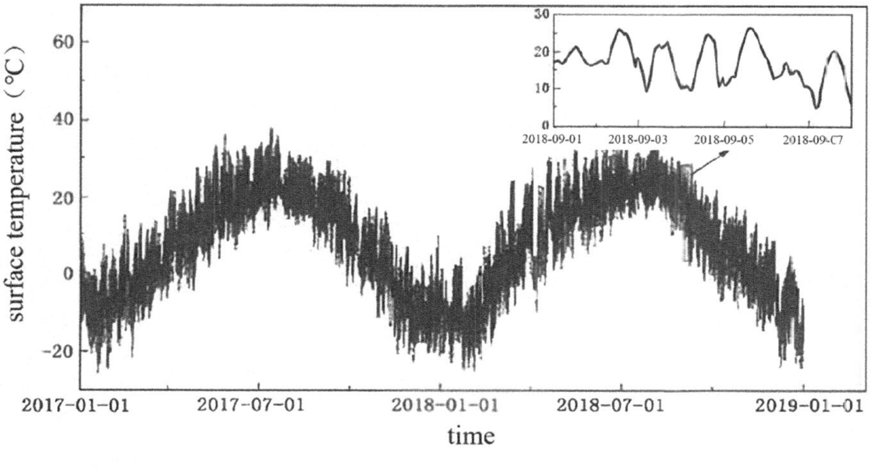

Fig. 6.13 Time-by-time variation curve of ground temperature in Datong, 2017–2018

many methods of fitting discrete time signals, including polynomial fitting. But when used for medium and long term discrete data fitting analysis, the fitted curves will diverge due to the presence of higher order terms. According to Fourier's theorem, any signal can be represented by an infinite series of frequency, amplitude and phase angle. By observing Figs. 6.12 and 6.13, it can be seen that the fluctuation of atmospheric pressure data has obvious annual periodicity, so the 1st order Fourier series fitting method is chosen to analyse its annual variation law, the specific Equation is as follows.

$$P(\mathrm{t}) = a_0 + a_1 \cos(\mathrm{t} \cdot \mathrm{w}) + \mathrm{b}_1 \sin(\mathrm{t} \cdot \mathrm{w}) \tag{6.5}$$

In the Equation, a_0, a_1, b_1—the Fourier series.

w—frequency.

The period T and amplitude A of the Fourier series fitting function are calculated as bellow.

$$T = \frac{1}{\mathrm{w}} \tag{6.6}$$

$$A = \sqrt{\mathrm{a}_1^2 + b_1^2} \tag{6.7}$$

The Fourier series fitting coefficients were obtained by performing 1st order Fourier series fitting on the atmospheric pressure and temperature time-by-time data from December 1, 2017 to December 31, 2018 in Datong area (Table 6.4).

Through the analysis of the sum of squares of deviations and adjusted sum of squares of deviations of the fitted equations of atmospheric pressure and temperature trends in Table 6.4, it can be seen that the annual variation law of atmospheric pressure and temperature can be better described using the 1st order Fourier series fitting. However, due to the dramatic daily fluctuations of atmospheric pressure and temperature, the sum of squared and mean squared errors in the simulation process are large. It is difficult to use the annual variation fitting curve to analyse the daily fluctuation pattern of atmospheric pressure and temperature. The Fourier fitted curves of atmospheric pressure and temperature are plotted according to the fitted equations (Figs. 6.14 and 6.15).

Table 6.4 Atmospheric pressure 1st order Fourier series fitting factors

Data type	Transformation fit coefficient				Variance	Sum of squares of deviations	Adjusted sum of squared deviations	Root mean square error
	a_0	a_1	b_1	w				
Atmospheric pressure	89,690	654.2	−5.352	0.00074	3.1×10^9	0.554	0.554	420.3
Temperature	7.59	−16.78	−1.594	0.0007176	7.3×10^9	0.7721	0.772	6.476

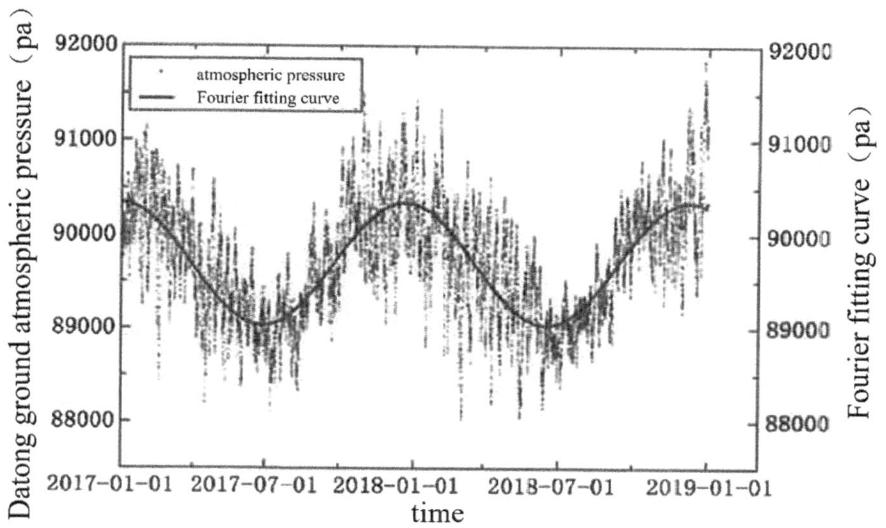

Fig. 6.14 Atmospheric pressure 1st order Fourier series fitting curve

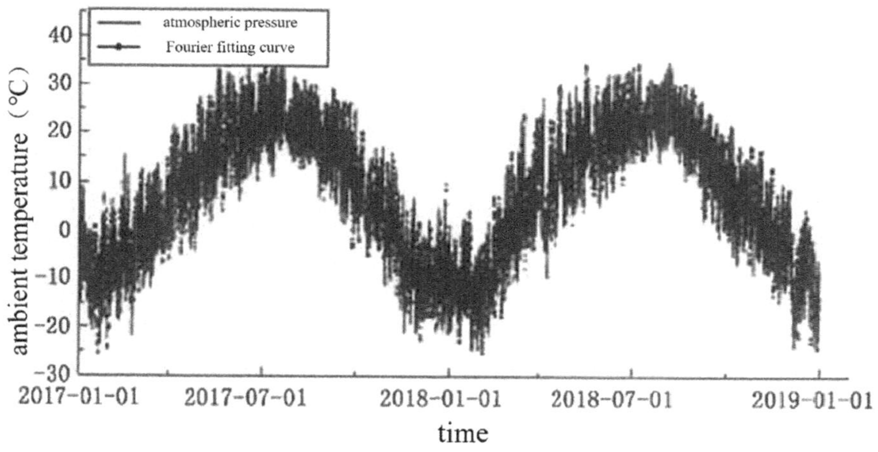

Fig. 6.15 Temperature 1st order Fourier series fitting graph

By analyzing Table 6.4 and Fig. 6.15, it can be seen that using the 1st order Fourier series fitting can better describe the annual variation law of atmospheric pressure, and the calculation result according to the period Equation shows that the annual period of atmospheric pressure is 353.78d, which is less different from 365d a year. Due to the violent nature of the time-by-time fluctuations of atmospheric pressure data, resulting in large errors in the sum variance and mean square error in the fitting process, it is difficult to analyse the daily fluctuation pattern of atmospheric pressure with this fitting Equation. According to the calculation of the amplitude Equation, the amplitude of the annual cycle fluctuation of atmospheric pressure in Datong area is 654.22 Pa, the mean value is 89,708.1 Pa, and the standard deviation is 629.4. Therefore, the time-by-time change of atmospheric pressure data fluctuates greatly and the data is very scattered.

Analysis of Table 6.4 and Fig. 6.15 shows that the annual cycle of temperature in the Datong area is 364.83d respectively, which would have been equal to a one-year cycle. The mean temperature in the Datong area is 7.6 °C, the amplitude of the annual cycle fluctuation is 16.83 °C, the mean value is 16.2 °C and the standard deviation is 13, so the time-by-time data fluctuation of temperature is small and its data distribution is more concentrated compared to the atmospheric pressure variation.

Through Pearson correlation analysis of the fluctuation curves of atmospheric pressure and temperature in Datong, the Pearsom correlation coefficient of atmospheric pressure and temperature in Datong is −0.752, and the correlation tests for them all shows a significance level of 0. Therefore, it is satisfied that atmospheric pressure and temperature are significantly correlated on a 0.01 level, and atmospheric pressure shows a significant linear correlation with temperature in terms of the annual cycle change pattern.

(2) Analysis of the pattern of daily cycle changes in atmospheric pressure and temperature

In order to accurately analyze the daily cycle characteristics of the time-by-time change data of atmospheric pressure throughout the year, it is necessary to eliminate the annual trend term of the original data to obtain the discrete points of atmospheric pressure with the annual cycle data removed (Fig. 6.16), and then by performing spectral analysis on its fast Fourier transform (Fig. 6.17), it can be seen that the discrete data periods of atmospheric pressure in the Datong area after eliminating the annual trend term are mainly 15.5d1d and 182.5d.

To verify the existence of a cycle in the spectrum analysis, a determination can be made with the help of the Fisher test. As Fisher has a cycle case, only logical means can be used for indirect analysis. Taking the significance level $\alpha = 0.05$, since $g_{0.05}$ (1000, 3) is greater than $g_{0.05}$ (100, 5), it follows that if the spectral density is less than $g_{0.05}$ (1000, 5). It means that the spectral density is similarly less than $g_{0.05}$ (1000, 3). The significance of the period was tested by calculating the spectral density through the Borie transform and finding the Fisher's critical value in the summation analysis, and the results are shown in Table 6.5.

The number of time by time data sampled for atmospheric pressure throughout the year is 17520, s = 17,520/2 = 8760. Since the critical values $g_{0.05}$ (8760, 1) < $g_{0.05}$ (1000, 1), $g_{0.05}$ (8760, 2) < $g_{0.05}$ (1000, 2), $g_{0.05}$ (8760, 3) < $g_{0.05}$ (1000, 5), the hypothesis test passes, i.e. there is a 15.5d, 1d and 182.5d period for atmospheric pressure. 1d and 182.5d periods with greater than 95% probability.

In order to accurately analyse the daily cycle characteristics of the time-by-time temperature change data throughout the year, the annual trend term of the original

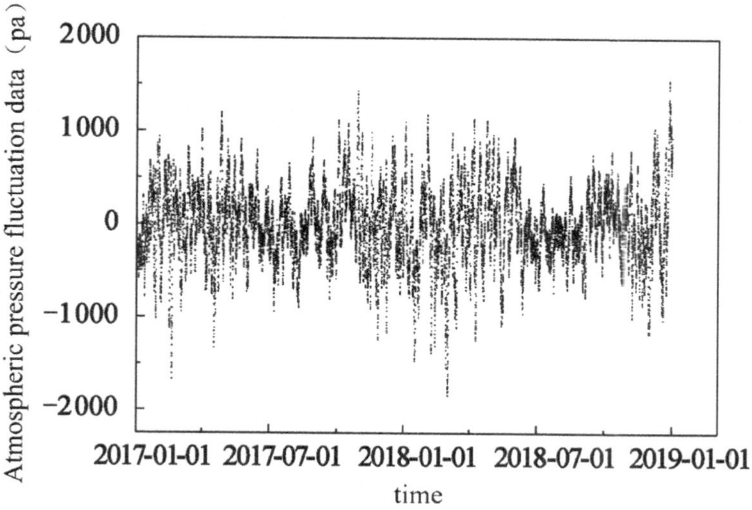

Fig. 6.16 Distribution of discrete points of the annual trend of atmospheric pressure removal

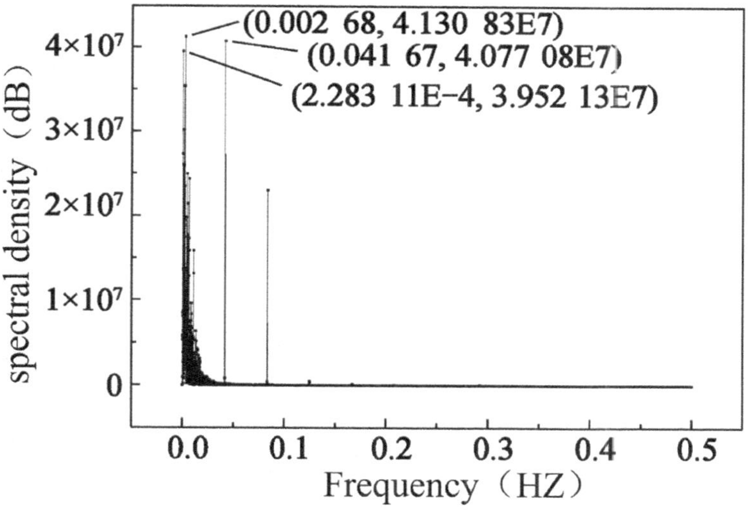

Fig. 6.17 Spectral analysis of atmospheric pressure

Table 6.5 Fisher test significance analysis table

Cycle time/d	Spectral density	Fisher threshold
15.5	0.0267	$g_{0.05}$ (1000, 1) = 0.00934
1	0.0263	$g_{0.05}$ (1000, 2) = 0.00790
182.5	0.0255	$g_{0.05}$ (1000, 5) = 0.00617

data needs to be eliminated to obtain the discrete points of the temperature de-annualized data (Fig. 6.18), and then by performing a spectral analysis of its fast Fourier transform (Fig. 6.19), it can be seen that the discrete data period of the atmospheric pressure in the Datong area after eliminating the annual trend term is mainly 1d.

To verify the existence of cycles in the spectral analysis, the determination can be made with the help of Fisher's harmonic analysis significance test. The spectral density of the daily cycle of temperature change $g_1 = 0.5063$ for the Datong area and $g_1 = 0.4679$ for the Shizuishan area. Taking the significance level $\alpha = 0.05$ and the daily cycle number $\gamma = 1$ for the Datong and Shizuishan areas, the Fisher's critical value table for the harmonic analysis significance test shows that $g_{0.05}$ (1000, 1) = 0.00984. Due to the length of sampling of temperature time by time data throughout the year of 17,520, s = 17,520/2 = 8760. The critical value $g_{0.05}$ (8760, 1) < $g_{0.05}$ (1000, 1), and the spectral density of temperature variation in Datong is much larger than the Fisher critical value. Therefore, the hypothesis test passes, i.e. the probability of the existence of a daily cycle in temperature is greater than 95%.

Atmospheric pressure and temperature data for July 2018 in the Datong area were extracted and time-by-time variation curves were plotted, as shown in Figs. 6.20 and 6.21.

The analysis shows that the temperature variation pattern shows an obvious daily periodicity, while the atmospheric pressure is less cyclical although it also fluctuates

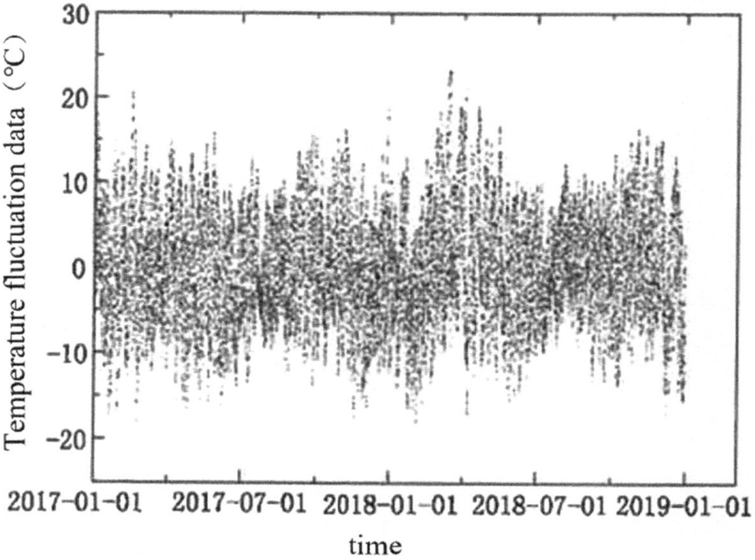

Fig. 6.18 Discrete point distribution of annual trend in temperature removal

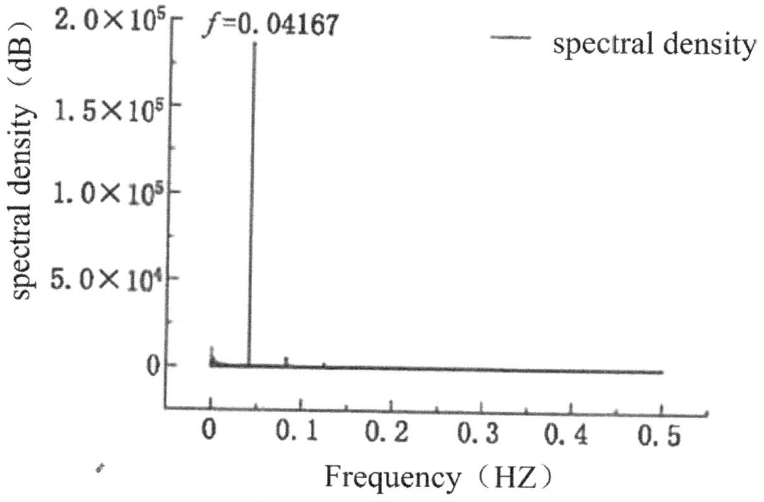

Fig. 6.19 Temperature spectrum analysis graph

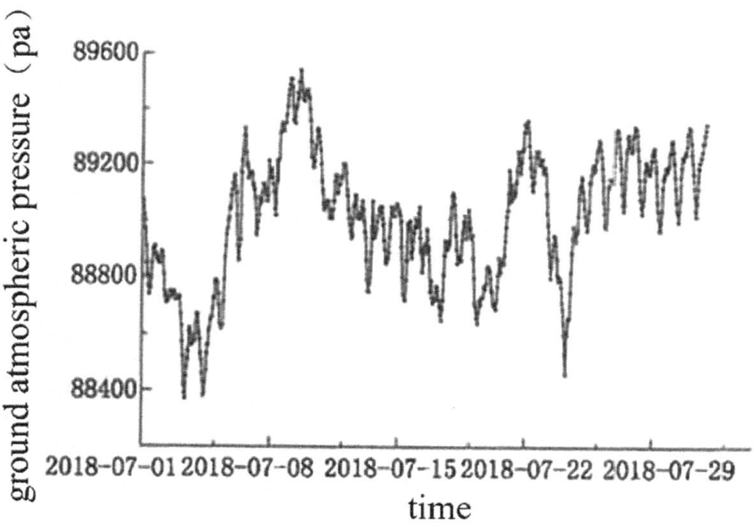

Fig. 6.20 Hour-by-hour variation curve of atmospheric pressure in July 2018

significantly from day to day. The SPSS software was used to correlate the time-by-time variation data of atmospheric pressure and temperature in Datong area, and the calculation results are shown in Table 6.6.

By comparing the standard deviation of atmospheric pressure and temperature changes, it can be seen that the overall variation in temperature is less different

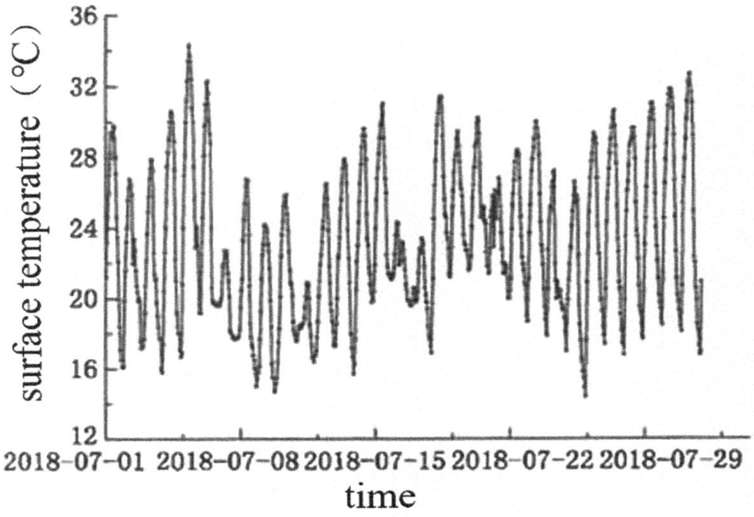

Fig. 6.21 Hour-by-hour temperature change curve for July 2018

Table 6.6 Correlation analysis between atmospheric pressure and temperature in the Datong area

	Atmospheric pressure/Pa	Temperature/°C
Average value	89,013	23.19
Standard deviation	230	4.4
Pearson's correlation coefficient	−0.26	
Bilateral Significance Test Results	0	

than that of atmospheric pressure, which has larger change data. The SPSS analysis shows that the time-by-time change data of atmospheric pressure and temperature in Datong and Shizuishan are significantly correlated at the 0.01 level. But due to their small Pearson correlation coefficients, the linear correlation between atmospheric pressure and temperature is poor. Therefore, the relationship between atmospheric pressure and temperature variation is significantly correlated but not significantly linearly correlated.

(3) Atmospheric pressure peak and trough distribution moments

Through statistical analysis of the two years of time-by-time data on atmospheric pressure, histograms of the peak and trough values of atmospheric pressure variation are plotted, as shown in Figs. 6.22 and 6.23 respectively, where the horizontal coordinates represent the moment in time (0:00–23:00) and the vertical coordinates represent the number of peak or trough values of atmospheric pressure occurring during the time period in question.

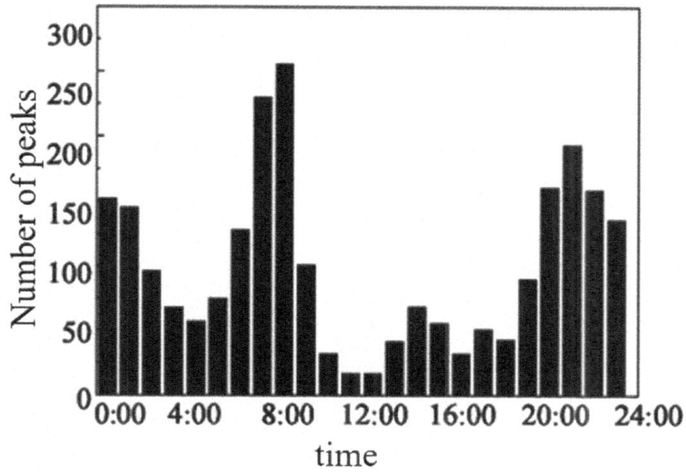

Fig. 6.22 Map of peak atmospheric pressure distribution over time

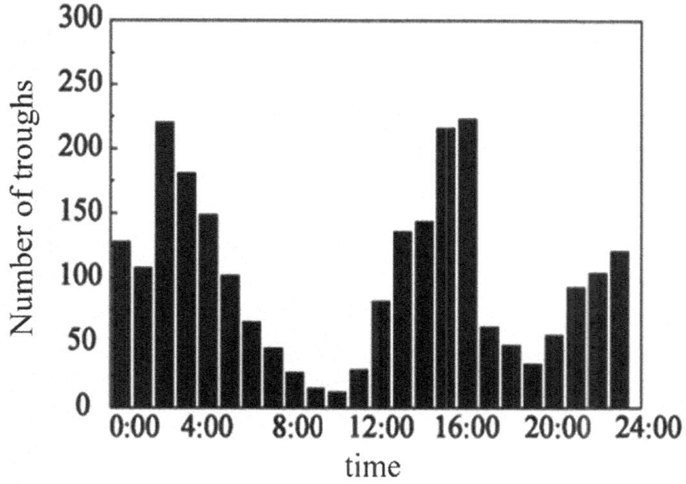

Fig. 6.23 Map of atmospheric pressure troughs over time

The analysis shows that the time period with the highest number of atmospheric pressure wave peaks is from 7:00–8:00, with the wave peaks accounting for 20.2% of the day's occurrences; next, the time periods with the highest number of wave troughs are 2:00, 15:00 and 16:00, accounting for 27.4%.

(4) Analysis of the rate of change of atmospheric pressure

Statistical analysis of the gas variation data was carried out by using Matlab Soft to plot statistical histograms of the number of different amplitude intensities and statistical histograms of the peak-to-trough transition times, as shown in Figs. 6.24 and 6.25 respectively.

By analyzing Fig. 6.24, it can be seen that the amplitude of the atmospheric pressure fluctuation process time by time is mostly concentrated within 500 Pa, among which less than 100 Pa is the most, and very few amplitudes exceed 1000 Pa.

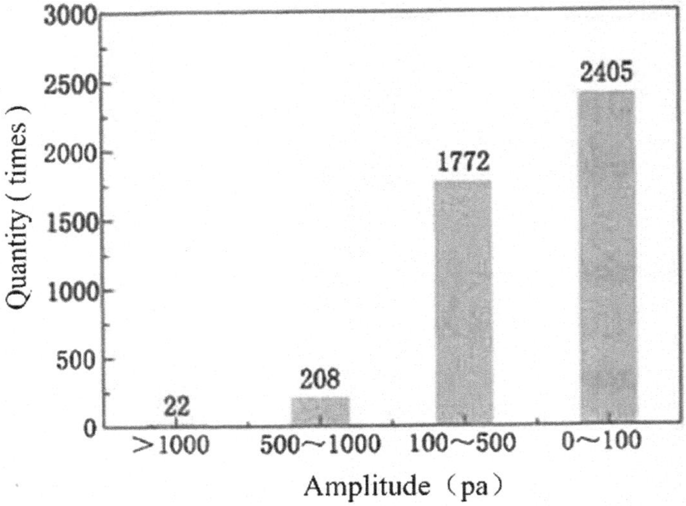

Fig. 6.24 Statistical histogram of the number of different amplitude intensities

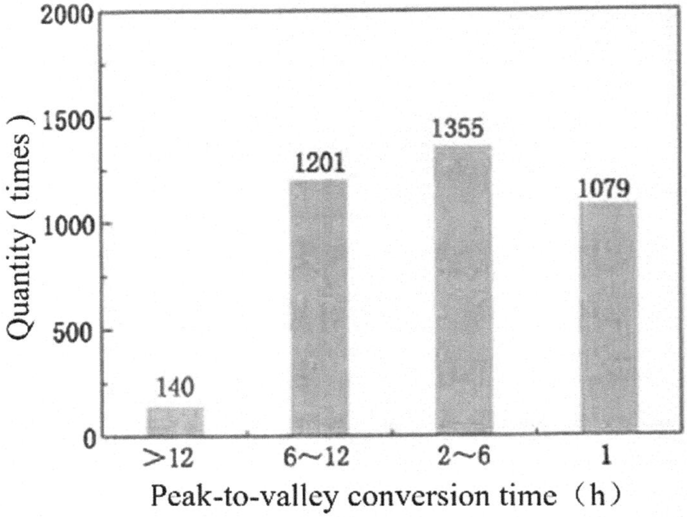

Fig. 6.25 Histogram of peak-to-trough transition time statistics

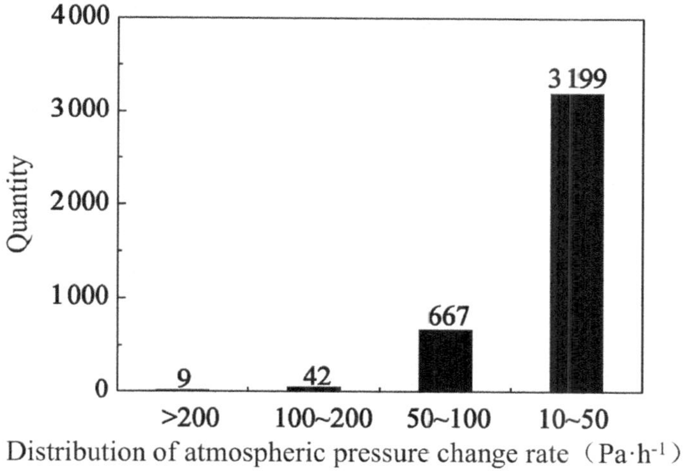

Fig. 6.26 Histogram of the number of daily fluctuation rates of atmospheric pressure

According to Fig. 6.25, it can be seen that the transition time of atmospheric pressure peak and trough is mainly concentrated in 2–6 h, which accounts for 68% of the total; 28.6% of the atmospheric waves have a peak-to-trough transition time of 1h. Due to the very short time, it can be considered as noise during the atmospheric pressure release. Therefore, this time does not constitute a valid fluctuation cycle. Combined with the analysis of the daily periodicity of the atmosphere, it can be seen that the atmospheric pressure fluctuates on a time-by-time basis with a daily periodicity. The peak-to-trough transition time is mainly concentrated around 6h (Fig. 6.25), so the shape of the daily atmospheric pressure change curve should be mainly bimodal.

A mathematical and statistical analysis of the daily fluctuation rate of atmospheric pressure (Fig. 6.26) shows that the rate of change of atmospheric pressure is mostly less than 100 Pa/h, with a few of them being greater than 100 Pa/h. This phenomenon is mainly caused by abnormal weather, such as cold wave outbreaks and thunderstorm processes.

6.4.2 Characteristics of the Propagation of Atmospheric Pressure Changes in Mines

There are many reasons for changes in underground air pressure. The first is the human factor, such as the change of fan speed, the sudden opening or closing of the dampers, the passing of mine cars, etc., all can cause changes in air pressure; the second is the change of the atmospheric environment. As we all know, the surface atmospheric pressure is changing all the time. Since human factors rarely change, the change of underground ventilation pressure depends to a large extent on the change of surface atmospheric pressure. Changes in surface atmospheric pressure lead to changes in pressure distribution in the underground tunnel, resulting in changes in the pressure difference between inside and outside the fire zone, causing air in the fire zone to leak out or outside air to flow into the fire zone, resulting in "respiration". This phenomenon is detrimental to suffocating the fire zone and preventing further development of the fire.

There are different time delays in the transfer of atmospheric pressure in different media. The length of the delay depends on the amount of resistance between the two points where the pressure is transmitted; the greater the resistance, the longer the time. According to the German scholar Plante's research, the propagation speed of atmospheric pressure in still air is the local speed of sound c. If the speed of the wind flow down-hole is v, the propagation speed of atmospheric pressure down-hole with the wind flow is c + v, and the speed against the wind flow is c-v in the down-hole roadway with limited distance and small wind flow speed. It can be considered that the pressure change of atmospheric pressure in the connected roadway down-hole is synchronized with the change trend of atmospheric pressure on the ground.

In studying the attenuation of the atmospheric pressure propagation process, the atmospheric pressure can be regarded as a proadway wave. According to the acoustics of the phase class knowledge can be known, when the proadway wave in the medium propagation, wave intensity will increase with the propagation distance and decrease, proadway over the eclipse vice mainly by the proadway wave absorption and scattering two parts caused. That is, part is caused by the viscosity of the medium, heat transfer and complex her image process, and it converts the ordered proadway wave into the medium's heat and internal energy. The other part is due to the scattering of energy from the proadway wave in the unevenly collapsed medium to other directions, which weakens the energy of the proadway wave in the original direction of propagation.

The attenuation of atmospheric pressure in the non-underground connecting roadway is very small (generally the attenuation ratio is around 4%), it can be considered that the atmospheric pressure on the surface and the atmospheric pressure in the underground roadway changes basically the same magnitude. But the atmospheric pressure in the propagation through the confinement wall to the mining area attenuation value is larger, the attenuation size is related to the degree of confinement, the material of the confinement wall, etc.

The criterion for determining the quality of a firewall (poor, better, good, excellent) is the time lag of the change in gas concentration within the containment zone caused by a change in atmospheric pressure. With poor quality firewall, changes in atmospheric pressure cause almost immediate changes in gas concentration within the containment zone. The greater the time lag, the better the quality of the firewall. As can be seen from Fig. 6.27a, when the atmospheric pressure increases, the CH_4 concentration decreases and the O_2 concentration increases, which means that atmospheric pressure changes have a strong influence on the gas concentration in the fire zone and that outside air leaks into the containment zone. This trend in the CH_4 and O concentrations indicates a poor containment effect. As can be seen from Fig. 6.27b, when the atmospheric pressure increases, the CH_4 concentration

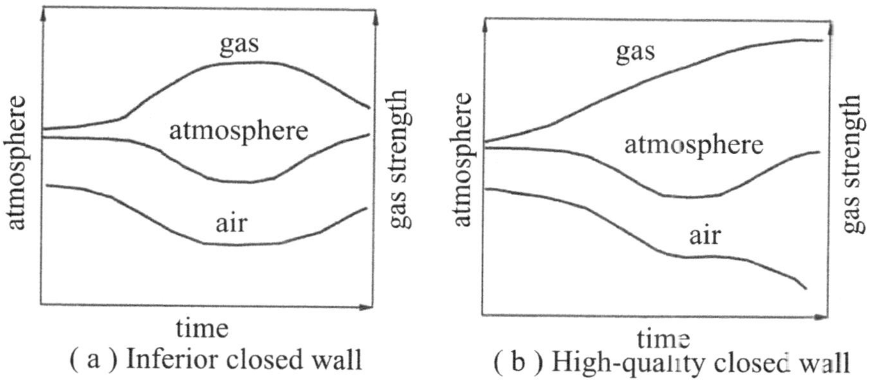

Fig. 6.27 The effect of changes in ground atmospheric pressure on air leakage through a confined wall

continues to increase and the O_2 concentration continues to decrease, indicating that the fire zone is tightly closed and the increase in atmospheric pressure does not increase air leakage. However, when the concentration of CH_4 is equal, if the atmospheric pressure increases, a small amount of air flows into the fire zone, resulting in CH_4 being forced out, the concentration of CH_4 in the fire zone decreases and the concentration of O_2 increases. This can also happen under high quality firewall conditions.

6.4.3 The Effect of Atmospheric Pressure on the Depth of the Sampling Tube in the Closed Fire Zone

The quality of the confinement wall in a mine's closed fire zone affects the amount of air leakage from the fire zone. For mine fire area management, the better the containment wall, the more effective the fire zone management and the shorter the management period. For fire zone gas sampling analysis, the better the quality of the confinement wall, the less mobile the gas in the closed fire zone, i.e. it is difficult to obtain a gas composition at the confinement wall that effectively reflects the true state of the fire zone through gas sampling. However, when the quality of the containment wall is very poor, it is difficult to reflect the true state of the fire zone due to the dilution of the gas concentration near the inside of the wall by air leakage. Therefore, in order to verify the true extinction of the fire zone before it is opened, it is sometimes necessary to control the air leakage (e.g. "oxygenation observation") and take samples to analyse the true gas composition in the vicinity of the fire source.

The airtightness of the confinement wall is an intrinsic factor in determining the air leakage in the fire zone, but the amount of air leakage is not consistent in different locations of the confinement wall, generally speaking, the amount of air leakage through the porous medium of the confinement wall itself is small, and its flow state is mostly laminar flow state. Soviet scholars believe that 70–95% of the air leakage from the confining wall exists at the perimeter of the contact between the confining wall and the coal and rock, and the flow of the air leakage is mostly in a transitional and turbulent state.

$$Q = kP\sqrt[n]{\frac{h}{9.8 \times \delta}} \tag{6.8}$$

In the Equation, Q—air leakage through the confining wall, m^3/s.

k—wind flow permeability coefficient, see Table 6.5

P—perimeter of the confining wall, m.

h—pressure difference between the two sides of the confining wall, Pa.

δ—thickness of the confining wall, m.

n—exponent, n = 1 for leaky laminar flow, n = 2 for leaky turbulent flow and $1 < n < 2$ for intermediate states.

The Soviet scholars M.A. Batrushev in the Donbass mine and Zeng-G. Molotkov in the Urals metal mine have carried out many measurements to prove that the confinement wall is built under the condition of intact gang rock, the value of n is 1.2–1.85; when there is fracture in the gang rock, the value of n is 1.5–1.9. When the confinement wall is constructed after 2–6 months, the confinement seal will be damaged and the air leakage flow will be changed from transitional flow to turbulent flow with index n equal to or close to 2. Therefore, when calculating the amount of air leakage from the confinement, the index n should be directly taken to be equal to 2.

According to the ideal gas equation of state, the main reason for the change of gas pressure in the closed fire zone of the mine is the change of gas mass, there are two main reasons for the change of gas mass in the closed fire zone. The first is the gushing out of gas in the confined area (mainly the gas and carbon dioxide in the coal seam). If the confinement is intact, the gas pressure in the confined area will gradually increase as the gas gushes out of the confined area. Secondly, the air leakage from the confining wall is caused by the change of pressure difference between inside and outside the confining wall. When the pressure outside the confinement wall is greater than the pressure inside the confinement wall, the outside world through the confinement wall to the enclosed area of wind leakage, which causes the closed fire zone pressure increases. When the pressure outside the confining wall is less than the pressure inside the confining wall, the air leakage from the enclosed area through the confining wall to the enclosed area will cause the pressure inside the closed fire zone to decrease.

For the mine fire area that has been confined for a period of time, it can be assumed that the gas gush in the confined area is low. While the mine confinement space is generally large, it can be assumed that the absolute pressure in the confined area is more constant during that time. Then, the pressure difference inside and outside the confinement wall is mainly caused by fluctuations in atmospheric pressure outside the confinement wall.

Table 6.7 Wind flow permeability coefficients for confined walls ($\times 10^{-5}$)

Type of group	Surrounding rocks	
	Complete	Cracked
Wooden board confinement	360	630
Filled and sealed	280	–
Wooden section confinement	260	430
Stone dense group	160	280
Brick dense mass and slag block confinement	120	215
Slag concrete dense mass and burl concrete confinement	90	165

The difference in pressure between the inside and outside of the confinement wall determines the amount of air leakage from the confinement wall. When the outside atmospheric pressure rises, this leads to an increase in air leakage from the outside into the fire zone. During this time, the gas composition within a certain area of the inner part of the confinement wall is affected by the dilution of the circular wind. Therefore, in order to ensure that the gas samples are taken in a way that reflects the true state of the fire zone, a non-sampling tube needs to be pre-buried into the wall to a certain depth L (i.e. the length at which atmospheric pressure "breathing" affects the displacement of the gas in the closed fire zone). The formula for calculating the length of sampling tube L in a closed fire zone is as follows.

$$L = \frac{\Delta T \cdot Q}{S} \tag{6.9}$$

In the Equation, L—the length of the sampling tube into the confining wall, m^3/s.

S—mine fire area closed tunnel section area, m^2;

ΔT—duration of air leakage, s.

Q—air leakage from the outside into the enclosed area, m^3/s.

According to the statistical results of the time by time data of atmospheric pressure in Datong area, it can be seen that the transition time of atmospheric pressure peaks and troughs in Datong area is mostly 6 h. The amplitude of atmospheric pressure fluctuations during the time by time data is mostly concentrated within 500 Pa, so it is assumed that half of this time period is the time when the external atmospheric pressure is greater than the air pressure in the enclosed area and the atmospheric pressure fluctuation is linear, then $\Delta T = 3 \times 60 \times 60 = 10800$ s, $h = 250$ Pa. For masonry confinement walls, according to Table 6.7, k is taken as 120×10^{-5}, assuming that the thickness of the confinement wall is 3 m, the tunnel section size is 3 m × 4 m (height x width) and the index n is equal to 2. Then, according to Eq. (6.8), the total air leakage from the outside environment through the confinement wall to the enclosed area in 3 h is 453.5 m^3. In order to prevent the outside atmospheric pressure fluctuations from interfering with the gas components in the enclosed area, the minimum length of the sampling tube that needs to be laid into the inner side of the enclosed fire zone should be greater than 37.8 m.

Reference

1. Lei B, Wu B. Theory and practice of gas analysis for mine fire relief. Beijing: Emergency Management Press; 2019.